Lecture Notes in Mathematics 2105

Editors-in-Chief:
J.-M. Morel, Cachan
B. Teissier, Paris

Advisory Board:
Camillo De Lellis (Zürich)
Mario di Bernardo (Bristol)
Alessio Figalli (Austin)
Davar Khoshnevisan (Salt Lake City)
Ioannis Kontoyiannis (Athens)
Gabor Lugosi (Barcelona)
Mark Podolskij (Aarhus)
Sylvia Serfaty (Paris and NY)
Catharina Stroppel (Bonn)
Anna Wienhard (Heidelberg)

T0155963

For further volumes:
http://www.springer.com/series/304

Siegfried Bosch

Lectures on Formal
and Rigid Geometry

 Springer

Siegfried Bosch
Westfälische Wilhelms-Universität
Mathematisches Institut
Münster, Germany

ISBN 978-3-319-04416-3   ISBN 978-3-319-04417-0 (eBook)
DOI 10.1007/978-3-319-04417-0
Springer Cham Heidelberg New York Dordrecht London

Lecture Notes in Mathematics ISSN print edition: 0075-8434
                            ISSN electronic edition: 1617-9692

Library of Congress Control Number: 2014944738

Mathematics Subject Classification (2010): 14G22, 14D15

Printed on acid-free paper

Springer is part of Springer Science+Business Media (www.springer.com)

Preface

This volume grew out of lectures that I gave on several occasions. First versions of the manuscript were prepared as handouts for students and later, in 2005, became a preprint of the Collaborative Research Center *Geometrical Structures in Mathematics* at the University of Münster.

The present Lecture Notes Volume is a revised and slightly expanded version of the earlier preprint. Although I kept the lecture-style presentation, I added more motivation on basic ideas as well as some fundamental examples. To make the text virtually self-contained, the theory of completed tensor products was included in a separate appendix.

It is a pleasure for me to express my gratitude to students, colleagues and, particularly, to M. Strauch for their valuable comments and suggestions. Also I would like to thank the referees for their constructive remarks which, finally, made the text more complete and easier to digest.

Münster, Germany   Siegfried Bosch
February 2014

Contents

Part II   Formal Geometry

Chapter 1
Introduction

Analytic Functions over Non-Archimedean Fields

Classical rigid geometry may be viewed as a theory of analytic functions over local fields or, more generally, over fields that are complete under a non-Archimedean absolute value; complete means that every Cauchy sequence is converging. For example, choosing a prime p, the field $\mathbb{Q}_p$ of p-adic numbers is such a field. To construct it, we start out from the field $\mathbb{Q}$ of rational numbers and complete it with respect to its p-adic absolute value $|\cdot|_p$, which is defined as follows: we set $|0|_p = 0$, and $|x|_p = p^{-r}$ for $x \in \mathbb{Q}^*$ with $x = p^r \frac{a}{b}$ where $a, b, r \in \mathbb{Z}$ and $p \nmid ab$. Then $|\cdot|_p$ exhibits the usual properties of an absolute value, as it satisfies the following conditions:

$$|x|_p = 0 \iff x = 0,$$

$$|xy|_p = |x|_p |y|_p,$$

$$|x + y|_p \leq \max\{|x|_p, |y|_p\}.$$

Furthermore, $|\cdot|_p$ extends to an absolute value on $\mathbb{Q}_p$ with the same properties. The third condition above is called the *non-Archimedean* triangle inequality, it is a sharpening of the usual *Archimedean* triangle inequality $|x + y| \leq |x| + |y|$.

This way, the field $\mathbb{Q}_p$ of p-adic numbers might be viewed as an analog of the field $\mathbb{R}$ of real numbers. There is also a p-adic analog $\mathbb{C}_p$ of the field $\mathbb{C}$ of complex numbers. Its construction is more complicated than in the Archimedean case. We first pass from $\mathbb{Q}_p$ to its algebraic closure $\mathbb{Q}_p^{\text{alg}}$. The theory of extensions of valuations and absolute values shows that there is a unique extension of $|\cdot|_p$ to this algebraic closure. However, as $\mathbb{Q}_p^{\text{alg}}$ is of infinite degree over $\mathbb{Q}_p$, we cannot conclude that $\mathbb{Q}_p^{\text{alg}}$ is complete again. In fact, it is not, and we have to pass from $\mathbb{Q}_p^{\text{alg}}$ to its completion. Fortunately, this completion remains algebraically closed; it is the field $\mathbb{C}_p$ we are looking for.

S. Bosch, *Lectures on Formal and Rigid Geometry*, Lecture Notes
in Mathematics 2105, DOI 10.1007/978-3-319-04417-0_1,
© Springer International Publishing Switzerland 2014

After the p-adic numbers had been discovered by Hensel in 1893, there were
several attempts to develop a theory of analytic functions over p-adic fields. At
the beginning, people were just curious about the question if there would exist a
reasonable analog of classical function theory over $\mathbb{C}$. However, later when algebraic
geometry had progressed so that applications to number theory were possible, a
good theory of analytic functions, say over $\mathbb{C}_p$, became sort of a necessity. To
explain this, let us look at a typical object of arithmetic algebraic geometry like
a scheme X of finite type over $\mathbb{Q}$ or $\mathbb{Z}$. Extending coefficients, we can derive from
it an $\mathbb{R}$-scheme $X_\mathbb{R}$ as well as a $\mathbb{Q}_p$-scheme $X_{\mathbb{Q}_p}$ for each prime p. There is a local–
global principle, which was already envisioned by Hensel when discovering p-adic
numbers. It says that, in many cases, problems over $\mathbb{Q}$ can be attacked by solving
them over $\mathbb{R}$ as well as over each field $\mathbb{Q}_p$. Some evidence for this principle is given
by the formula

$$\prod_{c \in P \cup \{\infty\}} |x|_c = 1 \quad \text{for} \quad x \in \mathbb{Q}^*$$

where P stands for the set of all primes and $|\cdot|_\infty$ is the usual Archimedean absolute
value on $\mathbb{Q}$. So, in the case we are looking at, we have to consider the schemes $X_\mathbb{R}$
and $X_{\mathbb{Q}_p}$ for each p. Sometimes it is desirable to leave the algebraic context and to
apply analytical methods. For example, extending coefficients from $\mathbb{R}$ to $\mathbb{C}$, we can
pass from $X_\mathbb{R}$ to $X_\mathbb{C}$ and then apply methods of classical complex analysis to $X_\mathbb{C}$. In
the same way it is desirable to develop analytic methods for handling the schemes
$X_{\mathbb{C}_p}$ obtained from $X_{\mathbb{Q}_p}$ by extending $\mathbb{Q}_p$ to $\mathbb{C}_p$.

There is a nice motivating example, due to J. Tate, showing that analytical methods in the non-Archimedean case can give new insight, when dealing with objects
of algebraic geometry. Let K be an algebraically closed field with a complete non-
Archimedean absolute value $|\cdot|$, which is assumed to be non-trivial in the sense that
there are elements $a \in K$ with $|a| \neq 0, 1$; for example, we may take $K = \mathbb{C}_p$.
Then, using ζ as a variable, look at the algebra

$$\mathcal{O}(K^*) = \left\{ \sum_{\nu \in \mathbb{Z}} c_\nu \zeta^\nu ;\ c_\nu \in K,\ \lim_{|\nu| \to \infty} |c_\nu| r^\nu = 0 \text{ for all } r > 0 \right\}$$

of all Laurent series that are globally convergent on K^*. Viewing $\mathcal{O}(K^*)$ as the
ring of analytic functions on K^*, we can construct its associated field of fractions
$\mathcal{M}(K^*) = Q(\mathcal{O}(K^*))$ and think of it as of the field of meromorphic functions
on K^*.

Now choose an element $q \in K^*$ with $|q| < 1$, and write $\mathcal{M}^q(K^*)$ for all
meromorphic functions that are invariant under multiplication by q on K^*, i.e.

$$\mathcal{M}^q(K^*) = \{ f \in \mathcal{M}(K^*) ;\ f(q\zeta) = f(\zeta) \}.$$

Tate made the observation that $\mathcal{M}^q(K^*)$ is an elliptic function field with a non-
integral j-invariant, i.e. with $|j| > 1$. Furthermore, he saw that the set of K-valued

points of the associated elliptic curve E_K coincides canonically with the quotient $K^*/q^{\mathbb{Z}}$. Elliptic curves that are obtained in this way have been called *Tate elliptic curves* since then. As quotients of type $K^*/q^{\mathbb{Z}}$ are not meaningful in the setting of algebraic geometry, Tate begun to develop a theory of so-called rigid analytic spaces where such quotients make sense; cf. [T]. In fact, the existence of an analytical isomorphism of type $E_K \simeq K^*/q^{\mathbb{Z}}$ is a characterizing condition for Tate elliptic curves. The construction of such quotients in terms of rigid spaces will be discussed in Sect. 9.2.

The nature of Tate elliptic curves becomes more plausible if we look at the classical complex case. Choose $\omega \in \mathbb{C} - \mathbb{R}$ and consider $\Gamma = \mathbb{Z} \oplus \mathbb{Z}\omega$ as a subgroup of the additive group of $\mathbb{C}$; it is called a lattice in $\mathbb{C}$. The quotient $\mathbb{C}/\Gamma$ makes sense as a Riemann surface; topologically it is a torus, like a life-belt. Furthermore, the field of meromorphic functions on $\mathbb{C}/\Gamma$ may be identified with the field of Γ-periodic meromorphic functions on $\mathbb{C}$. It is well-known that the set of isomorphism classes of Riemann surfaces of type $\mathbb{C}/\Gamma$ is in one-to-one correspondence with the set of isomorphism classes of elliptic curves over $\mathbb{C}$. In fact, if $\wp(z)$ denotes the Weierstraß $\wp$-function associated to the lattice Γ, we can consider the map

$$\mathbb{C} \longrightarrow \mathbb{P}^2_{\mathbb{C}}, \qquad z \longmapsto (\wp(z), \wp'(z), 1),$$

from $\mathbb{C}$ into the projective plane over $\mathbb{C}$. It factors through $\mathbb{C}/\Gamma$ and induces an isomorphism $\mathbb{C}/\Gamma \overset{\sim}{\longrightarrow} E_{\mathbb{C}}$ onto an elliptic curve $E_{\mathbb{C}} \subset \mathbb{P}^2_{\mathbb{C}}$. The defining equation of $E_{\mathbb{C}}$ in $\mathbb{P}^2_{\mathbb{C}}$ is given by the differential equation of the Weierstraß $\wp$-function. Thus, we see that $\mathbb{C}/\Gamma$ is, in fact, an algebraic object. Since the isomorphism $\mathbb{C}/\mathbb{Z} \overset{\sim}{\longrightarrow} \mathbb{C}^*$ provided by the exponential function induces an isomorphism $\mathbb{C}/\Gamma \overset{\sim}{\longrightarrow} \mathbb{C}^*/q^{\mathbb{Z}}$ for $q = e^{2\pi i \omega}$, we can also represent $E_{\mathbb{C}}$ as the quotient $\mathbb{C}^*/q^{\mathbb{Z}}$, which is the analog of what we have in the case of Tate elliptic curves.

Returning to the non-Archimedean case, one can prove that, just as in the classical complex case, isomorphism classes of elliptic curves correspond one-to-one to isomorphism classes of Riemann surfaces of genus 1 in the sense of rigid analytic spaces. However, among these precisely the elliptic curves with non-integral j-invariant are Tate elliptic; all others are said to have *good reduction*. Tate elliptic curves may be viewed as the correct analogs of complex tori. However, they can only be represented from the multiplicative point of view as quotients $K^*/q^{\mathbb{Z}}$, since the additive point of view, as used in the complex case, does not work. The reason is that the exponential function, if defined at all, does not converge well enough.

The discovery of Tate elliptic curves was only the beginning of a series of breathtaking further developments where rigid analytic spaces, or their equivalents, played a central role. Mumford generalized the construction of Tate elliptic curves to curves of higher genus [M1], as well as to abelian varieties of higher dimension [M2], obtaining the so-called *Mumford curves* in the first and *totally degenerate abelian varieties* in the second case. Sort of a reverse, Raynaud worked on the rigid analytic uniformization of abelian varieties and their duals over complete non-Archimedean fields [R2]. As a culminating point, the results of Mumford and Raynaud served as essential ingredients for the compactification of moduli spaces of abelian varieties in the book of Faltings and Chai [FC]. All this amply demonstrates the usefulness of analytic methods in the non-Archimedean case.

However, looking closer at the analytic methods themselves, we will see from the next section on that it is by no means clear how to define general analytic functions over non-Archimedean fields. There have been several attempts, among them Krasner's theory of analytical elements in dimension 1, but the only approach that really has survived, is the one chosen by Tate in his fundamental paper *Rigid analytic spaces* [T]. This theory was further developed by Grauert, Remmert, Kiehl, Gerritzen, and others, and is today referred to as *classical rigid geometry*. It will be the subject of the first part of these lectures and is described in detail in the book [BGR], Parts B and C.

From the beginning on it was quite clear that rigid geometry is much closer to algebraic geometry than to the methods from complex analysis. Therefore it is not surprising that rigid analytic spaces can be approached via so-called formal schemes, which are objects from formal algebraic geometry. This point of view was envisioned by Grothendieck, but has really been launched by Raynaud, who explained it in the lecture [R1]. The basic idea is to view a rigid space as the generic fiber of suitable formal R-schemes, so-called formal R-models, where R is the valuation ring of the base field under consideration. For a systematic foundation of this point of view see the papers [F I, F II, F III, F IV], as well as the monograph by Abbes [EGR].

Rigid geometry in terms of formal schemes will be dealt with in the second part of these lectures. In contrast to classical rigid geometry, this approach allows quite general objects as base spaces. But more important, one can apply a multitude of well-established and powerful techniques from algebraic geometry. As a simple example, the concept of (admissible) blowing-up on the level of formal schemes, as dealt with in Sect. 8.2, is well suited to replace the manual calculus involving rational subdomains, one of the corner stones of classical rigid geometry. Another striking example is the openness of flat morphisms in classical rigid geometry, see 9.4/2. The proof of this fact is unthinkable without passing to the formal point of view. One uses the existence of flat formal models for flat morphisms of classical rigid spaces, see 9.4/1, and then applies the openness of flat morphisms in algebraic geometry.

To complete the picture, let us have a brief look at the main further branches that grew out of classical rigid geometry, although these are beyond the scope of our lectures. The situation is a bit like in algebraic geometry over a field K where, in the early days, one has looked at points with values in an algebraic closure of K and then, on a more advanced level, has passed to points with values in more general domains over K. In fact, a classical rigid space X over a complete non-Archimedean field K consists of points with values in an algebraic closure of K. Furthermore, X carries a canonical topology inherited from the base field K which, however, is totally disconnected. It is a consequence of Tate's Acyclicity Theorem 4.3/10 that any reasonable notion of structure sheaf on X requires the selection of certain *admissible open subsets* of X and of certain *admissible open coverings* of the admissible open subsets of X by sets of the same type. Sheaves on X are then considered relative to this additional structure, referred to as a *Grothendieck topology*, that completely replaces the use of ordinary topologies. It is this concept of employing Grothendieck topologies that corresponds to Tate's use of the label *rigid* for his analytic spaces.

The need to consider a Grothendieck topology instead of an ordinary topology entails certain inconveniences. For example, there can exist non-zero abelian sheaves on a rigid space, although all their stalks are zero. This is a clear indication for the fact that there are not enough points on such a rigid space and that additional points should be included. The first one to pursue the idea of looking for more general points was Berkovich; see [Be1]. He started considering points with values in fields that are equipped with a non-Archimedean $\mathbb{R}$-targeted absolute value, also referred to as a rank 1 absolute value. Thereby one gets a Hausdorff topology with remarkable properties on the resulting rigid spaces that nowadays are referred to as *Berkovich spaces*. In view of their pleasant topological properties such spaces have become quite popular, although the construction of global Berkovich spaces by means of gluing local parts is not as natural as one would like; see [Be2]. A more rigorous approach to enlarge the point set underlying a rigid space was launched by Huber [H], who replaced rank 1 absolute values by those of arbitrary rank. Just as before, the resulting rigid spaces, called *Huber spaces*, come equipped with a true topology which, however, will be non-Hausdorff in general.

There is a totally different approach to the problem of setting up an appropriate scene on a classical rigid space. It is based on the formal point of view and remedies many of the shortcomings we have to accept otherwise. Namely, starting out from a classical rigid space X one considers the projective limit $\langle X \rangle = \varprojlim \overline{X}$ over all formal models $\overline{X}$ of X. This is the so-called *Zariski–Riemann space* associated to X, as suggested by Fujiwara [F] and Fujiwara–Kato in the forthcoming book [FK]. In a certain sense, the Zariski–Riemann space of X is equivalent to the Huber space associated to X, while the corresponding Berkovich space may be viewed as the biggest Hausdorff quotient of $\langle X \rangle$. This way we can say that the approach to rigid geometry through formal schemes, as presented in the second part of these lectures, is at the heart of all derivatives of classical rigid geometry, although the books of Berkovich [Be1] and of Huber [H] provide direct access to the corresponding theories without making use of methods from formal geometry. But let us point out that, in order to access any advanced branch in rigid geometry, a prior knowledge of Tate's classical theory is indispensable or, at least, highly advisable.

Part I
Classical Rigid Geometry

Chapter 2
Tate Algebras

2.1 Topology Induced from a Non-Archimedean Absolute Value

We start by recalling the definition of a non-Archimedean absolute value.

Definition 1. *Let K be a field. A map $|\cdot|: K \longrightarrow \mathbb{R}_{\geq 0}$ is called a* non-Archimedean *absolute value if for all $a,b \in K$ the following hold*:

(i) $|a| = 0 \Longleftrightarrow a = 0$,

(ii) $|ab| = |a||b|$,

(iii) $|a + b| \leq \max\{|a|,|b|\}$.

One can immediately verify trivialities such as $|1| = 1$ and $|-a| = |a|$. To an absolute value as above one can always associate a *valuation* of K. This is a map $v: K \longrightarrow \mathbb{R} \cup \{\infty\}$ satisfying the following conditions:

(i) $v(a) = \infty \Longleftrightarrow a = 0$

(ii) $v(ab) = v(a) + v(b)$

(iii) $v(a + b) \geq \min\{v(a), v(b)\}$

Just let $v(a) = -\log|a|$ for $a \in K$. This sets up a one-to-one correspondence between non-Archimedean absolute values and valuations, as we can pass from valuations v back to absolute values by setting $|a| = e^{-v(a)}$ for $a \in K$. Frequently, we will make no difference between absolute values and valuations, just saying that K is a field with a valuation. An absolute value $|\cdot|$ is called *trivial* if it assumes only the values $0, 1 \in \mathbb{R}$. It is called *discrete* if $|K^*|$ is discrete in $\mathbb{R}_{>0}$. Likewise a valuation is called trivial if $v(K^*) = \{0\}$, and discrete if $v(K^*)$ is discrete in $\mathbb{R}$. Unless stated otherwise, we will always assume that absolute values and valuations on fields are *non-trivial*.

In the following, let K be a field with a non-Archimedean absolute value $|\cdot|$. As usual, the absolute value gives rise to a distance function by setting $d(a, b) = |a-b|$

S. Bosch, *Lectures on Formal and Rigid Geometry*, Lecture Notes in Mathematics 2105, DOI 10.1007/978-3-319-04417-0_2, © Springer International Publishing Switzerland 2014

and, thus, to a topology on K. Furthermore, we can consider sequences or infinite series of elements in K and define their convergence as in the Archimedean case. In particular, K is called *complete* if every Cauchy sequence converges in K.

It has to be pointed out, however, that the non-Archimedean triangle inequality in Definition 1 (iii) has far reaching consequences.

Proposition 2. *Let $a,b \in K$ satisfy $|a| \neq |b|$. Then*

$$|a + b| = \max\{|a|,|b|\}.$$

Proof. Assume $|b| < |a|$. Then $|a + b| < |a|$ implies

$$|a| = |(a + b) - b| \leq \max\{|a + b|, |b|\} < |a|,$$

which is impossible. So we must have $|a+b| = |a| = \max\{|a|, |b|\}$ as claimed. $\square$

Lemma 3. *A series $\sum_{\nu=0}^{\infty} a_\nu$ of elements $a_\nu \in K$ is a Cauchy sequence if and only if the coefficients a_ν form a zero sequence, i.e. if and only if $\lim_{\nu \to \infty} |a_\nu| = 0$.*
   Hence, if K is complete, the series is convergent if and only if $\lim_{\nu \to \infty} |a_\nu| = 0$.

Proof. Choose $\varepsilon > 0$. We have to show there exists an integer $N \in \mathbb{N}$ such that $|\sum_{\nu=i}^{j} a_\nu| < \varepsilon$ for all $j \geq i \geq N$. As $\lim_{\nu \to \infty} |a_\nu| = 0$, we know there is an $N \in \mathbb{N}$ such that $|a_\nu| < \varepsilon$ for all $\nu \geq N$. But then, for any integers $j \geq i \geq N$, an iterated application of the non-Archimedean triangle inequality yields

$$\left| \sum_{\nu=i}^{j} a_\nu \right| \leq \max_{\nu=i...j} |a_\nu| < \varepsilon,$$

which had to be shown. $\square$

In terms of distances between points in K, the non-Archimedean triangle inequality implies

$$d(y,z) \leq \max\{d(x,y), d(x,z)\} \qquad \text{for } x, y, z \in K,$$

where this inequality is, in fact, an equality if $d(x,y)$ is different from $d(x,z)$; cf. Proposition 2. In particular, given any three points in K, there exists one of them such that the distances between it and the two remaining points coincide. In other words, any triangle in K is isosceles. Furthermore, we can conclude that each point of a disk in K can serve as its center. Thus, if an intersection of two disks is non-empty, we can choose a point of their intersection as common center, and we see that they are concentric.

For a center $a \in K$ and a radius $r \in \mathbb{R}_{>0}$ we can consider the disk around a without periphery

$$D^-(a,r) = \{x \in K ; d(x,a) < r\},$$

which is open as well as closed in K; it is symbolically referred to as the *open* disk around a with radius r. Similarly, we can consider the same disk with periphery, namely

$$D^+(a,r) = \{x \in K ; d(x,a) \leq r\}.$$

It is open and closed just as well and symbolically referred to as the *closed* disk around a of radius r. In addition, there is the periphery itself, namely

$$\partial D(a,r) = \{x \in K ; d(x,a) = r\}.$$

Certainly, it is closed, but it is also open since, due to Proposition 2, we have $D^-(x,r) \subset \partial D(a,r)$ for any $x \in \partial D(a,r)$. It is for this reason that the periphery $\partial D(0,1)$ is sometimes called the *unit tire* in K.

The preceding considerations show another peculiarity of the topology of K:

Proposition 4. *The topology of K is totally disconnected, i.e. any subset in K consisting of more than just one point is not connected.*

Proof. Consider an arbitrary subset $A \subset K$ consisting of at least two different points x, y. For $\delta = \frac{1}{2}d(x,y)$, set $A_1 = D^-(x,\delta) \cap A$ and $A_2 = A - A_1$. Then A_1 is relatively open and closed in A, and the same is true for A_2. Furthermore, A is the disjoint union of the non-empty open parts A_1 and A_2. Consequently, A cannot be connected with respect to the topology induced from K on A. $\square$

We may draw some conclusions from the latter observation. Writing $[0,1]$ for the unit interval in $\mathbb{R}$, there cannot exist non-constant continuous paths $[0,1] \longrightarrow K$. Consequently, there is no obvious way to define line integrals, and it is excluded that there exists a straightforward replacement for classical complex Cauchy theory, providing the link between holomorphic and analytic functions. In fact, the concept of holomorphic functions, defined through differentiability, and that of analytic functions, defined via convergent power series expansions, differs largely. On the other hand, it should be admitted that in certain contexts notions of integrals and also line integrals have been developed.

Without making it more precise, we mention that a definition of holomorphic functions via differentiability is not very rewarding. The class of such functions is very big and does not have good enough properties. So the only approach towards a reasonable "function theory" over non-Archimedean fields that might be left, is via analyticity, i.e. via convergent power series expansions. However, it is by no means clear how to proceed with analyticity. Let us call a function $f : U \longrightarrow K$

defined on some open subset $U \subset K$ *locally analytic* if it admits a convergent power series expansion locally around each point $x \in U$. Then we cannot expect any identity theorem for such functions, since non-empty open subsets of K are not connected. For example, let us look at the case where U is a disk $D^-(0, 1)$ or $D^+(0, 1)$. Choosing a radius r with $0 < r < 1$, we may write

$$U = \bigcup_{a \in U} D^-(a, r)$$

and get from it a partition of U into *disjoint* disks. On each of these disks, let's call them D_i, $i \in I$, we can consider an arbitrary convergent power series f_i. Then $f : U \longrightarrow K$ defined by $f|_{D_i} = f_i$ is locally analytic. In particular, we can take the f_i to be constant, and it follows that, on disks, locally analytic functions do not necessarily admit globally convergent power series expansions. This shows that locally analytic functions cannot enjoy reasonable global properties.

The basic principle of rigid analytic geometry is to require that analytic functions on disks admit globally convergent power series expansions. We will discuss the details of the precise definition in subsequent sections.

2.2   Restricted Power Series

As always, we consider a field K with a complete non-Archimedean absolute value that is non-trivial. Let $\overline{K}$ be its algebraic closure. We will use the results of Appendix A, namely that the absolute value of K admits a unique extension to $\overline{K}$ and that, although $\overline{K}$ itself might not be complete, this absolute value nevertheless is complete on each *finite* subextension of $\overline{K}/K$. For integers $n \geq 1$ let

$$\mathbb{B}^n(\overline{K}) = \left\{ (x_1, \ldots, x_n) \in \overline{K}^n ; |x_i| \leq 1 \right\}$$

be the unit ball in $\overline{K}^n$.

Lemma 1. *A formal power series*

$$f = \sum_{v \in \mathbb{N}^n} c_v \zeta^v = \sum_{v \in \mathbb{N}^n} c_{v_1 \ldots v_n} \zeta_1^{v_1} \ldots \zeta_n^{v_n} \in K[\![\zeta_1, \ldots, \zeta_n]\!]$$

converges on $\mathbb{B}^n(\overline{K})$ *if and only if* $\lim_{|v| \to \infty} |c_v| = 0$.

Proof. If f is convergent at the point $(1, \ldots, 1) \in \mathbb{B}^n(\overline{K})$, the series $\sum_v c_v$ is convergent, and we must have $\lim_{|v| \to \infty} |c_v| = 0$ by 2.1/3. Conversely, considering a point $x \in \mathbb{B}^n(\overline{K})$, there is a finite and, hence, complete subextension K' of $\overline{K}/K$ such that all components of x belong to K'. Then, if $\lim_{|v| \to \infty} |c_v| = 0$, we have $\lim_{|v| \to \infty} |c_v||x^v| = 0$, and $f(x)$ is convergent in $K' \subset \overline{K}$ by 2.1/3, again. $\square$

Definition 2. *The K-algebra $T_n = K\langle\zeta_1,\ldots,\zeta_n\rangle$ of all formal power series*

$$\sum_{\nu\in\mathbb{N}^n} c_\nu\zeta^\nu \in K[\![\zeta_1,\ldots,\zeta_n]\!], \qquad c_\nu \in K, \qquad \lim_{|\nu|\to\infty} |c_\nu| = 0,$$

thus, converging on $\mathbb{B}^n(\overline{K})$, *is called the* Tate algebra *of restricted, or* strictly convergent *power series. By convention we write* $T_0 = K$.

That T_n is, in fact, a K-algebra is easily checked. Also it is clear that the canonical map from T_n to the set of maps $\mathbb{B}^n(\overline{K}) \longrightarrow \overline{K}$ is a homomorphism of K-algebras. We define the so-called *Gauß norm* on T_n by setting

$$|f| = \max_\nu |c_\nu| \quad \text{for} \quad f = \sum_\nu c_\nu\zeta^\nu.$$

It satisfies the conditions of a K-algebra norm, i.e. for $c \in K$ and $f, g \in T_n$ we have

$$|f| = 0 \Longleftrightarrow f = 0,$$
$$|cf| = |c||f|,$$
$$|fg| = |f||g|,$$
$$|f + g| \leq \max\{|f|, |g|\},$$

where, strictly speaking, only the submultiplicativity $|fg| \leq |f||g|$ would be required for a K-algebra norm. In particular, it follows from the multiplicativity in the third line that T_n is an integral domain. The stated properties of the norm $|\cdot|$ are easy to verify, except possibly for the multiplicativity. Note first that we have $|fg| \leq |f||g|$ for trivial reasons. To show that this estimate is, in fact, an equality, we look at the *valuation ring* $R = \{a \in K ; |a| \leq 1\}$ of K; it is a subring of K with a unique maximal ideal $\mathfrak{m} = \{a \in K ; |a| < 1\}$. Thus, $k = R/\mathfrak{m}$ is a field, the *residue field* of K, and there is a canonical residue epimorphism $R \longrightarrow k$, which we will indicate by $a \longmapsto \tilde{a}$. Denoting by $R\langle\zeta_1,\ldots,\zeta_n\rangle$ the R-algebra of all restricted power series $f \in T_n$ having coefficients in R or, equivalently, with $|f| \leq 1$, the epimorphism $R \longrightarrow k$ extends to an epimorphism

$$\pi: R\langle\zeta_1,\ldots,\zeta_n\rangle \longrightarrow k[\zeta_1,\ldots,\zeta_n], \qquad \sum c_\nu\zeta^\nu \longmapsto \sum \tilde{c}_\nu\zeta^\nu.$$

For an element $f \in R\langle\zeta_1,\ldots,\zeta_n\rangle$ we will call $\tilde{f} = \pi(f)$ the *reduction* of f. Note that $\tilde{f} = 0$ if and only if $|f| < 1$. Now consider $f, g \in T_n$ with $|f| = |g| = 1$. Then f, g, and fg belong to $R\langle\zeta_1,\ldots,\zeta_n\rangle$, and we have

$$\pi(fg) = \tilde{f}\tilde{g} \neq 0,$$

since $k[\zeta_1,\ldots,\zeta_n]$ is an integral domain. But then we must have $|fg| = 1$.

For general $f, g \in T_n$ the assertion $|fg| = |f||g|$ holds for trivial reasons if f or g are constant. If both, f and g are non-constant, we can write $f = cf'$ and $g = dg'$ with $|f| = |c|$, $|g| = |d|$, and $|f'| = |g'| = 1$. Then

$$|fg| = |cdf'g'| = |cd||f'g'| = |c||d| = |f||g|,$$

which we had to show.

Proposition 3. T_n *is complete with respect to the Gauß norm. So it is a Banach K-algebra, i.e. a K-algebra that is complete under the given norm.*

Proof. Consider a series $\sum_{i=0}^{\infty} f_i$ with restricted power series $f_i = \sum_\nu c_{i\nu} \zeta^\nu \in T_n$ satisfying $\lim_{i \to \infty} f_i = 0$. Then, as $|c_{i\nu}| \leq |f_i|$, we have $\lim_{i \to \infty} |c_{i\nu}| = 0$ for all ν so that the limits $c_\nu = \sum_{i=0}^{\infty} c_{i\nu}$ exist. We claim that the series $f = \sum_\nu c_\nu \zeta^\nu$ is strictly convergent and that $f = \sum_{i=0}^{\infty} f_i$.

Choose $\varepsilon > 0$. As the f_i form a zero sequence, there is an integer N such that $|c_{i\nu}| < \varepsilon$ for all $i \geq N$ and all ν. Furthermore, as the coefficients of the series $f_0, \ldots, f_{N-1}$ form a zero sequence, almost all of these coefficients must have an absolute value smaller than ε. This implies that almost all of the absolute values $|c_{i\nu}|$ with arbitrary i and ν are smaller than ε and, hence, that the elements $c_{i\nu}$ form a zero sequence in K (under any ordering). Now, using the fact that the non-Archimedean triangle inequality generalizes for convergent series to an inequality of type

$$\left| \sum_{i=0}^{\infty} a_\nu \right| \leq \max_{i=0\ldots\infty} |a_\nu|,$$

we see immediately that the power series f belongs to T_n and that $f = \sum_{i=0}^{\infty} f_i$. $\square$

With the help of Proposition 3 we can easily characterize units in T_n.

Corollary 4. *A series $f \in T_n$ with $|f| = 1$ is a unit if and only if its reduction $\tilde{f} \in k[\zeta_1, \ldots, \zeta_n]$ is a unit, i.e. if and only if $\tilde{f} \in k^*$. More generally, an arbitrary series $f \in T_n$ is a unit if and only if $|f - f(0)| < |f(0)|$, i.e. if and only if the absolute value of the constant coefficient of f is strictly bigger than the one of all other coefficients of f.*

Proof. It is only necessary to consider elements $f \in T_n$ with Gauß norm 1. If f is a unit in T_n, it is also a unit in $R\langle \zeta_1, \ldots, \zeta_n \rangle$. Then $\tilde{f}$ is a unit in $k[\zeta_1, \ldots, \zeta_n]$ and, hence, in k^*. Conversely, if $\tilde{f} \in k^*$, the constant term $f(0)$ of f satisfies $|f(0)| = 1$, and we may even assume $f(0) = 1$. But then f is of type $f = 1 - g$ with $|g| < 1$, and $\sum_{i=0}^{\infty} g^i$ is an inverse of f. $\square$

Proposition 5 (Maximum Principle). *Let $f \in T_n$. Then $|f(x)| \leq |f|$ for all points $x \in \mathbb{B}^n(\overline{K})$, and there exists a point $x \in \mathbb{B}^n(\overline{K})$ such that $|f(x)| = |f|$.*

Proof. The first assertion is trivial. To verify the second one, assume $|f| = 1$ and consider again the canonical epimorphism $\pi \colon R\langle \zeta_1, \ldots, \zeta_n \rangle \longrightarrow k[\zeta_1, \ldots, \zeta_n]$. Then $\tilde{f} = \pi(f)$ is a non-trivial polynomial in n variables, and there exists a point $\tilde{x} \in \overline{k}^n$ with $\overline{k}$ the algebraic closure of k, such that $\tilde{f}(\tilde{x}) \neq 0$. The theory of valuations and absolute values shows that $\overline{k}$ may be interpreted as the residue field of $\overline{K}$, the algebraic closure of K. Writing $\overline{R}$ for the valuation ring of $\overline{K}$ and choosing a lifting $x \in \mathbb{B}^n(\overline{K})$ of $\tilde{x}$, we can consider the commutative diagram

$$
\begin{array}{ccc}
R\langle \zeta_1, \ldots, \zeta_n \rangle & \longrightarrow & k[\zeta_1, \ldots, \zeta_n] \\
\downarrow & & \downarrow \\
\overline{R} & \longrightarrow & \overline{k}
\end{array}
$$

where the first vertical map is evaluation at x and the second one evaluation at $\tilde{x}$. As $f(x) \in \overline{R}$ is mapped onto $\tilde{f}(\tilde{x}) \in \overline{k}$ and the latter is non-zero, we obtain $|f(x)| = 1 = |f|$, which had to be shown. $\square$

The Tate algebra T_n has many properties in common with the polynomial ring in n variables over K, as we will see. The key tool for proving all these properties is *Weierstraß theory*, which we will explain now and which is quite analogous to Weierstraß theory in the classical complex case. In particular, we will establish *Weierstraß division*, a division process similar to Euclid's division on polynomial rings. In Weierstraß theory the role of monic polynomials is taken over by so-called *distinguished* restricted power series, or later by so-called *Weierstraß polynomials*.

Definition 6. *A restricted power series $g = \sum_{\nu=0}^{\infty} g_\nu \zeta_n^\nu \in T_n$ with coefficients $g_\nu \in T_{n-1}$ is called ζ_n-distinguished of some order $s \in \mathbb{N}$ if the following hold:*

(i) *g_s is a unit in T_{n-1}.*
(ii) *$|g_s| = |g|$ and $|g_s| > |g_\nu|$ for $\nu > s$.*

In particular, if $g = \sum_{\nu=0}^{\infty} g_\nu \zeta_n^\nu$ satisfies $|g| = 1$, then g is ζ_n-distinguished of order s if and only if its reduction $\tilde{g}$ is of type

$$
\tilde{g} = \tilde{g}_s \zeta_n^s + \tilde{g}_{s-1} \zeta_n^{s-1} + \ldots + \tilde{g}_0 \zeta_n^0
$$

with a unit $\tilde{g}_s \in k^*$; use Corollary 4. Thereby we see that an arbitrary series $g \in T_n$ is ζ_n-distinguished of order 0 if and only if it is a unit. Furthermore, for $n = 1$, every non-zero element $g \in T_1$ is ζ_1-distinguished of some order $s \in \mathbb{N}$.

Lemma 7. *Given finitely many non-zero elements* $f_1, \ldots, f_r \in T_n$, *there is a continuous automorphism*

$$\sigma : T_n \longrightarrow T_n, \qquad \zeta_i \longmapsto \begin{cases} \zeta_i + \zeta_n^{\alpha_i} & \text{for } i < n \\ \zeta_n & \text{for } i = n \end{cases}$$

with suitable exponents $\alpha_1, \ldots, \alpha_{n-1} \in \mathbb{N}$ *such that the elements* $\sigma(f_1), \ldots, \sigma(f_r)$ *are* ζ_n-*distinguished.*[1] *Furthermore,* $|\sigma(f)| = |f|$ *for all* $f \in T_n$.

Proof. It is clear that we can define a continuous K-homomorphism σ of T_n by mapping the variables ζ_i as indicated in the assertion. Then

$$T_n \longrightarrow T_n, \qquad \zeta_i \longmapsto \begin{cases} \zeta_i - \zeta_n^{\alpha_i} & \text{for } i < n \\ \zeta_n & \text{for } i = n \end{cases}$$

defines an inverse σ^{-1} of σ, and we see that both homomorphisms are isomorphisms. As $|\sigma(f)| \leq |f|$ for all $f \in T_n$ and a similar estimate holds for σ^{-1}, we have, in fact, $|\sigma(f)| = |f|$ for all $f \in T_n$.

We start with the case where we are dealing with just one element $f \in T_n$. Assuming $|f| = 1$, we can consider the reduction $\tilde{f}$ of f, say $\tilde{f} = \sum_{\nu \in N} \tilde{c}_\nu \zeta^\nu$ where N is a finite subset of $\mathbb{N}^n$. Discarding all trivial terms of this sum, we may assume that N is minimal, i.e. that $\tilde{c}_\nu \neq 0$ for all $\nu \in N$. Now choose t greater than the maximum of all ν_i occurring as a component of some $\nu \in N$, and consider the automorphism σ of T_n obtained from $\alpha_1 = t^{n-1}, \ldots, \alpha_{n-1} = t$. Its reduction $\tilde{\sigma}$ on $k[\zeta_1, \ldots, \zeta_n]$ satisfies

$$\tilde{\sigma}(\tilde{f}) = \sum_{\nu \in N} \tilde{c}_\nu (\zeta_1 + \zeta_n^{\alpha_1})^{\nu_1} \ldots (\zeta_{n-1} + \zeta_n^{\alpha_{n-1}})^{\nu_{n-1}} \zeta_n^{\nu_n}$$

$$= \sum_{\nu \in N} \tilde{c}_\nu \zeta_n^{\alpha_1 \nu_1 + \ldots + \alpha_{n-1} \nu_{n-1} + \nu_n} + \tilde{g},$$

where $\tilde{g} \in k[\zeta_1, \ldots, \zeta_n]$ is a polynomial whose degree in ζ_n is strictly less than the maximum of all exponents $\alpha_1 \nu_1 + \ldots + \alpha_{n-1} \nu_{n-1} + \nu_n$ with ν varying over N. Due to the choice of $\alpha_1, \ldots, \alpha_{n-1}$, these exponents are pairwise different and, hence, their maximum s is assumed at a single index $\bar{\nu} \in N$. But then

$$\tilde{\sigma}(\tilde{f}) = \tilde{c}_{\bar{\nu}} \zeta_n^s + \text{a polynomial of degree } < s \text{ in } \zeta_n.$$

As $\tilde{c}_{\bar{\nu}} \neq 0$, it follows that $\sigma(f)$ is ζ_n-distinguished of order s.

---

[1] Later, in 3.1/20, we will see that homomorphisms of Tate algebras are automatically continuous.

The general case of finitely many non-zero elements $f_1, \ldots, f_r \in T_n$ is dealt with in the same way. One just has to choose t big enough such that it works for all f_i simultaneously.   □

The reason for considering distinguished elements in T_n is that there is Weierstraß division by such elements, which is the analog of Euclid's division on polynomial rings.

Theorem 8 (Weierstraß Division). *Let $g \in T_n$ be ζ_n-distinguished of some order s. Then, for any $f \in T_n$, there are a unique series $q \in T_n$ and a unique polynomial $r \in T_{n-1}[\zeta_n]$ of degree $r < s$ satisfying*

$$f = qg + r.$$

Furthermore, $|f| = \max(|q||g|, |r|)$.

Proof. Without loss of generality we may assume $|g| = 1$. First, let us consider an equation $f = qg + r$ of the required type. Then, clearly, $|f| \leq \max(|q||g|, |r|)$. If, however, $|f|$ is strictly smaller than the right-hand side, we may assume that $\max(|q||g|, |r|) = 1$. Then we would have $\tilde{q}\tilde{g} + \tilde{r} = 0$ with $\tilde{q} \neq 0$ or $\tilde{r} \neq 0$, and this would contradict Euclid's division in $k[\zeta_1, \ldots, \zeta_{n-1}][\zeta_n]$. Therefore we must have $|f| = \max(|q||g|, |r|)$, and uniqueness of the division formula is a consequence.

To verify the existence of the division formula, we write $g = \sum_{\nu=0}^{\infty} g_\nu \zeta_n^\nu$ with coefficients $g_\nu \in T_{n-1}$ where g_s is a unit and where $|g_\nu| < |g_s| = |g| = 1$ for $\nu > s$. Set $\varepsilon = \max_{\nu > s} |g_\nu|$ so that $\varepsilon < 1$. We want to show the following slightly weaker assertion:

(W) *For any $f \in T_n$, there exist $q, f_1 \in T_n$ and a polynomial $r \in T_{n-1}[\zeta_n]$ of degree $< s$ with*

$$f = qg + r + f_1,$$

$$|q|, |r| \leq |f|, \qquad |f_1| \leq \varepsilon |f|.$$

This is enough, since proceeding inductively and starting with $f_0 = f$, we obtain equations

$$f_i = q_i g + r_i + f_{i+1}, \qquad i \in \mathbb{N},$$

$$|q_i|, |r_i| \leq \varepsilon^i |f|, \qquad |f_{i+1}| \leq \varepsilon^{i+1} |f|,$$

and, hence, an equation

$$f = \left(\sum_{i=0}^{\infty} q_i\right) g + \left(\sum_{i=0}^{\infty} r_i\right)$$

as required. To verify the assertion (W), we may approximate f by a polynomial in $T_{n-1}[\zeta_n]$ and thereby assume $f \in T_{n-1}[\zeta_n]$. Furthermore, set $g' = \sum_{i=0}^{s} g_i \zeta_n^i$, where now g' is a polynomial in ζ_n that is ζ_n-distinguished of order s and satisfies $|g'| = 1$. Then Euclid's division in $T_{n-1}[\zeta_n]$ yields a decomposition

$$f = qg' + r$$

with an element $q \in T_n$ and a polynomial $r \in T_{n-1}[\zeta_n]$ of degree $< s$. As shown above, $|f| = \max(|q|, |r|)$. But from this we get

$$f = qg + r + f_1$$

with $f_1 = qg' - qg$. As $|g - g'| = \varepsilon$ and $|q| \leq |f|$, we have $|f_1| \leq \varepsilon |f|$, and we are done. $\square$

Corollary 9 (Weierstraß Preparation Theorem). *Let $g \in T_n$ be ζ_n-distinguished of order s. Then there exists a unique monic polynomial $\omega \in T_{n-1}[\zeta_n]$ of degree s such that $g = e\omega$ for a unit $e \in T_n$. Furthermore, $|\omega| = 1$ so that ω is ζ_n-distinguished of order s.*

Proof. Applying the Weierstraß division formula, we get an equation

$$\zeta_n^s = qg + r$$

with a series $q \in T_n$ and a polynomial $r \in T_{n-1}[\zeta_n]$ of degree $< s$ that satisfies $|r| \leq 1$. Writing $\omega = \zeta_n^s - r$, we see that $\omega = qg$ satisfies $|\omega| = 1$ and is ζ_n-distinguished of order s. To verify the existence of the asserted decomposition of g, we have to show that q is a unit in T_n. Assuming $|g| = |q| = 1$, we can look at the equation $\tilde{\omega} = \tilde{q}\tilde{g}$ obtained via reduction. Then both, $\tilde{\omega}$ and $\tilde{g}$, are polynomials of degree s in ζ_n, and it follows that $\tilde{q}$ is a unit in k^*, as $\tilde{\omega}$ is monic. But then q is a unit in T_n by Corollary 4.

To show uniqueness, start with a decomposition $g = e\omega$. Defining $r = \zeta_n^s - \omega$, we get

$$\zeta_n^s = e^{-1}g + r,$$

which by the uniqueness of Weierstraß division shows the uniqueness of e^{-1} and r and, hence, of e and ω. $\square$

Corollary 10. *The Tate algebra $T_1 = K\langle\zeta_1\rangle$ of restricted power series in a single variable ζ_1 is a Euclidean domain and, in particular, a principal ideal domain.*

Proof. Every non-zero element $g \in T_1$ is ζ_1-distinguished of a well-defined order $s \in \mathbb{N}$. Thus, in view of Weierstraß division, the map $T_1 - \{0\} \longrightarrow \mathbb{N}$ that associates to g its order s of being distinguished is a Euclidean function. It follows that T_1 is a Euclidean domain and, in particular, a principal ideal domain.   □

Monic polynomials $\omega \in T_{n-1}[\zeta_n]$ with $|\omega| = 1$, as occurring in Corollary 9, are called *Weierstraß polynomials* in ζ_n. So each ζ_n-distinguished element $f \in T_n$ is associated to a Weierstraß polynomial. Furthermore, if f is an arbitrary non-zero element in T_n, we can assume by Lemma 7 that the indeterminates $\zeta_1, \ldots, \zeta_n \in T_n$ are chosen in such a way that f is ζ_n-distinguished of some order s.

Corollary 11 (Noether Normalization). *For any proper ideal* $\mathfrak{a} \subsetneq T_n$, *there is a* K-*algebra monomorphism* $T_d \longrightarrow T_n$ *for some* $d \in \mathbb{N}$ *such that the composition* $T_d \longrightarrow T_n \longrightarrow T_n/\mathfrak{a}$ *is a finite monomorphism. The integer* d *is uniquely determined as the Krull dimension of* $T_n/\mathfrak{a}$.

Proof. Assuming $\mathfrak{a} \neq 0$, we can choose an element $g \neq 0$ in $\mathfrak{a}$. Furthermore, applying a suitable automorphism to T_n, we can assume by Lemma 7 that g is ζ_n-distinguished of some order $s \geq 0$. By Weierstraß division we know that any $f \in T_n$ is congruent modulo g to a polynomial $r \in T_{n-1}[\zeta_n]$ of degree $< s$. In other words, the canonical morphism $T_{n-1} \longrightarrow T_n \longrightarrow T_n/(g)$ is finite; in fact using the uniqueness of Weierstraß division, $T_n/(g)$ is a free T_{n-1}-module generated by the residue classes of $\zeta_n^0, \ldots, \zeta_n^{s-1}$.

Now consider the composition $T_{n-1} \longrightarrow T_n/(g) \longrightarrow T_n/\mathfrak{a}$ and write $\mathfrak{a}_1$ for its kernel. If $\mathfrak{a}_1 = 0$, we are done. Else we can proceed with $\mathfrak{a}_1$ and T_{n-1} in the same way as we did with $\mathfrak{a}$ and T_n. Then, as the composition of finite morphisms is finite again, we will get a finite monomorphism $T_d \longrightarrow T_n/\mathfrak{a}$ after finitely many steps.

Finally, it follows from commutative algebra, see [Bo], 3.3/6, that the Krull dimension of $T_n/\mathfrak{a}$ coincides with the one of T_d. However, the latter equals d, as we will see below in Proposition 17.   □

Looking at the proof of Corollary 11, it should be pointed out that the resulting monomorphism $T_d \longrightarrow T_n/\mathfrak{a}$ does not necessarily coincide with the canonical one sending $\zeta_i \in T_d$ to the residue class of ζ_i in $T_n/\mathfrak{a}$. In fact, this canonical morphism will, in general, be neither injective nor finite, as can be seen from simple examples.

Corollary 12. *Let* $\mathfrak{m} \subset T_n$ *be a maximal ideal. Then the field* $T_n/\mathfrak{m}$ *is finite over* K.

Proof. Using Noether normalization, there is a finite monomorphism $T_d \hookrightarrow T_n/\mathfrak{m}$ for a suitable $d \in \mathbb{N}$. As $T_n/\mathfrak{m}$ is a field, the same is true for T_d. So we must have $d = 0$ and, hence, $T_d = K$.   □

A direct consequence of Corollary 12 is the following:

Corollary 13. *The map*

$$\mathbb{B}^n(\overline{K}) \longrightarrow \mathrm{Max}\, T_n, \qquad x \longmapsto \mathfrak{m}_x = \{f \in T_n \,;\, f(x) = 0\},$$

from the unit ball in $\overline{K}^n$ *to the set of all maximal ideals in* T_n *is surjective.*

Proof. Evaluation of functions $f \in T_n$ at a point $x = (x_1, \ldots, x_n) \in \mathbb{B}^n(\overline{K})$ defines a continuous epimorphism $\varphi_x : T_n \longrightarrow K(x_1, \ldots, x_n)$. Thus its kernel, which equals $\mathfrak{m}_x$, is a maximal ideal in T_n.

Conversely, given any maximal ideal $\mathfrak{m} \subset T_n$, the field $K' = T_n/\mathfrak{m}$ is finite over K by Corollary 12, and we can choose an embedding $K' \hookrightarrow \overline{K}$. We claim that the resulting map $\varphi : T_n \longrightarrow \overline{K}$ is contractive in the sense that $|\varphi(a)| \leq |a|$ for all $a \in T_n$. Proceeding indirectly, we assume there is an element $a \in T_n$ with $|\varphi(a)| > |a|$. Then $a \neq 0$, and we may assume $|a| = 1$. Write $\alpha = \varphi(a)$, and let

$$p(\eta) = \eta^r + c_1 \eta^{r-1} + \ldots + c_r \in K[\eta]$$

be the minimal polynomial of α over K. If $\alpha_1, \ldots, \alpha_r$ denote the (not necessarily pairwise different) conjugates of α over K, we have

$$p(\eta) = \prod_{j=1}^{r} (\eta - \alpha_j).$$

All fields $K(\alpha_j)$ are canonically isomorphic to $K(\alpha)$. As K is complete and the absolute value of K extends uniquely to $K(\alpha)$, we get $|\alpha_j| = |\alpha|$ for all j. In particular, we have $|c_r| = |\alpha|^r$ and

$$|c_j| \leq |\alpha|^j < |\alpha|^r = |c_r| \qquad \text{for } j < r,$$

as $|\alpha| > 1$. Then, by Corollary 4, the expression $p(a) = a^r + c_1 a^{r-1} + \ldots + c_r$ is a unit in T_n and, consequently, it must be mapped under φ to a unit in $K' \subset \overline{K}$. On the other hand, the image $\varphi(p(a))$ is trivial, as it equals $\varphi(p(a)) = p(\alpha) = 0$. Thus, we obtain a contradiction and therefore have $|\varphi(a)| \leq |a|$ for all $a \in T_n$. From this it follows in particular, that $\varphi : T_n \longrightarrow \overline{K}$ is continuous. But then, setting $x_i = \varphi(\zeta_i)$ for $i = 1, \ldots, n$, it is clear that the point $x = (x_1, \ldots, x_n)$ belongs to $\mathbb{B}^n(\overline{K})$, that φ coincides with φ_x and, hence, that $\mathfrak{m} = \mathfrak{m}_{(x_1, \ldots, x_n)}$. $\square$

We want to end this section by deriving some standard properties of T_n.

Proposition 14. T_n *is Noetherian, i.e. each ideal* $\mathfrak{a} \subset T_n$ *is finitely generated.*

Proof. Proceeding by induction, we can assume that T_{n-1} is Noetherian. Now, consider a non-trivial ideal $\mathfrak{a} \subset T_n$. Then we can choose a non-zero element $g \in \mathfrak{a}$ that, using Lemma 7, can be assumed to be ζ_n-distinguished. By Weierstraß division, $T_n/(g)$ is a finite T_{n-1}-module and, hence, a Noetherian T_{n-1}-module, as T_{n-1} is Noetherian. Consequently, $\mathfrak{a}/(g)$ is finitely generated over T_{n-1} and, thus, $\mathfrak{a}$ is finitely generated on T_n. $\square$

Proposition 15. T_n *is factorial and, hence, normal, i.e. integrally closed in its field of fractions.*

Proof. Proceeding by induction, we may assume that T_{n-1} is factorial and, hence, by the Lemma of Gauß, that $T_{n-1}[\zeta_n]$ is factorial. Consider a non-zero element $f \in T_n$ that is not a unit. Again, by Lemma 7, we may assume that f is ζ_n-distinguished and, by Corollary 9, that f is, in fact, a Weierstraß polynomial. Now consider a factorization $f = \omega_1 \ldots \omega_r$ into prime elements $\omega_i \in T_{n-1}[\zeta_n]$. As f is a monic polynomial in ζ_n, we can assume the same for $\omega_1, \ldots, \omega_r$. Then, as $|\omega_i| \geq 1$, we have necessarily $|\omega_i| = 1$ for all i, since $|f| = 1$. So the ω_i are Weierstraß polynomials.

It remains to show that the ω_i, being prime in $T_{n-1}[\zeta_n]$, are prime in T_n as well. To verify this, it is enough to show for any Weierstraß polynomial $\omega \in T_{n-1}[\zeta_n]$ of some degree s that the canonical morphism

$$T_{n-1}[\zeta_n]/(\omega) \longrightarrow T_n/(\omega)$$

is an isomorphism. However this is clear, since both sides are free T_{n-1}-modules generated by the residue classes of $\zeta_n^0, \ldots, \zeta_n^{s-1}$, the left-hand side by Euclid's division and the right-hand side by Weierstraß' division. So T_n is factorial.

To see that this implies T_n being normal, consider an integral equation

$$\left(\frac{f}{g}\right)^r + a_1 \left(\frac{f}{g}\right)^{r-1} + \ldots + a_r = 0$$

for some fraction $\frac{f}{g} \in Q(T_n)$ of elements $f, g \in T_n$ and coefficients $a_i \in T_n$. Using the fact that T_n is factorial, we may assume that the gcd of f and g is 1. But then, since the equation

$$f^r + a_1 f^{r-1} g + \ldots + a_r g^r = 0$$

shows that any prime divisor of g must also divide f, it follows that g is a unit and, hence, that $\frac{f}{g} \in T_n$.   □

Proposition 16. *T_n is Jacobson, i.e. for any ideal $\mathfrak{a} \subset T_n$ its nilradical* rad $\mathfrak{a}$ *equals the intersection of all maximal ideals* $\mathfrak{m} \in$ Max T_n *containing $\mathfrak{a}$.*

Proof. One knows from commutative algebra that the nilradical rad $\mathfrak{a}$ of any ideal $\mathfrak{a} \subset T_n$ equals the intersection of all prime ideals in T_n containing $\mathfrak{a}$. So we have only to show that any prime ideal $\mathfrak{p} \subset T_n$ is an intersection of maximal ideals.

First, let us consider the case where $\mathfrak{p} = 0$. Let $f \in \bigcap_{\mathfrak{m} \in \text{Max } T_n} \mathfrak{m}$. Then, by Corollary 13, f vanishes at all points $x \in \mathbb{B}^n(\overline{K})$, and it follows $f = 0$ by Proposition 5.

Next assume that $\mathfrak{p}$ is not necessarily the zero ideal. Then, using Noether normalization as in Corollary 11, there is a finite monomorphism $T_d \hookrightarrow T_n/\mathfrak{p}$ for some $d \in \mathbb{N}$. One knows from commutative algebra that over each maximal ideal $\mathfrak{m} \subset T_d$ there is a maximal ideal $\mathfrak{m}' \subset T_n/\mathfrak{p}$ with $\mathfrak{m}' \cap T_d = \mathfrak{m}$. Thus, if $\mathfrak{q} \subset T_n/\mathfrak{p}$ is the intersection of all maximal ideals in $T_n/\mathfrak{p}$, we know $\mathfrak{q} \cap T_d = 0$. Now, if $\mathfrak{q}$ is non-zero, we can choose a non-zero element $f \in \mathfrak{q}$. Let

$$f^r + a_1 f^{r-1} + \ldots + a_r = 0$$

be an integral equation of minimal degree for f over T_d. Then $a_r \neq 0$. On the other hand, we see that

$$a_r = -f^r - a_1 f^{r-1} - \ldots - a_{r-1} f \in \mathfrak{q} \cap T_d = 0$$

and, hence, is trivial. Thus, we must have $f = 0$ and therefore $\mathfrak{q} = 0$, which concludes our proof. $\qquad \square$

Proposition 17. *Every maximal ideal* $\mathfrak{m} \subset T_n$ *is of height n and can be generated by n elements. In particular, the Krull dimension of T_n is n.*

Proof. Assume $n \geq 1$ and let $\mathfrak{m} \subset T_n$ be a maximal ideal. We claim that $\mathfrak{n} = \mathfrak{m} \cap T_{n-1}$ is a maximal ideal in T_{n-1}. Indeed, look at the derived injections $K \hookrightarrow T_{n-1}/\mathfrak{n} \hookrightarrow T_n/\mathfrak{m}$. Since $T_n/\mathfrak{m}$ is finite over K by Corollary 12, the same is true for $T_{n-1}/\mathfrak{n}$. Then it follows from commutative algebra that $T_{n-1}/\mathfrak{n}$ is a field. Therefore $\mathfrak{n}$ is a maximal ideal in T_{n-1}.

As the field $T_{n-1}/\mathfrak{n}$ is finite over K, it carries a unique complete absolute value $|\cdot|$ extending the one of K, just as is the case for $T_n/\mathfrak{m}$. Furthermore, we see from Corollary 13 and its proof that the projections $\varphi': T_{n-1} \longrightarrow T_{n-1}/\mathfrak{n}$ and $\varphi: T_n \longrightarrow T_n/\mathfrak{m}$ are contractive in the sense that $|\varphi'(a')| \leq |a'|$ and $|\varphi(a)| \leq |a|$ for all $a' \in T_{n-1}$ and $a \in T_n$. Therefore we can look at the following canonical commutative diagram of continuous K-algebra homomorphisms

$$
\begin{array}{ccc}
T_{n-1}\langle \zeta_n \rangle & =\!=\!=\!= & T_n \\
{\scriptstyle \varphi'} \downarrow & & \downarrow {\scriptstyle \varphi} \\
(T_{n-1}/\mathfrak{n})\langle \zeta_n \rangle & \xrightarrow{\ \pi\ } & T_n/\mathfrak{m}
\end{array}
\qquad (*)
$$

where π maps ζ_n onto its residue class in $T_n/\mathfrak{m}$. We claim that φ' is surjective and that its kernel is the ideal $\mathfrak{n} T_n$ generated by $\mathfrak{n}$ in $T_{n-1}\langle \zeta_n \rangle = T_n$. Then, since $(T_{n-1}/\mathfrak{n})\langle \zeta_n \rangle$ is an integral domain, it follows that $\mathfrak{n} T_n$ is a prime ideal in T_n.

For the surjectivity of φ' we need to know that any zero sequence in $T_{n-1}/\mathfrak{n}$ can be lifted to a zero sequence in T_{n-1}. This assertion can be derived from general arguments on affinoid K-algebras in Sect. 3.1, or it can be obtained by a direct argument as follows. Since $T_{n-1}/\mathfrak{n}$ is a finite-dimensional K-vector space, we can choose a K-basis $u_1, \ldots, u_r$ on it. Defining a norm on $T_{n-1}/\mathfrak{n}$ by setting

$$\left| \sum_{i=1}^{r} c_i u_i \right| = \max_{i=1,\ldots,r} |c_i|, \qquad c_i \in K,$$

we conclude from Theorem 1 of Appendix A that this norm is equivalent to the given absolute value on $T_{n-1}/\mathfrak{n}$. But then, choosing representatives of $u_1, \ldots, u_r$ in T_{n-1}, it is easy to lift zero sequences in $T_{n-1}/\mathfrak{n}$ to zero sequences in T_{n-1}. Therefore the homomorphism φ' of the above diagram will be surjective.

It remains to look at the kernel of φ'. It consists of all restricted power series $f = \sum_{\nu=0}^{\infty} f_\nu \zeta_n^\nu \in T_{n-1}\langle\zeta_n\rangle$ with coefficients $f_\nu \in \mathfrak{n}$. In particular, we see that the inclusion $\mathfrak{n} \cdot T_{n-1}\langle\zeta_n\rangle \subset \ker \varphi'$ is trivial. On the other hand, consider a restricted power series $f = \sum_{\nu=0}^{\infty} f_\nu \zeta_n^\nu \in T_{n-1}\langle\zeta_n\rangle$ such that $f_\nu \in \mathfrak{n}$ for all ν. We know from Proposition 14 that the ideal $\mathfrak{n} \subset T_{n-1}$ is finitely generated, but we need the stronger result 2.3/7 implying that there are generators $a_1, \ldots, a_r$ of $\mathfrak{n}$ with $|a_i| = 1$ such that for each $\nu \in \mathbb{N}$ there are elements $f_{\nu 1}, \ldots, f_{\nu r} \in T_{n-1}$ satisfying

$$|f_{\nu i}| \leq |f_\nu|, \qquad f_\nu = \sum_{i=1}^{r} f_{\nu i} a_i.$$

Then, for fixed i, the elements $f_{\nu i}$, $\nu \in \mathbb{N}$, form a zero sequence in T_{n-1}, and we see that

$$f = \sum_{\nu=0}^{\infty} f_\nu \zeta_n^\nu = \sum_{\nu=0}^{\infty} \left(\sum_{i=1}^{r} f_{\nu i} a_i\right) \zeta_n^\nu = \sum_{i=1}^{r} \left(\sum_{\nu=0}^{\infty} f_{\nu i} \zeta_n^\nu\right) a_i \in \mathfrak{n} \cdot T_{n-1}\langle\zeta_n\rangle.$$

Therefore $\ker \varphi' = \mathfrak{n} \cdot T_{n-1}\langle\zeta_n\rangle$, as desired.

Now it is easy to see that every maximal ideal $\mathfrak{m} \subset T_n$ is generated by n elements. Proceeding by induction on n, we may assume that $\mathfrak{n} = \mathfrak{m} \cap T_{n-1}$ is generated by $n - 1$ elements $a_1, \ldots, a_{n-1} \in T_{n-1}$ and, hence, that the same is true for the kernel of the surjection φ' in the above diagram $(*)$. Since $(T_{n-1}/\mathfrak{n})\langle\zeta_n\rangle$ is a principal ideal domain by Corollary 10, the kernel of the surjection π is generated by a single element. Lifting the latter to an element $a_n \in T_{n-1}\langle\zeta_n\rangle$, it follows that $\mathfrak{m}$ is generated by the n elements $a_1, \ldots, a_n$.

Next, to show ht $\mathfrak{m} = n$, we look at the strictly ascending chain $T_0 \subsetneq \ldots \subsetneq T_n$ and set $\mathfrak{n}_i = \mathfrak{m} \cap T_i$ for $i = 0, \ldots, n$. Then, since $\mathfrak{n}_{i-1} = \mathfrak{n}_i \cap T_{i-1}$ for $i = n, \ldots, 1$, we can conclude inductively as before that $\mathfrak{n}_i$ is a maximal ideal in T_i for all i. We want to show that

$$0 \subsetneq \mathfrak{n}_1 T_n \subsetneq \ldots \subsetneq \mathfrak{n}_{n-1} T_n \subsetneq \mathfrak{n}_n T_n = \mathfrak{m} \qquad (**)$$

is a strictly ascending chain of prime ideals in T_n. To do this, we look at diagrams of type $(*)$ and construct for each $i = 1, \ldots, n$ a canonical commutative diagram of continuous K-algebra homomorphisms by just adding variables:

$$
\begin{array}{ccc}
T_{i-1}\langle\zeta_i\rangle\langle\zeta_{i+1}, \ldots, \zeta_n\rangle & =\!=\!=\!=\!= & T_i\langle\zeta_{i+1}, \ldots, \zeta_n\rangle \\
\varphi_i' \downarrow & & \downarrow \varphi_i \\
(T_{i-1}/\mathfrak{n}_{i-1})\langle\zeta_i\rangle\langle\zeta_{i+1}, \ldots, \zeta_n\rangle & \xrightarrow{\pi_i} & (T_i/\mathfrak{n}_i)\langle\zeta_{i+1}, \ldots, \zeta_n\rangle
\end{array}
$$

As before, the maps φ_i' and φ_i are surjections with kernels generated by $\mathfrak{n}_{i-1}$, respectively $\mathfrak{n}_i$, while π_i is a surjection having a certain non-trivial kernel. From this we conclude that

$$\mathfrak{n}_{i-1} T_n = \ker \varphi_i' \subset \ker \varphi_i = \mathfrak{n}_i T_n$$

is a strict inclusion of prime ideals in T_n and, thus, that $(**)$ is, indeed, a strictly ascending chain of prime ideals. In particular, we see that $\operatorname{ht} \mathfrak{m} \geq n$. On the other hand, $\mathfrak{m}$ can be generated by n elements, as we have seen, and this implies $\operatorname{ht} \mathfrak{m} \leq n$ by Krull's Dimension Theorem; cf. [Bo], 2.4/6. Thus, we get $\operatorname{ht} \mathfrak{m} = n$, as desired.

$\square$

2.3   Ideals in Tate Algebras

We know already from 2.2/14 that all ideals $\mathfrak{a} \subset T_n$ are finitely generated. Considering such an ideal $\mathfrak{a} = (a_1, \dots, a_r)$, say with generators a_i of absolute value $|a_i| = 1$, we can ask if any $f \in \mathfrak{a}$ admits a representation $f = \sum_{i=1}^r f_i a_i$ with elements $f_i \in T_n$ satisfying $|f_i| \leq |f|$. If this is the case, we can easily deduce that $\mathfrak{a}$ is complete under the Gauß norm of T_n and, hence, that $\mathfrak{a}$ is closed in T_n. To establish these and other assertions, we will use a technique involving normed vector spaces.

Definition 1. *Let R be a ring. A* ring norm *on R is a map $|\cdot|: R \longrightarrow \mathbb{R}_{\geq 0}$ satisfying*

  (i) $|a| = 0 \Longleftrightarrow a = 0$,
 (ii) $|ab| \leq |a||b|$,
(iii) $|a + b| \leq \max\{|a|, |b|\}$,
 (iv) $|1| \leq 1$.

The norm is called multiplicative *if instead of* (ii) *we have*

(ii') $|ab| = |a||b|$.

We claim that, instead of condition (iv), we actually have $|1| = 1$ if R is non-zero. In fact, we have $|1| \leq |1|^2$ due to (ii) and, thus, $|1| = |1|^2$, since $|1| \leq 1$ by (iv). This implies $|1| = 1$ or $|1| = 0$. As $1 \neq 0$ and, hence, $|1| \neq 0$ by (i) if R is non-zero, we can conclude that $|1| = 1$ in this case.

Definition 2. *Let R be a ring with a multiplicative ring norm $|\cdot|$ such that $|a| \leq 1$ for all $a \in R$.*

  (i) *R is called a B-ring if*

$$\{a \in R ; |a| = 1\} \subset R^*.$$

(ii) *R is called* bald *if*

$$\sup\{|a|\;;\,a \in R \text{ with } |a| < 1\} < 1.$$

We want to show the following assertion:

Proposition 3. *Let K be a field with a valuation and R its valuation ring. Then the smallest subring $R' \subset R$ containing a given zero sequence $a_0, a_1, \ldots \in R$ is bald.*

Proof. The smallest subring $S \subset R$ equals either $\mathbb{Z}/p\mathbb{Z}$ for some prime p, or $\mathbb{Z}$. It is bald, since any valuation on the finite field $\mathbb{Z}/p\mathbb{Z}$ is trivial and since the ideal $\{a \in \mathbb{Z}\;;\,|a| < 1\} \subset \mathbb{Z}$ is principal. If there is an $\varepsilon \in \mathbb{R}$ such that $|a_n| \leq \varepsilon < 1$ for all $n \in \mathbb{N}$, we see for trivial reasons that $S[a_0, a_1, \ldots]$ is bald. Thus, it is enough to show that, for a bald subring $S \subset R$ and an element $a \in R$ of value $|a| = 1$, the ring $S[a]$ is bald. To do this, we may localize S by all elements of value 1 and thereby assume that S is a B-ring. Then S contains a unique maximal ideal m, and $\widetilde{S} = S/\mathfrak{m}$ is a field. If the reduction $\widetilde{a} \in k$ is transcendental over $\widetilde{S}$, it follows that $S[a]$ is bald for trivial reasons. Indeed, for any polynomial $p = \sum_{i=0}^{r} c_i \zeta^i \in S[\zeta]$, we have $|p(a)| < 1$ if and only if $\sum_{i=0}^{r} \widetilde{c}_i \widetilde{a}^i = 0$, i.e. if and only if $\widetilde{c}_i = 0$ for all i. Thus, $|p(a)| < 1$ implies

$$|p(a)| \leq \sup\{|c|\;;\,c \in S, |c| < 1\} < 1$$

and we are done.

It remains to consider the case where $\widetilde{a}$ is algebraic over $\widetilde{S}$. Choose a polynomial $g = \zeta^n + c_1 \zeta^{n-1} + \ldots + c_n \in S[\zeta]$ of minimal degree such that its reduction $\widetilde{g}$ annihilates $\widetilde{a}$ or, in equivalent terms, such that $|g(a)| < 1$. Let $\varepsilon < 1$ be the supremum of $|g(a)|$ and of all values $|c|$ for $c \in S$ and $|c| < 1$. Now consider a polynomial $f \in S[\zeta]$ with $|f(a)| < 1$; we want to show $|f(a)| \leq \varepsilon$. Using Euclid's division, we get a decomposition $f = qg + r$ with $q, r \in S[\zeta]$ and $\deg r < n = \deg g$. Since $|g(a)| \leq \varepsilon$, we may assume $f = r$. If all coefficients of r have value < 1, this value must be $\leq \varepsilon$ and we are done. On the other hand, if one of the coefficients of r has value 1, the reduction $\widetilde{r}$ of r is non-trivial. But then we have $\widetilde{r}(\widetilde{a}) = 0$, and this contradicts the definition of g, since $\deg \widetilde{r} < \deg \widetilde{g}$ and $\widetilde{g}$, annihilating $\widetilde{a}$, was chosen of minimal degree. Thus, $|f(a)| < 1$ implies $|f(a)| \leq \varepsilon$, and we are done.   □

Given a bald subring $R' \subset R$, for example as constructed in the situation of Proposition 3, we can localize R' by all elements of value 1 and thereby obtain a B-ring $R'' \subset R$ that contains R' and is bald. Furthermore, assuming R to be complete, we may even pass to the completion of R''. As the completion of a B-ring yields a B-ring again, we see that the smallest complete B-ring in R containing a given bald subring of R, will be bald again.

Next we want to look at vector spaces and norms on them. As a prototype, we can consider a Tate algebra T_n with its Gauß norm and forget about multiplication. In the following, let K be a field with a complete non-Archimedean valuation.

Definition 4. *Let V be a K-vector space. A* norm *on V is a map $|\cdot|: V \longrightarrow \mathbb{R}_{\geq 0}$, such that*

  (i) $|x| = 0 \Longleftrightarrow x = 0$,
 (ii) $|x + y| \leq \max\{|x|,|y|\}$,
(iii) $|cx| = |c||x|$ *for $c \in K$ and $x \in V$.*

Definition 5. *Let V be a complete normed K-vector space. A system $(x_\nu)_{\nu \in N}$ of elements in V, where N is finite or at most countable, is called a* (topological) orthonormal basis *of V if the following hold:*

  (i) $|x_\nu| = 1$ *for all $\nu \in N$.*
 (ii) *Each $x \in V$ can be written as a convergent series $x = \sum_{\nu \in N} c_\nu x_\nu$ with coefficients $c_\nu \in K$.*
(iii) *For each equation $x = \sum_{\nu \in N} c_\nu x_\nu$ as in (ii) we have $|x| = \max_{\nu \in N} |c_\nu|$. In particular, the coefficients c_ν in (ii) are unique.*

For example, the monomials $\zeta^\nu \in T_n$ form an orthonormal basis if we consider $T_n = K\langle \zeta \rangle$ as a normed K-vector space. For any normed K-vector space V, we will use the notations

$$V^\circ = \{x \in V \,;\, |x| \leq 1\}$$

for its "unit ball" and

$$\tilde{V} = V^\circ / \{x \in V \,;\, |x| < 1\}$$

for its reduction.

Theorem 6. *Let K be a field with a complete valuation and V a complete normed K-vector space with an orthonormal basis $(x_\nu)_{\nu \in N}$. Write R for the valuation ring of K, and consider a system of elements*

$$y_\mu = \sum_{\nu \in N} c_{\mu\nu} x_\nu \in V^\circ, \qquad \mu \in M,$$

where the smallest subring of R containing all coefficients $c_{\mu\nu}$ is bald. Then, if the residue classes $\tilde{y}_\mu \in \tilde{V}$ form a k-basis of $\tilde{V}$, the elements y_μ form an orthonormal basis of V.

Proof. The systems $(\tilde{x}_\nu)_{\nu\in N}$ and $(\tilde{y}_\mu)_{\mu\in M}$ form a k-basis of $\tilde{V}$. So M and N have the same cardinality, and M is at most countable. In particular, $(y_\mu)_{\mu\in M}$ is an orthonormal basis of a subspace $V' \subset V$. Now let S be the smallest complete B-ring in R containing all coefficients $c_{\mu\nu}$. Then S is bald by our assumption; let $\varepsilon = \sup\{|a| \,; \, a \in S, |a| < 1\}$. Setting

$$V'_S = \widehat{\sum_{\mu\in M}} Sy_\mu, \qquad V_S = \widehat{\sum_{\nu\in N}} Sx_\nu,$$

where $\widehat{\sum}$ means the completion of the usual sum, we have $V'_S \subset V_S$, and we claim that, in fact, $V'_S = V_S$. To verify this, let us first look at reductions. If $\mathfrak{m} \subset S$ denotes the unique maximal ideal, we set

$$\tilde{S} = S/\mathfrak{m}, \qquad V'_{\tilde{S}} = V'_S/\mathfrak{m}V'_S, \qquad V_{\tilde{S}} = V_S/\mathfrak{m}V_S.$$

Then $\tilde{S}$ is a subfield of the residue field k of R, and we have

$$\tilde{V}' = V'_{\tilde{S}} \otimes_{\tilde{S}} k, \qquad \tilde{V} = V_{\tilde{S}} \otimes_{\tilde{S}} k.$$

From $\tilde{V}' = \tilde{V}$ and $V'_{\tilde{S}} \subset V_{\tilde{S}}$ we get $V'_{\tilde{S}} = V_{\tilde{S}}$. The latter implies that, for any x_ν, there is an element $z_\nu \in V'_S$ satisfying $|x_\nu - z_\nu| \leq \varepsilon$. Then, more generally, for any $x \in V_S$, there is an element $z \in V'_S$ with $|z| = |x|$ and $|x - z| \leq \varepsilon|x|$. But then, as V'_S and V_S are complete, we get $V'_S = V_S$ by iteration. $\qquad\square$

Now we want to apply Theorem 6 to Tate algebras.

Corollary 7. *Let $\mathfrak{a}$ be an ideal in T_n. Then there are generators $a_1,\ldots,a_r$ of $\mathfrak{a}$ satisfying the following conditions:*

(i) *$|a_i| = 1$ for all i.*

(ii) *For each $f \in \mathfrak{a}$, there are elements $f_1,\ldots,f_r \in T_n$ such that*

$$f = \sum_{i=1}^{r} f_i a_i, \qquad |f_i| \leq |f|.$$

Proof. Let $\tilde{\mathfrak{a}}$ be the reduction of $\mathfrak{a}$, i.e. the image of $\mathfrak{a} \cap R\langle\zeta\rangle$ under the reduction map $R\langle\zeta\rangle \longrightarrow k[\zeta]$ where R is the valuation ring of K. Then $\tilde{\mathfrak{a}}$ is an ideal in the Noetherian ring $k[\zeta]$ and, hence, finitely generated, say by the residue classes $\tilde{a}_1,\ldots,\tilde{a}_r$ of some elements $a_1,\ldots,a_r \in \mathfrak{a}$ having norm equal to 1. As the elements $\zeta^\nu\tilde{a}_i, \nu \in \mathbb{N}^n, i = 1,\ldots,r$, generate $\tilde{\mathfrak{a}}$ as a k-vector space, we can find a system $(y_\mu)_{\mu\in M'}$ of elements of type $\zeta^\nu a_i \in \mathfrak{a}$ such that its residue classes form a k-basis of $\tilde{\mathfrak{a}}$. Adding monomials of type $\zeta^\nu, \nu \in \mathbb{N}^n$, we can enlarge the system to a system $(y_\mu)_{\mu\in M}$ such that its residue classes form a k-basis of $k[\zeta]$.

On the other hand, let us consider the system $(\zeta^\nu)_{\nu \in \mathbb{N}^n}$ of all monomials in T_n; it is an orthonormal basis of T_n and its reduction forms a k-basis of $k[\zeta]$. Now apply Proposition 3 and Theorem 6. To write the elements y_μ as (converging) linear combinations of the ζ^ν, we need only the coefficients of the series $a_1, \ldots, a_r$. As these form a zero sequence, we see that $(y_\mu)_{\mu \in M}$ is an orthonormal basis, the same being true for $(\zeta^\nu)_{\nu \in \mathbb{N}^n}$.

We want to show that the elements $a_1, \ldots, a_r$ have the required properties. Choose $f \in \mathfrak{a}$. Then, since $(y_\mu)_{\mu \in M}$ is an orthonormal basis of T_n, there is an equation $f = \sum_{\mu \in M} c_\mu y_\mu$ with certain coefficients $c_\mu \in K$ satisfying $|c_\mu| \le |f|$. Writing $f' = \sum_{\mu \in M'} c_\mu y_\mu$, the choice of the elements y_μ, $\mu \in M'$, implies that we can write $f' = \sum_{i=1}^r f_i a_i$ with certain elements $f_i \in T_n$ satisfying $|f_i| \le |f|$. In particular, $f' \in \mathfrak{a}$, and we are done if we can show $f = f'$. To justify the latter equality, we may replace f by

$$f - f' = \sum_{\mu \in M - M'} c_\mu y_\mu \in \mathfrak{a}$$

and thereby assume $c_\mu = 0$ for $\mu \in M'$. Then, if $f \ne 0$, there is an index $\mu \in M - M'$ with $c_\mu \ne 0$. Assuming $|f| = 1$, we would get a non-trivial equation $\tilde{f} = \sum_{\mu \in M - M'} \tilde{c}_\mu \tilde{y}_\mu$ for the element $\tilde{f} \in \tilde{\mathfrak{a}}$, which however, contradicts the construction of the elements $\tilde{y}_\mu$.   □

The proof shows more precisely that elements $a_1, \ldots, a_r \in \mathfrak{a}$ with $|a_i| \le 1$ satisfy the assertion of Corollary 7 as soon as the residue classes $\tilde{a}_1, \ldots, \tilde{a}_r$ generate the ideal $\tilde{\mathfrak{a}} \subset k[\zeta]$. Furthermore, the system $(y_\mu)_{\mu \in M'}$ is seen to be an orthonormal basis of $\mathfrak{a}$. The reason is that $(y_\mu)_{\mu \in M'}$ is part of an orthonormal basis of T_n and, as we have seen in the proof above, any convergent series $\sum_{\mu \in M'} c_\mu y_\mu$ with coefficients $c_\mu \in K$ gives rise to an element of $\mathfrak{a}$.

Corollary 8. *Each ideal $\mathfrak{a} \subset T_n$ is complete and, hence, closed in T_n.*

Proof. Choose generators $a_1, \ldots, a_r$ of $\mathfrak{a}$ as in Corollary 7. If $f = \sum_{\lambda=0}^\infty f_\lambda$ is convergent in T_n with elements $f_\lambda \in \mathfrak{a}$, there are equations $f_\lambda = \sum_{i=1}^r f_{\lambda i} a_i$ with coefficients $f_{\lambda i} \in T_n$ satisfying $|f_{\lambda i}| \le |f_\lambda|$. But then $f = \sum_{i=1}^r (\sum_{\lambda=0}^\infty f_{\lambda i}) a_i$ belongs to $\mathfrak{a}$ and we are done.   □

Corollary 9. *Each ideal $\mathfrak{a} \subset T_n$ is strictly closed, i.e. for each $f \in T_n$ there is an element $a_0 \in \mathfrak{a}$ such that*

$$|f - a_0| = \inf_{a \in \mathfrak{a}} |f - a|.$$

Proof. Going back to the proof of Corollary 7, we use the orthonormal basis $(y_\mu)_{\mu \in M}$ of T_n and write $f = \sum_{\mu \in M} c_\mu y_\mu$ with coefficients $c_\mu \in K$. As $M' \subset M$

is a subset such that $(y_\mu)_{\mu \in M'}$ is an orthonormal basis of $\mathfrak{a}$, the assertion of the corollary holds for $a_0 = \sum_{\mu \in M'} c_\mu y_\mu$.   □

For later use, we add a version of Corollary 7 that applies to modules:

Corollary 10. *Let $N \subset T_n^s$ be a T_n-submodule of a finite direct sum of T_n with itself, and consider on T_n^s the maximum norm derived from the Gauß norm of T_n. Then there are generators $x_1, \ldots, x_r$ of N as a T_n-module, satisfying the following conditions*:

(i) $|x_i| = 1$ *for all i.*
(ii) *For each $x \in N$, there are elements $f_1, \ldots, f_r \in T_n$ such that*

$$x = \sum_{i=1}^{r} f_i x_i, \qquad |f_i| \le |x|.$$

Proof. Proceeding as in the proof of Corollary 7, we consider the reduction map $(R\langle \zeta \rangle)^s \longrightarrow (k[\zeta])^s$ where R is the valuation ring of K. Writing $\tilde{N}$ for the image of $N \cap (R\langle \zeta \rangle)^s$, we see that $\tilde{N}$ is a $k[\zeta]$-submodule of $(k[\zeta])^s$ and, hence, finitely generated, since $k[\zeta]$ is Noetherian. Thus, we can choose elements $x_1, \ldots, x_r \in N$ of norm 1 such that their residue classes $\tilde{x}_1, \ldots, \tilde{x}_r$ generate $\tilde{N}$ as $k[\zeta]$-module. Consequently, there exists a system $(y_\mu)_{\mu \in M'}$ of elements of type $\zeta^\nu x_i \in N$ such that their residue classes form a k-basis of $\tilde{N}$. Let $e_1, \ldots, e_s$ be the "unit vectors" in T_n^s. Then it is possible to enlarge the system $(y_\mu)_{\mu \in M'}$ to a system $(y_\mu)_{\mu \in M}$, by adding elements of type $\zeta^\nu e_j$ in such a way that the residue classes of the y_μ, $\mu \in M$, form a k-basis of $k[\zeta]^s$. On the other hand, we have the canonical system $\mathfrak{Z} = (\zeta^\nu e_j)_{\nu \in \mathbb{N}^n, j=1,\ldots,s}$, which is an orthonormal basis of T_n^s and which induces a k-basis of $(k[\zeta])^s$.

In order to represent the elements x_i, $i = 1, \ldots, s$, in terms of the orthonormal basis $\mathfrak{Z}$, using converging linear combinations with coefficients in K, we need finitely many zero sequences in R, and the smallest subring $R' \subset R$ containing all these coefficients is bald by Proposition 3. Since the elements y_μ, $\mu \in M'$, are obtained from $x_1, \ldots, x_r$ by multiplication with certain monomials ζ^ν, $\nu \in \mathbb{N}^n$, we see that $y_\mu \in \hat{\sum}_{z \in \mathfrak{Z}} R'z$ for all $\mu \in M$. Thus, by Theorem 6, $(y_\mu)_{\mu \in M}$ is an orthonormal basis of T_n^s, and it follows as in the proof of Corollary 7 that $(y_\mu)_{\mu \in M'}$ is an orthonormal basis of N. Hence, $x_1, \ldots, x_r$ are as required.   □

Chapter 3
Affinoid Algebras and Their Associated Spaces

3.1 Affinoid Algebras

So far we have viewed the elements of T_n as functions $\mathbb{B}^n(\overline{K}) \longrightarrow \overline{K}$. If $\mathfrak{a} \subset T_n$ is an ideal, we can look at its zero set

$$V(\mathfrak{a}) = \{x \in \mathbb{B}^n(\overline{K}) \, ; \, f(x) = 0 \text{ for all } f \in \mathfrak{a}\}$$

and restrict functions on $\mathbb{B}^n(\overline{K})$ to $V(\mathfrak{a})$. Thereby we get a homomorphism vanishing on $\mathfrak{a}$ from T_n to the K-algebra of all maps $V(\mathfrak{a}) \longrightarrow \overline{K}$. Thus, we may interpret the quotient $A = T_n/\mathfrak{a}$ as an algebra of "functions" on $V(\mathfrak{a})$. However note that, as we will conclude later from the fact that T_n and, hence, A are Jacobson, an element $f \in A$ induces the zero function on $V(\mathfrak{a})$ if and only if f is nilpotent in A. The purpose of the present section is to study algebras of type $A = T_n/\mathfrak{a}$, which we call affinoid K-algebras.

Definition 1. *A K-algebra A is called an affinoid K-algebra if there is an epimorphism of K-algebras $\alpha\colon T_n \longrightarrow A$ for some $n \in \mathbb{N}$.*

We can consider the affinoid K-algebras as a category, together with K-algebra homomorphisms between them as morphisms. Let us mention right away that this category admits amalgamated sums:

Proposition 2. *Write $\mathfrak{A}$ for the category of affinoid K-algebras and consider two morphisms $R \longrightarrow A_1$ and $R \longrightarrow A_2$ in $\mathfrak{A}$ equipping A_1 and A_2 with the structure of R-algebras. Then there exists an R-object T together with R-morphisms $\sigma_1\colon A_1 \longrightarrow T$ and $\sigma_2\colon A_2 \longrightarrow T$ in $\mathfrak{A}$ fulfilling the universal property of an amalgamated sum:*
Given R-morphisms $\varphi_1\colon A_1 \longrightarrow D$ and $\varphi_2\colon A_2 \longrightarrow D$ in $\mathfrak{A}$, there exists a unique R-morphism $\varphi\colon T \longrightarrow D$ in $\mathfrak{A}$ such that the diagram

S. Bosch, *Lectures on Formal and Rigid Geometry*, Lecture Notes
in Mathematics 2105, DOI 10.1007/978-3-319-04417-0_3,
© Springer International Publishing Switzerland 2014

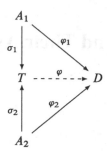

is commutative.

The R-algebra T, which is uniquely determined up to canonical isomorphism, is given by the completed tensor product $A_1 \hat{\otimes}_R A_2$ *of A_1 and A_2 over R; see the Appendix B and its Theorem 6 for details on such tensor products.*

A full discussion of completed tensor products would not be possible at this early stage since it requires residue norms on affinoid K-algebras as well as the continuity of homomorphisms between them, results to be proved only later in this section. Moreover, it is appropriate to consider completed tensor products within a more general setting, the one of normed modules, as is done in Appendix B.

We continue on a more elementary level by looking at some immediate consequences of the results 2.2/11, 2.2/14, and 2.2/16:

Proposition 3. *Let A be an affinoid K-algebra. Then:*

(i) *A is Noetherian.*
(ii) *A is Jacobson.*
(iii) *A satisfies Noether normalization, i.e. there exists a finite monomorphism $T_d \hookrightarrow A$ for some $d \in \mathbb{N}$.*

Proposition 4. *Let A be an affinoid K-algebra and $\mathfrak{q} \subset A$ an ideal whose nilradical is a maximal ideal in A. Then $A/\mathfrak{q}$ is of finite vector space dimension over K.*

Proof. Let $\mathfrak{m} = \operatorname{rad}\mathfrak{q}$. Applying Noether normalization, there is a finite monomorphism $T_d \hookrightarrow A/\mathfrak{q}$ for some $d \in \mathbb{N}$. However, we must have $d = 0$, since dividing out nilpotent elements yields a finite monomorphism $T_d \hookrightarrow A/\mathfrak{m}$. As $A/\mathfrak{m}$ is a field, the same must be true for T_d.   □

Affinoid K-algebras can easily be endowed with a topology (which is unique, as we will see later). Just note that for any epimorphism $\alpha \colon T_n \longrightarrow A$, the Gauß norm $|\cdot|$ of T_n induces a *residue norm* $|\cdot|_\alpha$ on A given by

$$|\alpha(f)|_\alpha = \inf_{a \in \ker \alpha} |f - a|.$$

We can say that $|\overline{f}|_\alpha$ for some $\overline{f} \in A$ is the infimum of all values $|f|$ with $f \in T_n$ varying over all inverse images of $\overline{f}$.

Proposition 5. *For an ideal* $\mathfrak{a} \subset T_n$, *view the quotient* $A = T_n/\mathfrak{a}$ *as an affinoid K-algebra via the projection map* $\alpha : T_n \longrightarrow T_n/\mathfrak{a}$. *The map* $|\cdot|_\alpha : T_n/\mathfrak{a} \longrightarrow \mathbb{R}_{\geq 0}$ *satisfies the following conditions*:

(i) $|\cdot|_\alpha$ *is a K-algebra norm, i.e. a ring norm and a K-vector space norm, and it induces the quotient topology of* T_n *on* $T_n/\mathfrak{a}$. *Furthermore,* $\alpha : T_n \longrightarrow T_n/\mathfrak{a}$ *is continuous and open.*

(ii) $T_n/\mathfrak{a}$ *is complete under* $|\cdot|_\alpha$.

(iii) *For any* $\overline{f} \in T_n/\mathfrak{a}$, *there is an inverse image* $f \in T_n$ *such that* $|\overline{f}|_\alpha = |f|$. *In particular, for any* $\overline{f} \in T_n/\mathfrak{a}$, *there is an element* $c \in K$ *with* $|\overline{f}|_\alpha = |c|$.

Proof. That $|\cdot|_\alpha$ is a K-algebra norm is easily verified; note that $|\overline{f}|_\alpha = 0$ implies $\overline{f} = 0$, since $\mathfrak{a}$ is closed in T_n by 2.3/8. It follows more generally from 2.3/9 that any $\overline{f} \in T_n/\mathfrak{a}$ admits an inverse image $f \in T_n$ with $|f| = |\overline{f}|_\alpha$. From this we see immediately that α maps an ε-neighborhood of $0 \in T_n$ onto an ε-neighborhood $0 \in T_n/\mathfrak{a}$ and, thus, is open. As it is continuous anyway, it induces the quotient topology on $T_n/\mathfrak{a}$. Finally, as we can lift Cauchy sequences in $T_n/\mathfrak{a}$ to Cauchy sequences in T_n, we see that $T_n/\mathfrak{a}$ is complete. $\square$

Viewing the elements f of an affinoid K-algebra $T_n/\mathfrak{a}$ as $\overline{K}$-valued functions on the zero set $V(\mathfrak{a}) \subset \mathbb{B}^n(\overline{K})$, we can introduce the supremum $|f|_{\sup}$ of all values that are assumed by f. The latter is finite, as can be seen from 2.2/5. However, to be independent of a special representation of an affinoid K-algebra A as a certain quotient $T_n/\mathfrak{a}$, we prefer to set for elements $f \in A$

$$|f|_{\sup} = \sup_{x \in \mathrm{Max}\,A} |f(x)|.$$

Here Max A is the spectrum of maximal ideals in A and, for any $x \in \mathrm{Max}\,A$, we write $f(x)$ for the residue class of f in A/x. The latter is a finite field extension of K by 2.2/12, and the value $|f(x)|$ is well-defined, since the valuation of K admits a unique extension to A/x. Usually $|\cdot|_{\sup}$ is called the *supremum norm* on A. However, to be more precise, it should be pointed out that, in the general case, $|\cdot|_{\sup}$ will only be a K-algebra *semi*-norm, which means that it satisfies the conditions of a norm, except for the condition that $|f|_{\sup} = 0$ implies $f = 0$. We start by listing some properties of the supremum norm that are more or less trivial.

Proposition 6. *The supremum norm is* power multiplicative, *i.e.* $|f^n|_{\sup} = |f|^n_{\sup}$ *for any element* f *of an affinoid K-algebra.*

Proposition 7. *Let* $\varphi : B \longrightarrow A$ *be a morphism between affinoid K-algebras. Then* $|\varphi(b)|_{\sup} \leq |b|_{\sup}$ *for all* $b \in B$.

Proof. If $\mathfrak{m}$ is a maximal ideal in A, the quotient $A/\mathfrak{m}$ is finite over K by 2.2/12. Thus, writing $\mathfrak{n} = \varphi^{-1}(\mathfrak{m})$ we get finite maps $K \hookrightarrow B/\mathfrak{n} \hookrightarrow A/\mathfrak{m}$ and we see that $\mathfrak{n}$ is a maximal ideal in B. As $|b(\mathfrak{n})| = |\varphi(b)(\mathfrak{m})|$, we are done. $\square$

Proposition 8. *On a Tate algebra T_n, the supremum norm $|\cdot|_{\sup}$ coincides with the Gauß norm $|\cdot|$.*

Proof. It follows from the Maximum Principle 2.2/5 that

$$|f| = \sup\{|f(x)| \,;\, x \in \mathbb{B}^n(\overline{K})\}$$

for any $f \in T_n$. To $x \in \mathbb{B}^n(\overline{K})$ we can always associate the maximal ideal of T_n given by $\mathfrak{m}_x = \{h \in T_n \,;\, h(x) = 0\}$, as we have seen in 2.2/13. Then evaluation at x yields an embedding $T_n/\mathfrak{m}_x \hookrightarrow \overline{K}$, and we see that $f(\mathfrak{m}_x) = f(x)$ and, hence, $|f(\mathfrak{m}_x)| = |f(x)|$. Since $x \longmapsto \mathfrak{m}_x$ defines a surjection $\mathbb{B}^n(\overline{K}) \longrightarrow \operatorname{Max} T_n$ by 2.2/13, we are done. $\square$

Proposition 9. *Let A be an affinoid K-algebra with a residue norm $|\cdot|_\alpha$ corresponding to some K-algebra epimorphism $\alpha\colon T_n \longrightarrow A$. Then $|f|_{\sup} \leq |f|_\alpha$ for all $f \in A$. In particular, $|f|_{\sup}$ is finite.*

Proof. Consider a maximal ideal $\mathfrak{m} \subset A$ and its inverse image $\mathfrak{n} = \alpha^{-1}(\mathfrak{m}) \subset T_n$. Fixing an element $f \in A$, let $g \in T_n$ be an inverse image satisfying $|f|_\alpha = |g|$. Then

$$\left|f(\mathfrak{m})\right| = \left|g(\mathfrak{n})\right| \leq \left|g\right| = \left|f\right|_\alpha$$

and, hence, $|f|_{\sup} \leq |f|_\alpha$. $\square$

Proposition 10. *Let A be an affinoid K-algebra. Then, for $f \in A$, the following are equivalent:*

(i) $|f|_{\sup} = 0$.
(ii) f *is nilpotent.*

Proof. Condition (i) is equivalent to $f \in \bigcap_{\mathfrak{m} \in \operatorname{Max} A} \mathfrak{m}$. As A is Jacobson by Proposition 3, the ideal $\bigcap_{\mathfrak{m} \in \operatorname{Max} A} \mathfrak{m}$ equals the nilradical of A. Thus, (i) is equivalent to (ii). $\square$

Next we want to relate the supremum norm $|\cdot|_{\sup}$ to residue norms $|\cdot|_\alpha$ on affinoid K-algebras A. We need some preparations.

Lemma 11. *For any polynomial*

$$p(\zeta) = \zeta^r + c_1\zeta^{r-1} + \ldots + c_r = \prod_{j=1}^{r}(\zeta - \alpha_j)$$

in $K[\zeta]$ *with zeros* $\alpha_1, \ldots, \alpha_r \in \overline{K}$, *one has*

$$\max_{j=1\ldots r} |\alpha_j| = \max_{i=1\ldots r} |c_i|^{\frac{1}{i}}.$$

Proof. As c_i equals the ith elementary symmetric function of the zeros $\alpha_1, \ldots, \alpha_r$, up to sign, we get

$$|c_i|^{\frac{1}{i}} \leq \max_{j=1\ldots r} |\alpha_j|$$

for $i = 1, \ldots, r$. On the other hand, assume that $|\alpha_j|$ is maximal precisely for $j = 1, \ldots, r'$. Then $|c_{r'}| = |\alpha_1| \ldots |\alpha_{r'}|$ and, hence,

$$|c_{r'}|^{\frac{1}{r'}} = \max_{j=1\ldots r} |\alpha_j|$$

so that we are done. $\square$

If $p = \zeta^r + c_1 \zeta^{r-1} + \ldots + c_r$ is a monic polynomial with coefficients c_i in a normed (or semi-normed) ring A, we call

$$\sigma(p) = \max_{j=1\ldots r} |c_i|^{\frac{1}{i}}$$

the *spectral value* of p. Thus we can say that, in the situation of Lemma 11, the spectral value $\sigma(p)$ equals the maximal value of the zeros of p. The assertion of Lemma 11 is true more generally if the coefficients of p and the zeros of p belong to a normed ring A, whose absolute value is *multiplicative*. Without the latter property we still have $\sigma(p) \leq \max_{j=1\ldots r} |\alpha_j|$ for any polynomial $p = \prod_{j=1}^{r}(\zeta - \alpha_j)$, as can also be seen from the next lemma.

Lemma 12. *Let A be a normed (or semi-normed) ring and let $p, q \in A[\zeta]$ be monic polynomials. Then the spectral value satisfies $\sigma(pq) \leq \max(\sigma(p), \sigma(q))$.*

Proof. Let $p = \sum_{i=0}^{m} a_i \zeta^{m-i}$ and $q = \sum_{j=0}^{n} a_j \zeta^{n-j}$ with $a_0 = b_0 = 1$. Then

$$pq = \sum_{\lambda=0}^{m+n} c_\lambda \zeta^{m+n-\lambda}, \qquad c_\lambda = \sum_{i+j=\lambda} a_i b_j.$$

Now $|a_i| \leq \sigma(p)^i$ for $i = 0, \ldots, m$ and $|b_j| \leq \sigma(q)^j$ for $j = 0, \ldots, n$. Thus,

$$|c_\lambda| \leq \max_{i+j=\lambda} |a_i||b_j| \leq \max_{i+j=\lambda} \sigma(p)^i \sigma(q)^j \leq \max(\sigma(p), \sigma(q))^\lambda$$

for all λ, and we see that $\sigma(pq) \leq \max(\sigma(p), \sigma(q))$. $\square$

Lemma 13. *Let $T_d \hookrightarrow A$ be a finite monomorphism into some K-algebra A. Let $f \in A$ and assume that A, as a T_d-module, is torsion-free.*

(i) *There is a unique monic polynomial $p_f = \zeta^r + a_1\zeta^{r-1} + \ldots + a_r \in T_d[\zeta]$ of minimal degree such that $p_f(f) = 0$. More precisely, p_f generates the kernel of the T_d-homomorphism*

$$T_d[\zeta] \longrightarrow A, \qquad \zeta \longmapsto f.$$

(ii) *Fixing a maximal ideal $x \in \operatorname{Max} T_d$, let $y_1, \ldots, y_s \in \operatorname{Max} A$ be those maximal ideals that restrict to x on T_d. Then*

$$\max_{j=1\ldots s} |f(y_j)| = \max_{i=1\ldots r} |a_i(x)|^{\frac{1}{i}}.$$

(iii) *The supremum norm of f is given by*

$$|f|_{\sup} = \max_{i=1\ldots r} |a_i|_{\sup}^{\frac{1}{i}}.$$

Proof. First note that A/y, for any $y \in \operatorname{Max} A$, is finite over K, due to the fact that A is finite over T_d. Therefore the values $|f(y)|$ and $|f|_{\sup}$ are well-defined for $f \in A$, even without knowing that A is, in fact, an affinoid K-algebra.

Starting with assertion (i), let us write $F = Q(T_d)$ for the field of fractions of T_d and $F(A) = A \otimes_{T_d} F$ for the F-algebra obtained from A. Since A is torsion-free over T_d, there is a commutative diagram of inclusions:

$$
\begin{array}{ccc}
T_d & \longrightarrow & A \\
\downarrow & & \downarrow \\
F & \longrightarrow & F(A)
\end{array}
$$

Now consider the F-homomorphism $F[\zeta] \longrightarrow F(A)$, given by $\zeta \longmapsto f$. Its kernel is generated by a unique monic polynomial $p_f \in F[\zeta]$, and we claim that $p_f \in T_d[\zeta]$. To justify this, observe that there is a monic polynomial $h \in T_d[\zeta]$ satisfying $h(f) = 0$, since A is finite and, hence, integral over T_d. Then p_f divides h in $F[\zeta]$, but also in $T_d[\zeta]$, due to the lemma of Gauß, which we can apply as T_d is factorial by 2.2/15. But then, by a similar argument, p_f must divide any polynomial $h \in T_d[\zeta]$ satisfying $h(f) = 0$. Consequently, p_f generates the kernel of $T_d[\zeta] \longrightarrow A, \zeta \longmapsto f$.

Next, let us look at assertion (ii). The theory of integral ring extensions (or a direct argument) shows that the restriction of maximal ideals yields surjections

$$\operatorname{Max} A \longrightarrow \operatorname{Max} T_d[f] \longrightarrow \operatorname{Max} T_d.$$

Thus, we may replace A by $T_d[f]$ and thereby assume $A = T_d[f]$. Now look at the field $L = T_d/x$, which is finite over K by 2.2/12. Writing $\overline{f}$ for the residue class of f in $A/(x)$ and $\overline{p}_f$ for the residue class of p_f in $L[\zeta]$, we obtain a finite morphism $L \longrightarrow A/(x) = L[\zeta]/(\overline{p}_f)$. Let $\alpha_1, \ldots, \alpha_r$ be the zeros of $\overline{p}_f$ in some algebraic closure of L. Then the kernels of the canonical L-morphisms

$$A/(x) = L[\overline{f}] \longrightarrow L[\alpha_i], \qquad \overline{f} \longmapsto \alpha_i,$$

(which might not be pairwise different) are just the maximal ideals of $A/(x)$ and, thus, coincide with the residue classes of the maximal ideals $y_1, \ldots, y_s \in \mathrm{Max}\, A$ lying over x. Using Lemma 11, we get

$$\max_{j=1\ldots s} |f(y_j)| = \max_{i=1\ldots r} |\alpha_i| = \max_{i=1\ldots r} |a_i|^{\frac{1}{i}}$$

and we are done.

Finally, assertion (iii) is a consequence of (ii). $\qquad\qquad\qquad\qquad\qquad\qquad$ $\square$

We need a slight generalization of Lemma 13 (iii).

Lemma 14. *Let* $\varphi\colon B \longrightarrow A$ *be a finite homomorphism of affinoid K-algebras. Then, for any* $f \in A$, *there is an integral equation*

$$f^r + b_1 f^{r-1} + \ldots + b_r = 0$$

with coefficients $b_j \in B$ *such that* $|f|_{\mathrm{sup}} = \max_{i=1\ldots r} |b_i|_{\mathrm{sup}}^{\frac{1}{i}}$.

Proof. Let us start with the case where A is an integral domain. Using Noether normalization 2.2/11, there is a morphism $T_d \longrightarrow B$ for some $d \in \mathbb{N}$ inducing a finite monomorphism $T_d \hookrightarrow B/\ker\varphi$. Then the resulting morphism $T_d \longrightarrow A$ is a finite monomorphism and, since A is an integral domain, it does not admit T_d-torsion. Applying Lemma 13 (iii), there is an integral equation $f^r + a_1 f^{r-1} + \ldots + a_r = 0$ with coefficients $a_i \in T_d$ satisfying $|f|_{\mathrm{sup}} = \max_{i=1\ldots r} |a_i|_{\mathrm{sup}}^{\frac{1}{i}}$. Replacing each a_i by its image b_i in B we obtain an integral equation $f^r + b_1 f^{r-1} + \ldots + b_r = 0$ of f over B. As $|b_i|_{\mathrm{sup}} \leq |a_i|_{\mathrm{sup}}$ by Proposition 7, we get $|f|_{\mathrm{sup}} \geq \max_{i=1\ldots r} |b_i|_{\mathrm{sup}}^{\frac{1}{i}}$. However, the integral equation of f over B shows that this inequality must, in fact, be an equality. Indeed, there exists an index i such that

$$|f^r|_{\mathrm{sup}} \leq |b_i f^{r-i}|_{\mathrm{sup}} \leq |b_i|_{\mathrm{sup}} |f|_{\mathrm{sup}}^{r-i},$$

and it follows $|f|_{\mathrm{sup}} \leq |b_i|_{\mathrm{sup}}^{\frac{1}{i}}$.

Next we consider the general case where A is not necessarily an integral domain. As A is Noetherian by Proposition 3, it contains only finitely many minimal prime ideals, say $\mathfrak{p}_1, \ldots, \mathfrak{p}_s$, and we can interpret Max A as the union of the sets Max $A/\mathfrak{p}_j$, $j = 1, \ldots, s$. Thus, we have $|f|_{\sup} = \max_{j=1\ldots s} |f_j|_{\sup}$, writing f_j for the residue class of f in $A/\mathfrak{p}_j$.

Now look at the induced maps $B \longrightarrow A \longrightarrow A/\mathfrak{p}_j$. As we have seen at the beginning, there are monic polynomials $q_1, \ldots, q_s \in B[\zeta]$ such that $q_j(f_j) = 0$ and $|f_j|_{\sup} = \sigma(q_j)$ where $\sigma(q_j)$ is the spectral value of q_j. The product $q_1 \ldots q_s$ assumes a value at f that is nilpotent in A. Thus, there is a certain power q of $q_1 \ldots q_s$ such that $q(f) = 0$ in A, and we have

$$|f|_{\sup} = \max_{j=1\ldots s} |f_j|_{\sup} = \max_{j=1\ldots s} \sigma(q_j) \geq \sigma(q)$$

by Lemma 12. However, as above, the equation $q(f) = 0$ shows that this inequality is, in fact, an equality.   □

There are some important consequences of Lemmata 13 and 14.

Theorem 15 (Maximum Principle). *For any affinoid K-algebra A and any $f \in A$, there exists a point $x \in$ Max A such that $|f(x)| = |f|_{\sup}$.*

Proof. As in the proof of Lemma 14, we consider the minimal prime ideals $\mathfrak{p}_1, \ldots, \mathfrak{p}_s$ of A. Writing f_j for the residue class of f in $A/\mathfrak{p}_j$, there is an index j satisfying $|f|_{\sup} = |f_j|_{\sup}$. Hence, we may replace A by $A/\mathfrak{p}_j$ and thereby assume that A is an integral domain. But then we can apply Noether normalization 2.2/11 to get a finite monomorphism $T_d \hookrightarrow A$, and derive the Maximum Principle for A with the help of Lemma 13 from the Maximum Principle 2.2/5 for Tate algebras. In fact, if

$$f^r + a_1 f^{r-1} + \ldots + a_r = 0$$

is the integral equation of minimal degree for f over T_d, we have

$$\max_{j=1\ldots s} |f(y_j)| = \max_{i=1\ldots r} |a_i(x)|^{\frac{1}{i}}$$

for any $x \in$ Max T_d and the points $y_1, \ldots, y_s \in$ Max A restricting to x; cf. Lemma 13 (ii). Then, by applying the Maximum Principle 2.2/5 to the product $a_1 \ldots a_r \in T_d$, we can find a point $x \in$ Max T_d such that

$$|a_1(x)| \ldots |a_r(x)| = |(a_1 \ldots a_r)(x)| = |a_1 \ldots a_r| = |a_1| \ldots |a_r|.$$

It follows $|a_i(x)| = |a_i|$ for all i and, hence, if $y_1, \ldots, y_s$ are the points of Max A restricting to x,

$$\max_{j=1\ldots s} |f(y_j)| = \max_{i=1\ldots r} |a_i(x)|^{\frac{1}{i}} = \max_{i=1\ldots r} |a_i|^{\frac{1}{i}} = |f|_{\sup}.$$

Therefore f assumes its supremum at one of the points $y_1, \ldots, y_s$.   □

Proposition 16. *Let* f *be an element of some affinoid* K-*algebra* A. *Then there is an integer* $n > 0$ *such that* $|f|_{\text{sup}}^n \in |K|$.

Proof. Use Noether normalization 2.2/11 in conjunction with Lemma 14 and the fact that the Gauß norm on Tate algebras assumes values in $|K|$. □

Using the methods developed in the proofs of Lemmata 13 and 14 in a direct way, one can even show the existence of an integer $n \in \mathbb{N}$ satisfying $|f|_{\text{sup}}^n \in |K|$ for *all* $f \in A$ simultaneously.

Theorem 17. *Let* A *be an affinoid* K-*algebra and let* $|\cdot|_\alpha$ *be a residue norm on* A. *Then for any* $f \in A$, *the following are equivalent*:

  (i) $|f|_{\text{sup}} \leq 1$
  (ii) *There is an integral equation* $f^r + a_1 f^{r-1} + \ldots + a_r = 0$ *with coefficients* $a_i \in A$ *satisfying* $|a_i|_\alpha \leq 1$.
  (iii) *The sequence* $|f^n|_\alpha, n \in \mathbb{N}$, *is bounded; we say,* f *is* power bounded (*with respect to* $|\cdot|_\alpha$).

In particular, the notion of power boundedness is independent of the residue norm under consideration.

Proof. Let $\alpha \colon T_n \longrightarrow A$ be the epimorphism that we use to define the residue norm $|\cdot|_\alpha$ on A. By Noether normalization 2.2/11, there is a monomorphism $T_d \hookrightarrow T_n$ such that the resulting morphism $T_d \hookrightarrow T_n \longrightarrow A$ is a finite monomorphism. Then, by Lemma 14, any $f \in A$ with $|f|_{\text{sup}} \leq 1$ satisfies an integral equation

$$f^r + a_1 f^{r-1} + \ldots + a_r = 0$$

with coefficients $a_i \in T_d$ where $|a_i|_{\text{sup}} = |a_i| \leq 1$. As $T_d \hookrightarrow T_n$ is contractive with respect to the supremum norm by Proposition 7 and, hence, with respect to Gauß norms, the images $\bar{a}_i \in A$ of a_i satisfy $|\bar{a}_i|_\alpha \leq 1$, and the implication from (i) to (ii) is clear.

Next, let us assume (ii). Writing $A^\circ = \{g \in A \; ; \; |g|_\alpha \leq 1\}$, condition (ii) says that f is integral over A°. But then $A^\circ[f]$ is a finite A°-module, and it follows that the sequence $|f^n|_\alpha, n \in \mathbb{N}$, must be bounded.

Finally, that (iii) implies (i), follows from the fact that $|f|_{\text{sup}}^n = |f^n|_{\text{sup}} \leq |f^n|_\alpha$; use Propositions 6 and 9. □

Corollary 18. *Let* A *be an affinoid* K-*algebra and let* $|\cdot|_\alpha$ *be a residue norm on* A. *Then for any* $f \in A$, *the following are equivalent*:

  (i) $|f|_{\text{sup}} < 1$
  (ii) *The sequence* $|f^n|_\alpha, n \in \mathbb{N}$, *is a zero sequence; we say,* f *is* topologically nilpotent *with respect to* $|\cdot|_\alpha$.

In particular, the notion of topological nilpotency is independent of the residue norm under consideration.

Proof. The assertion follows from Proposition 10 if $|f|_{\sup} = 0$. So let us assume $0 < |f|_{\sup} < 1$. Then, by Proposition 16, there is an integer $r > 0$ such that $|f^r|_{\sup} \in |K^*|$; let $c \in K^*$ with $|f^r|_{\sup} = |c|$ so that $|c| < 1$ and $|c^{-1} f^r|_{\sup} = 1$. As $c^{-1} f^r$ is power bounded with respect to $|\cdot|_\alpha$ by Theorem 17, say $|c^{-n} f^{rn}|_\alpha \leq M$ for $n \in \mathbb{N}$ and some $M \in \mathbb{R}$, we see that $|f^{rn}|_\alpha \leq c^n M$ and, hence, that f^r is topologically nilpotent. But then f itself is topologically nilpotent, and we see that assertion (i) implies (ii). Conversely, assume $\lim_{n \to \infty} |f^n|_\alpha = 0$. Then we must have $|f|_{\sup} < 1$ since

$$|f|_{\sup}^n = |f^n|_{\sup} \leq |f^n|_\alpha$$

by Proposition 9.   □

We are now in a position to show that all residue norms on an affinoid K-algebra A are equivalent, i.e. they induce the same topology on A. In particular, this stresses again the fact that the notions of power boundedness as in Theorem 17 and of topological nilpotency as in Corollary 18 are independent of the residue norm under consideration.

Lemma 19. *Let A be an affinoid K-algebra and consider elements $f_1, \ldots, f_n \in A$.*

(i) *Assume there is a K-morphism $\varphi \colon K\langle \zeta_1, \ldots, \zeta_n \rangle \longrightarrow A$ such that $\varphi(\zeta_i) = f_i$, $i = 1, \ldots, n$. Then $|f_i|_{\sup} \leq 1$ for all i.*

(ii) *Conversely, if $|f_i|_{\sup} \leq 1$ for all i, there exists a unique K-morphism $\varphi \colon K\langle \zeta_1, \ldots, \zeta_n \rangle \longrightarrow A$ such that $\varphi(\zeta_i) = f_i$ for all i. Furthermore, φ is continuous with respect to the Gauß norm on T_n and with respect to any residue norm on A.*

Proof. As $|\zeta_i|_{\sup} = |\zeta_i| = 1$, assertion (i) follows from Proposition 7. To verify (ii), fix a residue norm $|\cdot|_\alpha$ on A and define φ by setting

$$\varphi\left(\sum_{\nu \in \mathbb{N}^n} c_\nu \zeta_1^{\nu_1} \ldots \zeta_n^{\nu_n} \right) = \sum_{\nu \in \mathbb{N}^n} c_\nu f_1^{\nu_1} \ldots f_n^{\nu_n}.$$

Due to Theorem 17, $|f_i|_{\sup} \leq 1$ implies that the f_i are power bounded with respect to any residue norm on A. From this we see immediately that φ is well-defined and unique as a *continuous* morphism mapping ζ_i to f_i. Thus, it remains to prove that, apart from φ, there cannot exist any further K-morphism $\varphi' \colon K\langle \zeta_1, \ldots, \zeta_n \rangle \longrightarrow A$ mapping ζ_i to f_i. Let us first consider the case where A, as a K-vector space, is of finite dimension over K. We show that, in this case, any K-morphism $\varphi' \colon K\langle \zeta_1, \ldots, \zeta_n \rangle \longrightarrow A$ is continuous. As is known for finite dimensional vector spaces over complete fields, any K-vector space norm on A induces the product topology in the sense that any isomorphism of K-vector spaces $A \xrightarrow{\sim} K^d$, where $d = \dim_K A$, is a homeomorphism; see Theorem 1 of Appendix A. Now viewing $T_n / \ker \varphi'$ as an affinoid K-algebra with canonical residue norm, it is enough to show that the induced morphism $T_n / \ker \varphi' \hookrightarrow A$ is continuous. However, the latter is clear, since linear forms $V \longrightarrow K$ on a finite dimensional normed vector space V are continuous if V carries the product topology. Thus, φ' is continuous.

To deal with the general case, consider two K-morphisms $\varphi, \varphi' : T_n \longrightarrow A$, both mapping ζ_i to f_i. Then, choosing a maximal ideal $\mathfrak{m} \subset A$ and some integer $r > 0$, we know from Proposition 4 that $A / \mathfrak{m}^r$ is of finite vector space dimension over K. Hence, by what we have seen before, the induced maps $T_n \longrightarrow A / \mathfrak{m}^r$ are continuous and, thus, coincide. Therefore it is enough to show that any $f \in A$ satisfying $f \equiv 0 \mod \mathfrak{m}^r$ for all $\mathfrak{m} \in \mathrm{Max}\, A$ and all $r > 0$ will be trivial. To do this, apply Krull's Intersection Theorem (see for example 7.1/2) to all localizations $A_{\mathfrak{m}}, \mathfrak{m} \in \mathrm{Max}\, A$. It states that $\bigcap_{r \in \mathbb{N}} \mathfrak{m}^r A_{\mathfrak{m}} = 0$. Therefore the image of f in any localization $A_{\mathfrak{m}}$ is trivial and, thus, f itself must be trivial. $\square$

Proposition 20. *Any morphism $B \longrightarrow A$ between affinoid K-algebras is continuous with respect to any residue norms on A and B. In particular, all residue norms on an affinoid K-algebra are equivalent.*

Proof. Choose an epimorphism $T_n \longrightarrow B$ and consider the resulting composition $T_n \longrightarrow B \longrightarrow A$. By Lemma 19 the latter is continuous with respect to any residue norm on A. But then also $B \longrightarrow A$ is continuous. $\square$

Alternatively, one can derive Proposition 20 from the Closed Graph Theorem and the Open Mapping Theorem for Banach spaces (i.e. complete normed vector spaces); see [EVT] for these results of functional analysis. The Open Mapping Theorem can further be used to show that the supremum norm $|\cdot|_{\sup}$ on any *reduced* affinoid K-algebra A is equivalent to all possible residue norms. However, also this result can be obtained in a more direct way, using (sophisticated, though) techniques of affinoid K-algebras. One shows that, after replacing K by a suitable finite extension, the R-algebra $\{f \in A \,;\, |f|_{\sup} \leq 1\}$, divided by its nilradical, is finite over $\{f \in A \,;\, |f|_\alpha \leq 1\}$ for any residue norm $|\cdot|_\alpha$ on A.

Let us add that, although affinoid K-algebras have been defined as quotients of Tate algebras without taking into account any topology, their handling nevertheless requires the use of a residue norm or topology. Otherwise, convergence will not be defined, and we run already into troubles when we want to give explicit constructions of simple things such as a morphism $T_n \longrightarrow A$ from a Tate algebra T_n into some affinoid K-algebra A.

We end this section by an example underlining the usefulness of Proposition 20.

Example 21. Consider an affinoid K-algebra A and on it the topology given by any residue norm. Then, for a set of variables $\xi = (\xi_1, \ldots, \xi_n)$, the K-algebra

$$A\langle\xi\rangle = \left\{ \sum_{\nu\in\mathbb{N}^n} a_\nu \xi^\nu \in A[\![\xi]\!] \, ; a_\nu \in A, \lim_{\nu\in\mathbb{N}^n} a_\nu = 0 \right\}$$

is well-defined, independently of the chosen residue norm on A. It is called the algebra of *restricted power series* in ξ with coefficients in A. We can even show that $A\langle\xi\rangle$ is an affinoid K-algebra. Just choose an epimorphism $\alpha\colon K\langle\zeta\rangle \longrightarrow A$ for a set of variables $\zeta = (\zeta_1,\ldots,\zeta_m)$ and extend it to a morphism of K-algebras

$$\tilde{\alpha}\colon T_{m+n} = K\langle\zeta,\xi\rangle \longrightarrow A\langle\xi\rangle,$$

$$\sum_{\mu\in\mathbb{N}^m} \left(\sum_{\nu\in\mathbb{N}^n} a_{\mu,\nu}\zeta^\nu \right)\xi^\mu \longmapsto \sum_{\mu\in\mathbb{N}^m} \alpha\left(\sum_{\nu\in\mathbb{N}^n} a_{\mu,\nu}\zeta^\nu \right)\xi^\mu,$$

which is, in fact, an epimorphism. The corresponding residue norm coincides with the Gauß norm on $A\langle\xi\rangle$ that is derived from the residue norm via α on A:

$$\left| \sum_{\nu\in\mathbb{N}^n} a_\nu\xi^\nu \right|_{\tilde{\alpha}} = \max_{\nu\in\mathbb{N}^n} |a_\nu|_\alpha$$

3.2 Affinoid Spaces

Let A be an affinoid K-algebra. As we have seen, the elements of A can be viewed as "functions" on Max A, the spectrum of maximal ideals of A. To be more specific, let us define $f(x)$ for $f \in A$ and $x \in$ Max A as the residue class of f in A/x. Embedding A/x into an algebraic closure $\overline{K}$ of K, the value $f(x) \in \overline{K}$ is defined up to conjugation over K, whereas the absolute value $|f(x)|$ is well-defined, as it is independent of the chosen embedding $A/x \hookrightarrow \overline{K}$.

In the following we will write Sp A for the set Max A together with its K-algebra of "functions" A and call it the *affinoid K-space* associated to A. Frequently, we will use Sp A also in the sense of Max A and talk about the *spectrum* of A. Usually, points in Sp A will be denoted by letters $x, y, \ldots$, and the corresponding maximal ideals in A by $\mathfrak{m}_x, \mathfrak{m}_y, \ldots$. One might ask, why we restrict ourselves to maximal ideals instead of considering the spectrum of all prime ideals in A, as is done in algebraic geometry. There is a simple reason for this. In the next section, we will introduce a certain process of localization for affinoid K-algebras, more precisely, of complete localization, since we do not want to leave the context of affinoid K-algebras. Similarly as in algebraic geometry, this localization process is used in order to endow affinoid K-spaces with the structure of a ringed space. As only maximal ideals behave well with respect to localization in this sense, we must restrict ourselves to spectra of maximal ideals. For example, considering such a localization $A \longrightarrow A_S$ and a (non-maximal) prime ideal $\mathfrak{q} \subset A_S$, it can happen that there is no prime ideal $\mathfrak{p} \subset A$ satisfying $\mathfrak{q} = \mathfrak{p} A_S$; see 3.3/22 for a detailed discussion of such a phenomenon.

The Zariski topology on an affinoid K-space $\mathrm{Sp}\, A$ can be defined as usual. For any ideal $\mathfrak{a} \subset A$ we consider its zero set

$$V(\mathfrak{a}) = \{x \in \mathrm{Sp}\, A ;\ f(x) = 0 \text{ for all } f \in \mathfrak{a}\} = \{x \in \mathrm{Sp}\, A ;\ \mathfrak{a} \subset \mathfrak{m}_x\}$$

and call it a *Zariski closed* subset of $\mathrm{Sp}\, A$.

Lemma 1. *Let A be an affinoid K-algebra, and consider ideals $\mathfrak{a}, \mathfrak{b} \subset A$ as well as a family $(\mathfrak{a}_i)_{i \in I}$ of ideals in A.*

(i) $\mathfrak{a} \subset \mathfrak{b} \Longrightarrow V(\mathfrak{a}) \supset V(\mathfrak{b})$.
(ii) $V(\sum_{i \in I} \mathfrak{a}_i) = \bigcap_{i \in I} V(\mathfrak{a}_i)$.
(iii) $V(\mathfrak{a}\mathfrak{b}) = V(\mathfrak{a}) \cup V(\mathfrak{b})$.

The *proof* of (i) and (ii) is straightforward. So it remains to look at (iii). We have $V(\mathfrak{a}) \cup V(\mathfrak{b}) \subset V(\mathfrak{a}\mathfrak{b})$ by (i). To show the converse, consider a point $x \in \mathrm{Sp}\, A$ that is neither in $V(\mathfrak{a})$, nor in $V(\mathfrak{b})$. So there are elements $f \in \mathfrak{a}$ and $g \in \mathfrak{b}$ such that $f(x) \neq 0$ and $g(x) \neq 0$. Then $f, g \notin \mathfrak{m}_x$ and, hence, $fg \notin \mathfrak{m}_x$, since $\mathfrak{m}_x$ is a prime ideal. So $fg(x) \neq 0$, which implies $x \notin V(\mathfrak{a}\mathfrak{b})$. $\qquad\Box$

Assertions (ii) and (iii) show that there really is a topology on $\mathrm{Sp}\, A$, namely the Zariski topology, whose closed sets are just the sets of type $V(\mathfrak{a})$. Also note that, for any epimorphism $\alpha \colon T_n \longrightarrow A$, the map $\mathrm{Sp}\, A \longrightarrow \mathrm{Sp}\, T_n$, $\mathfrak{m} \longmapsto \alpha^{-1}(\mathfrak{m})$, yields a homeomorphism with respect to Zariski topologies between $\mathrm{Sp}\, A$ and the Zariski closed subset $V(\ker \alpha) \subset \mathrm{Sp}\, T_n$.

Proposition 2. *Let A be an affinoid K-algebra. Then the sets*

$$D_f = \{x \in \mathrm{Sp}\, A ;\ f(x) \neq 0\}, \qquad f \in A,$$

form a basis of the Zariski open subsets of $\mathrm{Sp}\, A$.

Proof. First, the sets D_f are Zariski open, since they are the complements of the Zariski closed sets $V(f)$. Next, consider an ideal $\mathfrak{a} = (f_1, \ldots, f_r) \subset A$. Then $V(\mathfrak{a}) = \bigcap_{i=1}^r V(f_i)$ by Lemma 1(ii), and its complement equals the union of the open sets D_{f_i}, $i = 1, \ldots, r$. $\qquad\Box$

As usual, we can associate to any subset $Y \subset \mathrm{Sp}\, A$ the ideal

$$\mathrm{id}(Y) = \{f \in A ;\ f(y) = 0 \text{ for all } y \in Y\} = \bigcap_{y \in Y} \mathfrak{m}_y.$$

Clearly $Y \subset Y'$ implies $\mathrm{id}(Y) \supset \mathrm{id}(Y')$. We want to show that the maps $V(\cdot)$ and $\mathrm{id}(\cdot)$ are inverse to each other in a certain sense.

Proposition 3. *Let A be an affinoid K-algebra and $Y \subset \operatorname{Sp} A$ a subset. Then $V(\operatorname{id}(Y))$ equals the closure of Y in $\operatorname{Sp} A$ with respect to the Zariski topology. In particular, if Y is Zariski closed, we have $V(\operatorname{id}(Y)) = Y$.*

Proof. Writing $\mathfrak{a} = \operatorname{id}(Y)$, we have $V(\operatorname{id}(Y)) = \bigcap_{f \in \mathfrak{a}} V(f)$ by Lemma 1 (ii). On the other hand, the closure $\overline{Y}$ of Y equals the intersection of all closed sets $Y' \subset \operatorname{Sp} A$ containing Y. Since, again by Lemma 1 (ii), any such Y' may be written as an intersection of sets of type $V(g)$, we get

$$\overline{Y} = \bigcap_{g \in A, Y \subset V(g)} V(g) = \bigcap_{f \in \mathfrak{a}} V(f) = V(\operatorname{id}(Y)).$$

$\square$

Theorem 4 (Hilbert's Nullstellensatz). *Let A be an affinoid K-algebra and $\mathfrak{a} \subset A$ an ideal. Then*

$$\operatorname{id}(V(\mathfrak{a})) = \operatorname{rad} \mathfrak{a}.$$

Proof. We have

$$\operatorname{id}(V(\mathfrak{a})) = \operatorname{id}(\{x \in \operatorname{Sp} A \,;\, \mathfrak{a} \subset \mathfrak{m}_x\}) = \bigcap_{\mathfrak{a} \subset \mathfrak{m}_x} \mathfrak{m}_x,$$

and the intersection on the right-hand side equals the nilradical of $\mathfrak{a}$, since A is Jacobson; cf. 3.1/3. $\square$

Corollary 5. *For any affinoid K-algebra A, the maps $V(\cdot)$ and $\operatorname{id}(\cdot)$ define mutually inverse bijections between the set of reduced ideals in A and the set of Zariski closed subsets of $\operatorname{Sp} A$.*

Corollary 6. *Consider a set of functions $f_i, i \in I$, of an affinoid K-algebra A. The following are equivalent:*

 (i) *The f_i have no common zeros on $\operatorname{Sp} A$.*
 (ii) *The f_i generate the unit ideal in A.*

As in algebraic geometry, a non-empty subset $Y \subset \operatorname{Sp} A$ is called *irreducible* if Y (endowed with the topology induced from the Zariski topology on $\operatorname{Sp} A$) cannot be written as a union $Y_1 \cup Y_2$ of two proper relatively closed subsets $Y_1, Y_2 \subsetneq Y$. One shows that, under the bijection of Corollary 5, the irreducible Zariski closed subsets of $\operatorname{Sp} A$ correspond precisely to the prime ideals in A. Furthermore, as affinoid K-algebras are Noetherian, any Zariski closed subset $Y \subset \operatorname{Sp} A$ admits a unique decomposition into finitely many irreducible closed subsets.

Finally, let us point out that any morphism $\sigma: B \longrightarrow A$ of affinoid K-algebras induces an associated map

$$^a\sigma: \operatorname{Sp} A \longrightarrow \operatorname{Sp} B, \qquad \mathfrak{m} \longmapsto \sigma^{-1}(\mathfrak{m}).$$

We have used this fact already implicitly in Sect. 3.1. Note that $\sigma^{-1}(\mathfrak{m}) \subset B$ is maximal, since we have a chain of injections

$$K \hookrightarrow B/\sigma^{-1}(\mathfrak{m}) \hookrightarrow A/\mathfrak{m}$$

and since $A/\mathfrak{m}$ is a field that is finite over K. The map $^a\sigma: \operatorname{Sp} A \longrightarrow \operatorname{Sp} B$ (together with its inducing homomorphism σ) will be called a *morphism of affinoid K-spaces*, more precisely, the morphism of affinoid K-spaces associated to $\sigma: B \longrightarrow A$. Frequently, we will write $\varphi: \operatorname{Sp} A \longrightarrow \operatorname{Sp} B$ for a morphism of affinoid K-spaces and $\varphi^*: B \longrightarrow A$ for the inherent morphism of affinoid K-algebras. In fact, φ^* may be interpreted as pulling back functions from $\operatorname{Sp} B$ to $\operatorname{Sp} A$ via composition with φ, as for any $x \in \operatorname{Sp} A$ the commutative diagram

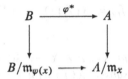

implies

$$\varphi^*(g)(x) = g(\varphi(x))$$

for all $g \in B$.

The affinoid K-spaces together with their morphisms form a category, which can be interpreted as the opposite of the category of affinoid K-algebras. Since the latter category admits amalgamated sums, see 3.1/2 and Theorem 6 of Appendix B, we can conclude:

Proposition 7. *For two affinoid K-spaces over a third one Z, the fiber product $X \times_Z Y$ exists as an affinoid K-space.*

3.3 Affinoid Subdomains

The Zariski topology on an affinoid K-space is quite coarse. In the present section we want to introduce a finer one that is directly induced from the topology of K. We can think of an affinoid K-space $\operatorname{Sp} A$ as of a Zariski closed subspace of $\operatorname{Sp} T_n$ for some $n \in \mathbb{N}$, and the latter can be identified with the unit ball $\mathbb{B}^n(K)$, at least if K

is algebraically closed. Thereby we see that the topology of the affine n-space K^n gives rise to a topology on $\operatorname{Sp} A$ that, as we will see, is independent of the particular embedding $\operatorname{Sp} A \hookrightarrow \operatorname{Sp} T_n$; it will be referred to as the *canonical* topology of $\operatorname{Sp} A$. If K is not necessarily algebraically closed, we can proceed similarly by viewing $\operatorname{Sp} T_n$ as the quotient of $\mathbb{B}^n(\overline{K})$ by the action of $\operatorname{Aut}_K(\overline{K})$, providing $\operatorname{Sp} T_n$ with the quotient topology.

To give a more rigorous approach, consider an affinoid K-space $X = \operatorname{Sp} A$ and set

$$X(f;\varepsilon) = \{x \in X \; ; \; |f(x)| \leq \varepsilon\}.$$

for $f \in A$ and $\varepsilon \in \mathbb{R}_{>0}$.

Definition 1. *For any affinoid K-space $X = \operatorname{Sp} A$, the topology generated by all sets of type $X(f;\varepsilon)$ with $f \in A$ and $\varepsilon \in \mathbb{R}_{>0}$ is called the* canonical *topology of X.*

Thus, a subset $U \subset X$ is open with respect to the canonical topology if and only if it is a union of finite intersections of sets of type $X(f;\varepsilon)$. Writing $X(f) = X(f;1)$ for any $f \in A$ and $X(f_1,\ldots,f_r) = X(f_1) \cap \ldots \cap X(f_r)$ for $f_1,\ldots,f_r \in A$, we can even say:

Proposition 2. *For any affinoid K-space $X = \operatorname{Sp} A$, the canonical topology is generated by the system of all subsets $X(f)$ with f varying over A. In particular, a subset $U \subset \operatorname{Sp} A$ is open if and only if it is a union of sets of type $X(f_1,\ldots,f_r)$ for elements $f_1,\ldots,f_r \in A, r \in \mathbb{N}$.*

Proof. For any $f \in A$, the function $|f|: \operatorname{Sp} A \longrightarrow \mathbb{R}_{\geq 0}$ assumes values in $|\overline{K}|$. Therefore, if $\varepsilon \in \mathbb{R}_{>0}$, we can write

$$X(f;\varepsilon) = \bigcup_{\varepsilon' \in |\overline{K}^*|, \, \varepsilon' \leq \varepsilon} X(f;\varepsilon').$$

For $\varepsilon' \in |\overline{K}^*|$ we can always find an element $c \in K^*$ and an integer $s > 0$ such that $\varepsilon'^s = |c|$; see for example Theorem 3 of Appendix A. But then

$$X(f;\varepsilon') = X(f^s;\varepsilon'^s) = X(c^{-1}f^s)$$

and we are done. $\square$

We want to establish a basic lemma that will enable us to derive the openness of various types of sets.

Lemma 3. *For an affinoid K-space $X = \operatorname{Sp} A$, consider an element $f \in A$ and a point $x \in \operatorname{Sp} A$ such that $\varepsilon = |f(x)| > 0$. Then there is an element $g \in A$ satisfying*

$g(x) = 0$ *such that* $|f(y)| = \varepsilon$ *for all* $y \in X(g)$. *In particular,* $X(g)$ *is an open neighborhood of* x *contained in* $\{y \in X \; ; |f(y)| = \varepsilon\}$.

Proof. Let $\mathfrak{m}_x \subset A$ be the maximal ideal corresponding to x and write $\overline{f}$ for the residue class of f in $A/\mathfrak{m}_x$. Furthermore, let

$$P(\zeta) = \zeta^n + c_1 \zeta^{n-1} + \ldots + c_n \in K[\zeta]$$

be the minimal polynomial of $\overline{f}$ over K and let

$$P(\zeta) = \prod_{i=1}^{n} (\zeta - \alpha_i)$$

be its product decomposition with zeros $\alpha_i \in \overline{K}$. Then, choosing an embedding $A/\mathfrak{m}_x \hookrightarrow \overline{K}$, we have $\varepsilon = |f(x)| = |\overline{f}| = |\alpha_i|$ for all i by the uniqueness of the valuation on $\overline{K}$.

Now consider the element $g = P(f) \in A$. Then $g(x) = P(f(x)) = 0$ and we claim:

$$y \in X \text{ with } |g(y)| < \varepsilon^n \quad \Longrightarrow \quad |f(y)| = \varepsilon$$

In fact, assume $|f(y)| \neq \varepsilon$ for some $y \in X$ satisfying $|g(y)| < \varepsilon^n$. Then, choosing an embedding $A/\mathfrak{m}_y \hookrightarrow \overline{K}$, we have

$$\left| f(y) - \alpha_i \right| = \max(|f(y)|, |\alpha_i|) \geq |\alpha_i| = \varepsilon$$

for all i and, thus,

$$|g(y)| = |P(f(y))| = \prod_{i=1}^{n} |f(y) - \alpha_i| \geq \varepsilon^n,$$

which contradicts the choice of y. Therefore, if $c \in K^*$ satisfies $|c| < \varepsilon^n$, we have $|f(y)| = \varepsilon$ for all $y \in X(c^{-1}g)$. $\qquad\qquad\square$

As a direct consequence of Lemma 3, we can state:

Proposition 4. *Let* Sp A *be an affinoid* K*-space. Then, for* $f \in A$ *and* $\varepsilon \in \mathbb{R}_{>0}$, *the following sets are open with respect to the canonical topology:*

$$\{x \in \text{Sp } A; f(x) \neq 0\}$$

$$\{x \in \text{Sp } A; |f(x)| \leq \varepsilon\}$$

$$\{x \in \text{Sp } A; |f(x)| = \varepsilon\}$$

$$\{x \in \text{Sp } A; |f(x)| \geq \varepsilon\}$$

Proposition 5. *Let $X = \operatorname{Sp} A$ be an affinoid K-space, and let $x \in X$ correspond to the maximal ideal $\mathfrak{m}_x \subset A$. Then the sets $X(f_1, \ldots, f_r)$ for $f_1, \ldots, f_r \in \mathfrak{m}_x$ and variable r form a basis of neighborhoods of x.*

Proposition 6. *Let $\varphi^*: A \longrightarrow B$ be a morphism of affinoid K-algebras, and let $\varphi: \operatorname{Sp} B \longrightarrow \operatorname{Sp} A$ be the associated morphism of affinoid K-spaces. Then, for $f_1, \ldots, f_r \in A$, we have*

$$\varphi^{-1}\big((\operatorname{Sp} A)(f_1, \ldots, f_r)\big) = (\operatorname{Sp} B)\big(\varphi^*(f_1), \ldots, \varphi^*(f_r)\big).$$

In particular, φ is continuous with respect to the canonical topology.

Proof. Each $y \in \operatorname{Sp} B$ gives rise to a commutative diagram

$$
\begin{array}{ccc}
A & \xrightarrow{\ \varphi^*\ } & B \\
\downarrow & & \downarrow \\
A/\mathfrak{m}_{\varphi(y)} & \longrightarrow & B/\mathfrak{m}_y
\end{array}
$$

with a monomorphism in the lower row. As we may embed the latter into $\overline{K}$, we see that $|f(\varphi(y))| = |\varphi^*(f)(y)|$ holds for any $f \in A$. This implies

$$\varphi^{-1}\big((\operatorname{Sp} A)(f)\big) = (\operatorname{Sp} B)\big(\varphi^*(f)\big)$$

and, hence, forming intersections, we are done.   □

Next we want to introduce certain special open subsets of affinoid K-spaces that, themselves, have a structure of affinoid K-space again.

Definition 7. *Let $X = \operatorname{Sp} A$ be an affinoid K-space.*

 (i) *A subset in X of type*

$$X(f_1, \ldots, f_r) = \{x \in X \,;\, |f_i(x)| \leq 1\}$$

for functions $f_1, \ldots, f_r \in A$ is called a Weierstraß domain *in X.*
 (ii) *A subset in X of type*

$$X(f_1, \ldots, f_r, g_1^{-1}, \ldots, g_s^{-1}) = \{x \in X \,;\, |f_i(x)| \leq 1, |g_j(x)| \geq 1\}$$

for functions $f_1, \ldots, f_r, g_1, \ldots, g_s \in A$ is called a Laurent domain *in X.*
(iii) *A subset in X of type*

$$X\left(\frac{f_1}{f_0}, \ldots, \frac{f_r}{f_0}\right) = \{x \in X \,;\, |f_i(x)| \leq |f_0(x)|\}$$

for functions $f_0, \ldots, f_r \in A$ without common zeros is called a rational domain *in X.*

Note that the condition in (iii), namely that $f_0, \ldots, f_r$ have no common zero on $\mathrm{Sp}\, A$, is equivalent to the fact that these functions generate the unit ideal in A.

Lemma 8. *The domains of Definition 7 are open in $X = \mathrm{Sp}\, A$ with respect to the canonical topology. The Weierstraß domains form a basis of this topology.*

Proof. The openness of Weierstraß and Laurent domains can be read from the assertion of Lemma 3. In the case of a rational domain the same is true, as for any $x \in X\left(\frac{f_1}{f_0}, \ldots \frac{f_r}{f_0}\right)$ we must have $f_0(x) \neq 0$, due to the fact that the f_i are not allowed to have a common zero on X. $\qquad\square$

Let us point out that the condition in Definition 7 (iii), namely that the elements $f_0, \ldots, f_r \in A$ have no common zeros on X, is necessary to assure that sets of type $X\left(\frac{f_1}{f_0}, \ldots \frac{f_r}{f_0}\right)$ are open in $X = \mathrm{Sp}\, A$. For example, look at $X = \mathrm{Sp}\, T_1 = \mathrm{Sp}\, K\langle \zeta_1 \rangle$ and choose a constant $c \in K$ such that $0 < |c| < 1$. Then the set

$$\left\{ x \in X \; ; \; |\zeta_1(x)| \leq |c\zeta_1(x)| \right\}$$

consists of a single point, namely the one given by the maximal ideal $(\zeta_1) \subset T_1$. However, in view of Proposition 5, such a point cannot define an open subset in X.

The domains introduced in Definition 7 are important examples of more general subdomains, whose definition we will give now.

Definition 9. *Let $X = \mathrm{Sp}\, A$ be an affinoid K-space. A subset $U \subset X$ is called an* affinoid subdomain *of X if there exists a morphism of affinoid K-spaces $\iota: X' \longrightarrow X$ such that $\iota(X') \subset U$ and the following universal property holds:*

Any morphism of affinoid K-spaces $\varphi: Y \longrightarrow X$ satisfying $\varphi(Y) \subset U$ admits a unique factorization through $\iota: X' \longrightarrow X$ via a morphism of affinoid K-spaces $\varphi': Y \longrightarrow X'$.

Lemma 10. *In the situation of Definition 9, let us write $X = \mathrm{Sp}\, A$ and $X' = \mathrm{Sp}\, A'$, and let $\iota^*: A \longrightarrow A'$ be the K-morphism corresponding to ι. Then the following hold:*

(i) *ι is injective and satisfies $\iota(X') = U$. Hence, it induces a bijection of sets $X' \xrightarrow{\sim} U$.*

(ii) *For any $x \in X'$ and $n \in \mathbb{N}$, the map ι^* induces an isomorphism of affinoid K-algebras $A/\mathfrak{m}_{\iota(x)}^n \xrightarrow{\sim} A'/\mathfrak{m}_x^n$.*

(iii) *For $x \in X'$ we have $\mathfrak{m}_x = \mathfrak{m}_{\iota(x)} A'$.*

Proof. Choosing a point $y \in U$, we get a commutative diagram

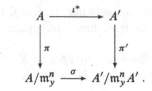

Then $\operatorname{Sp} A/\mathfrak{m}_y^n$ is a one-point space that is mapped by ${}^a\pi$ onto the point $y \in U$, and it follows from the universal property of ι or ι^* that π factors through $\iota^*\colon A \longrightarrow A'$ via a unique K-morphism $\alpha\colon A' \longrightarrow A/\mathfrak{m}_y^n$. Now insert α into the above diagram:

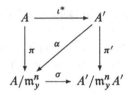

The upper triangle will be commutative, and we claim that the same holds for the lower triangle, i.e. that $\pi' = \sigma \circ \alpha$. To justify this, note that the map of affinoid K-spaces associated to $\sigma \circ \pi$ has image $y \in U$, too. As $\sigma \circ \pi$ factors through $\iota^*\colon A \longrightarrow A'$ via both, π' and $\sigma \circ \alpha$, the uniqueness part of the universal property of ι^* yields $\pi' = \sigma \circ \alpha$.

Now the surjectivity of π' implies the surjectivity of σ. Furthermore, α is surjective since π is surjective, and we have $\ker \pi' = \mathfrak{m}_y^n A' \subset \ker \alpha$. Thus, σ must be injective and, hence, bijective. For $n = 1$ we see that the ideal $\mathfrak{m}_y A'$ is maximal in A'. Thus, the fiber of ι over y is non-empty and consists of precisely one point $x \in X'$ where $\mathfrak{m}_x = \mathfrak{m}_y A'$. This shows (i) and (iii). Then we get (ii) from the bijectivity of σ and from the fact that $\mathfrak{m}_x = \mathfrak{m}_y A' = \mathfrak{m}_{\iota(x)} A'$.   □

When dealing with affinoid subdomains in the sense of Definition 9, we will use Lemma 10 (i) and always identify the subset $U \subset X$ with the set of X'. We thereby get a structure of affinoid K-space on any affinoid subdomain $U \subset X$, and this structure is unique up to canonical isomorphism. In fact, we can talk about the affinoid subdomain $X' \hookrightarrow X$. Such a subdomain is called *open* in X if it is open with respect to the canonical topology. Later in Proposition 19 we will see that any affinoid subdomain $X' \hookrightarrow X$ is open in X.

We now want to show that the domains listed in Definition 7 define open affinoid subdomains in the sense of Definition 9.

Proposition 11. *For any affinoid K-space $X = \operatorname{Sp} A$, Weierstraß, Laurent, and rational domains in X are examples of open affinoid subdomains. These are called special affinoid subdomains.*

Proof. First, it follows from Lemma 8 that Weierstraß, Laurent, and rational domains are open in X. To show that they satisfy the defining condition of affinoid subdomains, we start with a Weierstraß domain $X(f) \subset X$ where f stands for a tuple of functions $f_1, \ldots, f_r \in A$. Let $A\langle \zeta_1, \ldots, \zeta_r \rangle$ be the affinoid K-algebra of restricted power series in the variables $\zeta_1, \ldots, \zeta_r$ over A, the topology of A being provided by some residue norm; see Example 3.1/21. Then consider

$$A\langle f \rangle = A\langle f_1, \ldots, f_r \rangle = A\langle \zeta_1, \ldots, \zeta_r \rangle / (\zeta_i - f_i ; i = 1, \ldots, r).$$

as an affinoid K-algebra. There is a canonical morphism of affinoid K-algebras $\iota^*: A \longrightarrow A\langle f \rangle$ and, associated to it, a morphism between affinoid K-spaces $\iota: \mathrm{Sp}\, A\langle f \rangle \longrightarrow X$. We claim that ι has image in $X(f)$ and that all other morphisms of affinoid K-spaces $\varphi: Y \longrightarrow X$ with $\operatorname{im} \varphi \subset X(f)$ admit a unique factorization through ι.

To check this consider a morphism of affinoid K-spaces $\varphi: Y \longrightarrow X$ and let it correspond to a morphism of affinoid K-algebras $\varphi^*: A \longrightarrow B$. Then, for any $y \in Y$, we get

$$\left| \varphi^*(f_i)(y) \right| = \left| f_i(\varphi(y)) \right|, \qquad i = 1, \ldots, r,$$

by looking at the inclusion $A/\mathfrak{m}_{\varphi(y)} \hookrightarrow B/\mathfrak{m}_y$ between finite extensions of K, as induced from φ^*. Therefore, $\varphi(Y) \subset X(f)$ is equivalent to $|\varphi^*(f_i)|_{\sup} \leq 1$ for all i. Since $\iota^*(f_i)$ equals the residue class of ζ_i in $A\langle f \rangle$, we have $|f_i|_{\sup} \leq 1$ by 3.1/9. Thus, it follows that ι has image in $X(f)$, and it remains to show the following universal property for ι^*:

Each morphism of affinoid K-algebras $\varphi^*: A \longrightarrow B$ with $|\varphi^*(f_i)|_{\sup} \leq 1$ for all i admits a unique factorization through $\iota^*: A \longrightarrow A\langle f \rangle$.

However, this is easy to do. Given such a morphism $\varphi^*: A \longrightarrow B$, we can extend it to a morphism $A\langle \zeta \rangle \longrightarrow B$ by mapping ζ_i to $\varphi^*(f_i)$ for all i. Then the elements $\zeta_i - f_i$ belong to the kernel, and we get an induced morphism $A\langle f \rangle \longrightarrow B$ that is a factorization of $\varphi^*: A \longrightarrow B$ through $\iota^*: A \longrightarrow A\langle f \rangle$. That this factorization is unique follows from the fact that the image of A is dense in $A\langle f \rangle$.

Next, let us look at the case of a Laurent domain $X(f, g^{-1}) \subset X$ where we use tuples $f = (f_1, \ldots, f_r)$ and $g = (g_1, \ldots, g_s)$ of elements of A. Then look at the affinoid K-algebra

$$A\langle f, g^{-1} \rangle = A\langle f_1, \ldots, f_r, g_1^{-1}, \ldots, g_s^{-1} \rangle$$
$$= A\langle \zeta_1, \ldots, \zeta_r, \xi_1, \ldots, \xi_s \rangle / (\zeta_i - f_i, 1 - g_j \xi_j ; i = 1, \ldots, r ; j = 1, \ldots, s).$$

There is a canonical morphism of affinoid K-algebras $\iota^*: A \longrightarrow A\langle f, g^{-1} \rangle$ and, associated to it, a morphism of affinoid K-spaces $\iota: \mathrm{Sp}\, A\langle f, g^{-1} \rangle \longrightarrow \mathrm{Sp}\, A$. Similarly as before, a morphism of affinoid K-spaces $\varphi: Y \longrightarrow X$ corresponding to a morphism of affinoid K-algebras $\varphi^*: A \longrightarrow B$ has image in $X(f, g^{-1})$ if and only if

$$\left| \varphi^*(f_i)(y) \right| \leq 1, \qquad \left| \varphi^*(g_j)(y) \right| \geq 1, \qquad \text{for all } y \in Y, \text{ all } i \text{ and } j.$$

Of course, the inequalities involving the f_i are equivalent to $|\varphi^*(f_i)|_{\sup} \leq 1$, whereas the ones on the g_j may be replaced by the condition that the $\varphi^*(g_j)$ are units in B satisfying $|\varphi^*(g_j)^{-1}|_{\sup} \leq 1$ for all j. Now consider the map ι in place of φ and use a bar to indicate residue classes in $A\langle f, g^{-1}\rangle$. Then one concludes from

$$\overline{\zeta}_i - \iota^*(f_i) = 0, \qquad |\overline{\zeta}_i|_{\sup} \leq 1, \qquad i = 1, \ldots, r,$$

$$\iota^*(g_j)\overline{\xi}_j = 1, \qquad |\overline{\xi}_i|_{\sup} \leq 1, \qquad j = 1, \ldots, s,$$

similarly as before that ι has image in $X(f, g^{-1})$. Thus, it remains to show:

Each morphism of affinoid K-algebras $\varphi^*: A \longrightarrow B$, where $|\varphi^*(f_i)|_{\sup} \leq 1$, $i = 1, \ldots, r$, and $\varphi^*(g_j)$ is a unit in B with $|\varphi^*(g_j)^{-1}|_{\sup} \leq 1$, $j = 1, \ldots, s$, admits a unique factorization through $\iota^*: A \longrightarrow A\langle f, g^{-1}\rangle$.

So consider a morphism $\varphi^*: A \longrightarrow B$ with the properties listed above. We can extend it to a morphism $A\langle \zeta, \xi\rangle \longrightarrow B$ by mapping ζ_i to $\varphi^*(f_i)$ and ξ_j to $\varphi^*(g_j)^{-1}$. As the kernel contains all elements $\zeta_i - f_i$ and all elements $1 - g_j\xi_j$, we get an induced map $A\langle f, g^{-1}\rangle \longrightarrow B$ that is a factorization of φ^* through ι^*. The latter is unique, as the image of $A[g^{-1}]$ is dense in $A\langle f, g^{-1}\rangle$.

Finally, let us look at a rational domain $X\left(\frac{f}{f_0}\right) \subset X$ where we have written $f = (f_1, \ldots, f_r)$ and where $f_0, \ldots, f_r \in A$ have no common zero on $\mathrm{Sp}\, A$. We set

$$A\left\langle \frac{f}{f_0}\right\rangle = A\left\langle \frac{f_1}{f_0}, \ldots, \frac{f_r}{f_0}\right\rangle = A\langle \zeta_1, \ldots, \zeta_r\rangle / (f_i - f_0\zeta_i \; ; \; i = 1, \ldots, r)$$

and consider the canonical morphism of affinoid K-algebras $\iota^*: A \longrightarrow A\langle\frac{f}{f_0}\rangle$, as well as its associated morphism of affinoid K-spaces $\iota: \mathrm{Sp}\, A\langle\frac{f}{f_0}\rangle \longrightarrow \mathrm{Sp}\, A$.

Next, let $\varphi: Y \longrightarrow X$ be any morphism of affinoid K-spaces with corresponding morphism of affinoid K-algebras $\varphi^*: A \longrightarrow B$. Then φ maps Y into $X\left(\frac{f}{f_0}\right)$ if and only if we have

$$\left|\varphi^*(f_i)(y)\right| \leq \left|\varphi^*(f_0)(y)\right|, \qquad \text{for all } y \in \mathrm{Sp}\, B \text{ and all } i. \qquad (*)$$

As $f_0, \ldots, f_r$ generate the unit ideal in A, the same is true for their images in B, and we see that $(*)$ is equivalent to

$$\varphi^*(f_0) \in B^*, \qquad \left|\varphi^*(f_i) \cdot \varphi^*(f_0)^{-1}\right|_{\sup} \leq 1, \qquad \text{for all } i, \qquad (**)$$

where B^* is the group of units in B. Now consider the map ι in place of φ and use a bar to indicate residue classes in $A\langle\frac{f}{f_0}\rangle$. Then one concludes property $(*)$ for ι^* in place of φ^* from

$$\iota^*(f_i) = \iota^*(f_0)\overline{\xi}_i, \qquad |\overline{\xi}_i|_{\sup} \leq 1, \qquad i = 1, \ldots, r,$$

and it follows that ι has image in $X\left(\frac{f}{f_0}\right)$. As above, it remains to show that $\iota^*\colon A \longrightarrow A\langle\frac{f}{f_0}\rangle$ satisfies the following universal property:

Each morphism of affinoid K-algebras $\varphi^*\colon A \longrightarrow B$ with $(**)$ admits a unique factorization through $\iota^*\colon A \longrightarrow A\langle\frac{f}{f_0}\rangle$.

To verify this, start with a morphism $\varphi^*\colon A \longrightarrow B$ satisfying $(**)$ and extend it to a morphism $A\langle\zeta_1,\ldots,\zeta_r\rangle \longrightarrow B$ by mapping ζ_i to $\varphi^*(f_i)\cdot\varphi^*(f_0)^{-1}$. As the kernel contains all elements $f_i - f_0\zeta_i$, we get an induced morphism $A\langle\frac{f}{f_0}\rangle \longrightarrow B$ that is a factorization of φ^* through ι^*. The latter is unique as the image of $A[f_0^{-1}]$ is dense in $A\langle\frac{f}{f_0}\rangle$. $\qquad\square$

Proposition 12 (Transitivity of Affinoid Subdomains). *For an affinoid K-space X, consider an affinoid subdomain $V \subset X$, and an affinoid subdomain $U \subset V$. Then U is an affinoid subdomain in X as well.*

Proof. Consider a morphism of affinoid K-spaces $\varphi\colon Y \longrightarrow X$ having image in U. Then, as $U \subset V$ and V is an affinoid subdomain of X, there is a unique factorization $\varphi'\colon Y \longrightarrow V$ of φ through $V \hookrightarrow X$. Furthermore, φ' admits a unique factorization $\varphi''\colon Y \longrightarrow U$ through $U \hookrightarrow V$, as U is an affinoid subdomain of V. Then, of course, φ'' is a factorization of φ through $U \hookrightarrow V \hookrightarrow X$ that, using the uniqueness of factorizations through $U \hookrightarrow V$ and $V \hookrightarrow X$, is easily seen to be unique. $\qquad\square$

Proposition 13. *Let $\varphi\colon Y \longrightarrow X$ be a morphism of affinoid K-spaces and let $X' \hookrightarrow X$ be an affinoid subdomain. Then $Y' = \varphi^{-1}(X')$ is an affinoid subdomain of Y, and there is a unique morphism of affinoid K-spaces $\varphi'\colon Y' \longrightarrow X'$ such that the diagram*

$$
\begin{array}{ccc}
Y' & \xrightarrow{\ \varphi'\ } & X' \\
\big\downarrow & & \big\downarrow \\
Y & \xrightarrow{\ \varphi\ } & X
\end{array}
$$

is commutative. In fact, the diagram is cartesian in the sense that it characterizes Y' as the fiber product of Y and X' over X.

If X' is Weierstraß, Laurent, or rational in X, the corresponding fact is true for Y' as an affinoid subdomain of Y. More specifically, if $\varphi^*\colon A \longrightarrow B$ is the morphism of affinoid K-algebras associated to $\varphi\colon Y \longrightarrow X$, and if

$$
f = (f_1,\ldots,f_r), \qquad g = (g_1,\ldots,g_s), \qquad h = (h_0,\ldots,h_t)
$$

are tuples of elements in A, such that the h_i generate the unit ideal in A, then

$$\varphi^{-1}\big(X(f)\big) = Y\big(\varphi^*(f)\big),$$

$$\varphi^{-1}\big(X(f,g^{-1})\big) = Y\big(\varphi^*(f),\varphi^*(g)^{-1}\big),$$

$$\varphi^{-1}\Big(X\Big(\frac{h}{h_0}\Big)\Big) = Y\Big(\frac{\varphi^*(h)}{\varphi^*(h_0)}\Big).$$

Proof. In the case where X' is a general affinoid subdomain of X, we use the fact 3.2/7 that the category of affinoid K-spaces admits fiber products or, equivalently, that the category of affinoid K-algebras admits amalgamated sums; cf. Theorem 6 of Appendix B. Relying on the existence of the fiber product $Y \times_X X'$, it is easy to see that the first projection $p: Y \times_X X' \longrightarrow Y$ defines $\varphi^{-1}(X')$ as an affinoid subdomain in Y. Just look at the commutative diagram

$$
\begin{array}{ccc}
Y \times_X X' & \longrightarrow & X' \\
{\scriptstyle p}\big\downarrow & & \big\downarrow \\
Y & \xrightarrow{\ \varphi\ } & X\,.
\end{array}
$$

It shows that p maps $Y \times_X X'$ into $\varphi^{-1}(X')$. Furthermore, consider a morphism of affinoid K-spaces $\psi: Z \longrightarrow Y$ having image in $\varphi^{-1}(X')$. Then the composition $\varphi \circ \psi: Z \longrightarrow X$ factors through $X' \hookrightarrow X$, and the universal property of fiber products yields a unique factorization of ψ via $p: Y \times_X X' \longrightarrow Y$. Thus, p defines $Y' = Y \times_X X'$ as an affinoid subdomain of Y and we have $Y' = \varphi^{-1}(X')$ by Lemma 10 (i).

The second projection $\varphi': Y' = Y \times_X X' \longrightarrow X'$ is a morphism making the diagram mentioned in the assertion commutative. That φ' is uniquely determined by this property follows from the universal property of X' as an affinoid subdomain of X.

If $\varphi^*: A \longrightarrow B$ is the morphism of affinoid K-algebras corresponding to $\varphi: Y \longrightarrow X$, we have for any $y \in Y$ a commutative diagram

$$
\begin{array}{ccc}
A & \xrightarrow{\ \varphi^*\ } & B \\
\big\downarrow & & \big\downarrow \\
A/\mathfrak{m}_{\varphi(y)} & \longrightarrow & B/\mathfrak{m}_y
\end{array}
$$

with an injection of finite field extensions of K in the lower row. It follows $|f(\varphi(y))| = |\varphi^*(f)(y)|$ for any $f \in A$. As in Proposition 6, one deduces the stated identities for the inverse of Weierstraß, Laurent, and rational domains. In the case of rational domains we use the fact that the images $\varphi^*(h_0), \dots, \varphi^*(h_t)$ will generate the unit ideal in B as soon as the elements $h_0, \dots, h_t$ generate the unit ideal in A. □

The morphism of affinoid K-algebras $\varphi'^*\colon A' \longrightarrow B'$ that in the setting of Proposition 13 is associated to $\varphi'\colon Y' \longrightarrow X'$, is obtained from $\varphi^*\colon A \longrightarrow B$ by tensoring the latter with A' over A, of course in the sense of completed tensor products. If X' is a special affinoid subdomain, φ'^* can be described in more explicit terms. For example, for a Weierstraß domain $X' = X(f)$, the map φ'^* is obtained via the canonical commutative diagram

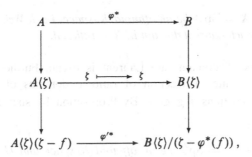

and there are similar diagrams for Laurent and rational domains.

Proposition 14. *Let X be an affinoid K-space and let $U, V \subset X$ be affinoid subdomains. Then $U \cap V$ is an affinoid subdomain of X. If U and V are Weierstraß, resp. Laurent, resp. rational domains, the same is true for $U \cap V$.*

Proof. Let $\varphi\colon U \hookrightarrow X$ be the morphism defining U as an affinoid subdomain of X. Then $U \cap V = \varphi^{-1}(V)$ and we see that $U \cap V$ is an affinoid subdomain of U by Proposition 13. Hence, by Proposition 12, $U \cap V$ is an affinoid subdomain of X.

Next let us consider the case where U and V are rational subdomains of X, say

$$U = X\left(\frac{f_1}{f_0}, \ldots, \frac{f_r}{f_0}\right), \qquad V = X\left(\frac{g_1}{g_0}, \ldots, \frac{g_s}{g_0}\right)$$

with functions f_i, g_j satisfying $(f_0, \ldots, f_r) = (1)$, as well as $(g_0, \ldots, g_s) = (1)$. The product of both ideals is the unit ideal again and we see that the functions $f_i g_j$, $i = 0, \ldots, r$, $j = 0, \ldots, s$ have no common zero on X. Therefore

$$W = X\left(\frac{f_i g_j}{f_0 g_0}; i = 0, \ldots, r; j = 0, \ldots, s\right)$$

is a well-defined rational subdomain in X, and we claim that it equals the intersection $U \cap V$. Clearly, we have $U \cap V \subset W$ since for any $x \in X$ the inequalities $|f_i(x)| \leq |f_0(x)|$, $i = 0, \ldots, r$ and $|g_j(x)| \leq |g_0(x)|$, $j = 0, \ldots, s$ imply $|(f_i g_j)(x)| \leq |(f_0 g_0)(x)|$ for all i, j. Conversely, consider a point $x \in X$ such that $|(f_i g_j)(x)| \leq |(f_0 g_0)(x)|$ for all i, j. Then, as the $f_i g_j$ have no common zero on X, we must have $(f_0 g_0)(x) \neq 0$ and, hence, $f_0(x) \neq 0$ and $g_0(x) \neq 0$. But then the inequalities

$$\left|(f_i g_0)(x)\right| \le \left|(f_0 g_0)(x)\right|, \qquad i = 1, \dots, r,$$

imply $|f_i(x)| \le |f_0(x)|$ for all i and, hence $x \in U$. Similarly, we get $x \in V$ and therefore $x \in U \cap V$ so that we have $W \subset U \cap V$ and, hence, $W = U \cap V$.

Finally, that the intersection of Laurent or Weierstraß domains in X is of the same type again is trivial.   □

Corollary 15. *Let $X = \operatorname{Sp} A$ be an affinoid K-space. Each Weierstraß domain in X is Laurent, and each Laurent domain in X is rational.*

Proof. That Weierstraß domains are Laurent is trivial. Furthermore, a Laurent domain in X is a finite intersection of rational domains of type $X\left(\frac{f}{1}\right)$ and $X\left(\frac{1}{g}\right)$ for suitable functions $f, g \in A$. By Proposition 14, such an intersection is rational.   □

Proposition 16. *Let $X = \operatorname{Sp} A$ be an affinoid K-space and $U \subset X$ a rational subdomain. Then there is a Laurent domain $U' \subset X$ such that U is contained in U' as a Weierstraß domain.*

Proof. Let $U = \operatorname{Sp} A' = X\left(\frac{f_1}{f_0}, \dots, \frac{f_r}{f_0}\right)$ with functions f_i having no common zero on X. Then, as $|f_i(x)| \le |f_0(x)|$ for all i, we must have $f_0(x) \ne 0$ for all $x \in U$. Consequently, the restriction $f_0|_U$ of f_0 to U, which is meant as the image of f in A' via the morphism $A \longrightarrow A'$ given by the affinoid subdomain $\operatorname{Sp} A' \hookrightarrow \operatorname{Sp} A$, is a unit in A'. Applying the Maximum Principle 3.1/15 to $(f_0|_U)^{-1}$, there is a constant $c \in K^*$ such that $|c f_0(x)| \ge 1$ for all $x \in U$. But then, setting $U' = X\left((c f_0)^{-1}\right)$, we have $U \subset U'$ and, in fact,

$$U = U'\left(f_1|_{U'} \cdot (f_0|_{U'})^{-1}, \dots, f_r|_{U'} \cdot (f_0|_{U'})^{-1}\right)$$

where $f_0|_{U'}$ is a unit on U'. So U is a Weierstraß domain in U' and U' is a Laurent domain in X, as claimed.   □

Proposition 17 (Transitivity of Special Affinoid Subdomains). *Let X be an affinoid K-space, V a Weierstraß (resp. rational) domain in X, and U a Weierstraß (resp. rational) domain in V. Then U is a Weierstraß (resp. rational) domain in X. In view of Proposition 16, the assertion does not extend to Laurent domains.*

Proof. Let $X = \operatorname{Sp} A$. Starting with the case of Weierstraß domains, let us write $V = X(f)$ and $U = V(g)$ for a tuple f of functions in A and a tuple g of functions in $A\langle f \rangle$, the affinoid K-algebra of V. As the image of A is dense in $A\langle f \rangle$ and as we may subtract from g a tuple of supremum norm ≤ 1 without changing $U = V(g)$ (use the non-Archimedean triangle inequality), we may assume that g is (the restriction of) a tuple of functions in A. But then we can write $U = X(f, g)$ and we are done.

It remains to look at the case of rational domains. So let $V = X\left(\frac{f_1}{f_0}, \ldots, \frac{f_r}{f_0}\right)$ with functions $f_0, \ldots, f_r \in A$ having no common zero on X. Using the fact that U is a Weierstraß domain in a Laurent domain of V, cf. Proposition 16, as well as the fact that the intersection of finitely many rational domains is a rational domain again, cf. Proposition 14, it is enough to consider the cases where $U = V(g)$ or $U = V(g^{-1})$ with a single function g in $A\langle \frac{f_1}{f_0}, \ldots, \frac{f_r}{f_0}\rangle$, the affinoid algebra of V. As the image of $A[f_0^{-1}]$ is dense in this algebra and as we may subtract from g a function of supremum norm < 1 without changing $V(g)$ or $V(g^{-1})$, we may assume that there is an integer $n \in \mathbb{N}$ such that $f_0^n g$ extends to a function $g' \in A$. Then, as f_0 has no zero on V, we have

$$V(g) = V \cap \{x \in X \; ; \; |g'(x)| \le |f_0^n(x)|\},$$

$$V(g^{-1}) = V \cap \{x \in X \; ; \; |g'(x)| \ge |f_0^n(x)|\}.$$

Now applying the Maximum Principle 3.1/15 to $f_0^{-n}|_V$, we see that there is a constant $c \in K^*$ such that $|f_0^n(x)| \ge |c|$ for all $x \in V$. But then we can write

$$V(g) = V \cap X\left(\frac{g'}{f_0^n}, \frac{c}{f_0^n}\right), \qquad V(g^{-1}) = V \cap X\left(\frac{f_0^n}{g'}, \frac{c}{g'}\right),$$

and it follows from Proposition 14 that $V(g)$ and $V(g^{-1})$ are rational subdomains of X. $\qquad\qquad\square$

Using Proposition 5, we can conclude from Proposition 17 in conjunction with Corollary 15 that, for any Weierstraß, Laurent, or rational subdomain U of a rigid K-space X, the canonical topology of X restricts to the canonical topology of U; furthermore, U is open in X by Lemma 8. We want to generalize this to arbitrary affinoid subdomains of rigid K-spaces.

Lemma 18. *Let $\varphi: Y \longrightarrow X$ be a morphism of affinoid K-spaces with associated morphism of affinoid K-algebras $\varphi^*: A \longrightarrow B$, and let $x \in X$ be a point corresponding to a maximal ideal $\mathfrak{m} \subset A$.*

(i) *Assume that φ^* induces a surjection $A/\mathfrak{m} \longrightarrow B/\mathfrak{m}B$. Then there is a special affinoid subdomain $X' \hookrightarrow X$ containing x such that the resulting morphism $\varphi': Y' \longrightarrow X'$ induced from φ on $Y' = \varphi^{-1}(X')$ is a closed immersion in the sense that the corresponding morphism of affinoid K-algebras $\varphi'^*: A' \longrightarrow B'$ is surjective.*

(ii) *Assume that φ^* induces isomorphisms $A/\mathfrak{m}^n \xrightarrow{\sim} B/\mathfrak{m}^n B$ for all $n \in \mathbb{N}$. Then there is a special affinoid subdomain $X' \hookrightarrow X$ containing x such that the resulting morphism $\varphi': Y' \longrightarrow X'$ induced from φ on $Y' = \varphi^{-1}(X')$ is an isomorphism.*

Proof. We start with a general remark. Since $A/\mathfrak{m}$ is a field, we see that the surjection $A/\mathfrak{m} \longrightarrow B/\mathfrak{m}B$ is either an isomorphism or the zero mapping. Hence, $\mathfrak{m}B$ is either a maximal ideal in B or the unit ideal. Using this observation in conjunction with Lemma 10 and Propositions 13 and 17, we see that we may replace X without loss of generality by a special affinoid subdomain $X' \subset X$ containing x and Y by $Y' = \varphi^{-1}(X')$.

In the situation of (i) we choose affinoid generators $b_1, \ldots, b_n$ of B over A. Thereby we mean power bounded elements $b_i \in B$ giving rise to a surjection

$$\Phi^* : A\langle \zeta_1, \ldots, \zeta_r \rangle \longrightarrow B, \qquad \zeta_i \longmapsto b_i, \qquad i = 1, \ldots, r,$$

extending φ^*. Note that such generators exist, since $A \neq 0$ and since B, as an affinoid K-algebra, admits an epimorphism $T_r \longrightarrow B$ for some $r \in \mathbb{N}$. Let $m_1, \ldots, m_s$ generate the maximal ideal $\mathfrak{m} \subset A$. Then, as φ^* induces a surjection $A/\mathfrak{m} \longrightarrow B/\mathfrak{m}B$, there are elements $a_i \in A$ and $c_{ij} \in B$, $i = 1, \ldots, r$, $j = 1, \ldots, s$, such that

$$b_i - \varphi^*(a_i) = \sum_{j=1}^{s} c_{ij} m_j, \qquad i = 1, \ldots, r. \tag{$*$}$$

Choosing a residue norm $|\cdot|$ on A, we consider on $A\langle \zeta_1, \ldots, \zeta_r \rangle$ the natural (Gauß) norm derived from $|\cdot|$ and on B the residue norm via Φ^*. Furthermore, for any Weierstraß domain $X(f) \subset X$, we can consider the morphism $\varphi' : Y(\varphi^*(f)) \longrightarrow X(f)$ induced from φ, as well as the resulting commutative diagram

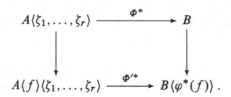

Going back to the explicit construction of $A\langle f \rangle$ and $B\langle \varphi^*(f) \rangle$ in the proof of Proposition 11, we get residue norms on the algebras in the lower row such that all morphisms of the diagram are contractive. Furthermore, Φ'^* is surjective, just as Φ^* is.

Adjusting norms via constants in K^* on the right-hand sides of the equations $(*)$, we can assume $|c_{ij}| \leq 1$ for all i, j. As explained in the beginning, we may replace X by a special affinoid subdomain $X' \subset X$ containing x. For example, we may take $X' = X(c^{-1}m_1, \ldots, c^{-1}m_s)$ for some $c \in K$, $0 < |c| < 1$ and thereby assume that

$$|b_i - \varphi^*(a_i)| \leq |c| < 1, \qquad i = 1, \ldots, r.$$

Then, as $|b_i| \leq 1$, we have $|\varphi^*(a_i)| \leq 1$ and, in particular, $|\varphi^*(a_i)|_{\text{sup}} \leq 1$ for all i. Now, if the fiber $\varphi^{-1}(x)$ is empty, there is an equation $\sum_{i=1}^r m_i g_i = 1$ with certain elements $g_i \in B$, and we can take $|c|$ small enough such that $\psi^{-1}(X')$ is empty. The assertion of (i) is trivial in this case. On the other hand, if the fiber $\varphi^{-1}(x)$ is non-empty, it consists of a single point $y \in Y$. Then we have

$$\left|a_i(x)\right| = \left|a_i(\varphi(y))\right| = \left|\varphi^*(a_i)(y)\right| \leq \left|\varphi^*(a_i)\right|_{\text{sup}} \leq 1$$

for all i, and we can, in fact, replace X by $X(a_1, \ldots, a_r, c^{-1}m_1, \ldots, c^{-1}m_s)$, thereby assuming

$$\left|a_i\right| \leq 1, \qquad \left|b_i - \varphi^*(a_i)\right| \leq |c| < 1, \qquad i = 1, \ldots, r.$$

Now by 3.1/5, the estimates above say that we can approximate every element $b \in B$ with $|b| \leq |c|^t$ for some $t \in \mathbb{N}$ by an element of type $\varphi^*(a)$ with $a \in A$ such that

$$\left|a\right| \leq |c|^t, \qquad \left|b - \varphi^*(a)\right| \leq |c|^{t+1} < |c|^t.$$

A standard limit argument shows then, that $\varphi^* : A \longrightarrow B$ is surjective.

It remains to verify (ii). As the assumption of (ii) includes the one of (i), we may assume that $\varphi^* : A \longrightarrow B$ is surjective. Furthermore, we get

$$\ker \varphi^* \subset \bigcap_{n \in \mathbb{N}} \mathfrak{m}^n.$$

By Krull's Intersection Theorem (see 7.1/2), there is an element $f \in A$ of type $f = 1 - m$ for some $m \in \mathfrak{m}$ such that f annihilates the kernel $\ker \varphi^*$. Since $A \longrightarrow A\langle f^{-1} \rangle$ factors through $A[f^{-1}]$, the kernel of φ^* is contained in the kernel of $A \longrightarrow A\langle f^{-1} \rangle$. Thus, there is a canonical diagram

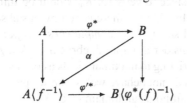

where the square is commutative, as well as the upper triangle, and where φ^* and φ'^* are surjective. But then, using the surjectivity of φ^*, also the lower triangle is commutative. Now consider the morphisms

$$A \xrightarrow{\varphi^*} B \xrightarrow{\alpha} A\langle f^{-1} \rangle$$

whose composition equals the canonical morphism $A \longrightarrow A\langle f^{-1} \rangle$. In other words, the canonical morphism $\operatorname{Sp} A\langle f^{-1} \rangle \hookrightarrow \operatorname{Sp} A$ factors through $\operatorname{Sp} B$. By restriction to inverse images over $\operatorname{Sp} A\langle f^{-1} \rangle \subset \operatorname{Sp} A$, we get morphisms

$$A\langle f^{-1} \rangle \longrightarrow B\langle \varphi^*(f)^{-1} \rangle \longrightarrow A\langle f^{-1} \rangle$$

whose composition is the identity. As $A\langle f^{-1}\rangle \longrightarrow B\langle \varphi^*(f)^{-1}\rangle$ is still surjective and necessarily injective, it is an isomorphism. Thus, $\varphi: Y \longrightarrow X$ restricts to an isomorphism $\varphi': Y(\varphi^*(f)^{-1}) \xrightarrow{\sim} X(f^{-1})$, and as $x \in X(f^{-1})$, we are done. $\square$

Proposition 19. *Let $U \longrightarrow X$ be a morphism of rigid K-spaces defining U as an affinoid subdomain of X. Then U is open in X, and the canonical topology of X restricts to the one of U.*

Proof. By Lemma 10 (ii), the morphism $U \longrightarrow X$ satisfies the conditions of Lemma 18 (ii). $\square$

To characterize the structure of general affinoid subdomains in more precise terms, we cite already at this place the following result:

Theorem 20 (Gerritzen–Grauert). *Let X be an affinoid K-space and $U \subset X$ an affinoid subdomain. Then U is a finite union of rational subdomains of X.*

A more general version of this theorem will be proved in Sect. 4.2; see 4.2/10 and 4.2/12. However, it should be noted that, in general, a finite union of affinoid subdomains of X, even of Weierstraß domains, does not yield an affinoid subdomain again.

To end this section, we want to explain why it is not advisable to consider the spectrum of *all* prime ideals of a given affinoid K-algebra as the point set of its associated affinoid K-space, as is the rule when dealing with affine schemes in algebraic geometry. A first observation shows for a prime ideal $\mathfrak{p}$ of some affinoid K-algebra A that its residue field $K_{\mathfrak{p}}$, i.e. the field of fractions of $A/\mathfrak{p}$, will in general be of infinite degree over K. In this case $K_{\mathfrak{p}}$ cannot be viewed as an affinoid K-algebra since, otherwise, $K_{\mathfrak{p}}$ would be finite over K by Noether Normalization 3.1/3 (iii). In addition, there is no obvious absolute value on $K_{\mathfrak{p}}$ that extends the one of K and satisfies the *completeness property*. So, in particular, it will not be possible to consider affinoid algebras over $K_{\mathfrak{p}}$. Another, may be more convincing reason for restricting to maximal ideals as points, consists in the fact that non-maximal prime ideals do not behave well when we pass back and forth from an affinoid K-space X to an affinoid subdomain $U \subset X$.

To exhibit such a behavior, let $A \longrightarrow A'$ be the morphism of affinoid K-algebras corresponding to an affinoid subdomain $U \hookrightarrow X$. For a prime ideal $\mathfrak{p} \subset A$ we can consider the Zariski closed subset $Y = V(\mathfrak{p}) \subset X$. Then we see with the help of Lemma 10 that the restriction of Y to U equals the Zariski closed subset $Y \cap U = V(\mathfrak{p}A')$ of U. If $\mathfrak{p}$ is a maximal ideal in A corresponding to a point $x \in U$ then $\mathfrak{p}A'$ is maximal in A'. However, for a non-maximal prime ideal $\mathfrak{p} \subset A$, the ideals $\mathfrak{p}A'$ or $\mathrm{rad}(\mathfrak{p}A')$ do not need to be prime, even if $V(\mathfrak{p}) \cap U \neq \emptyset$. Just look at the following example. Let $X = \mathrm{Sp}\, T_1$ be the unit disk with coordinate function ζ (the variable of T_1) and consider the Weierstraß subdomain

$$U = \left\{ x \in X \; ; \; \left| \zeta(x) \cdot (\zeta(x) - 1) \right| \leq \varepsilon \right\} \subset X$$

for some value $\varepsilon = |c| < 1$ where $c \in K$. Then U is the disjoint union of the Weierstraß subdomains

$$U_1 = \{x \in X \, ; \, |\zeta(x)| \leq \varepsilon\}, \qquad U_2 = \{x \in X \, ; \, |\zeta(x) - 1| \leq \varepsilon\}$$

in X. Looking at corresponding affinoid K-algebras, one can conclude by a direct approximation argument or, more easily, by applying Tate's Theorem 4.3/1 (to be proved in Chap. 4) that $T_1\langle c^{-1}\zeta(\zeta - 1)\rangle$, the affinoid K-algebra of U, is the direct product of two integral domains, namely the affinoid K-algebras corresponding to U_1 and U_2:

$$T_1\langle c^{-1}\zeta(\zeta - 1)\rangle \simeq T_1\langle c^{-1}\zeta\rangle \times T_1\langle c^{-1}(\zeta - 1)\rangle.$$

Therefore, working with full prime spectra instead of maximal spectra, we see that the "generic" point of X, which corresponds to the zero-ideal in T_1, gives rise to two different points on U, namely the "generic" point of U_1 *and* the "generic" point of U_2.

This is the first problem we encounter when dealing with affinoid subdomains in terms of full prime spectra instead of spectra of maximal ideals. But worse than that, it can happen that there exist prime ideal points in an affinoid subdomain $U \subset X$ that are not visible at all on X. In terms of the corresponding morphism of affinoid K-algebras $A \longrightarrow A'$ this means that there can exist non-maximal prime ideals $\mathfrak{p}' \subset A'$ such that the prime ideal $\mathfrak{p} = \mathfrak{p}' \cap A \subset A$ does not satisfy $\mathfrak{p}A' = \mathfrak{p}'$. Interpreting this phenomenon on the level of Zariski closed subsets, we can start with $Y' = V(\mathfrak{p}') \subset U$ and see from 3.2/3 that $Y = V(\mathfrak{p} \cap A)$ is the Zariski closure of Y' in X. Then, indeed, it can happen that the restriction $Y \cap U$ is strictly bigger than Y' and, thus, that there is no Zariski closed subset in X that restricts to Y' on U. To give an example we first need to show:

Example 21. *Assume that the valuation on K is not discrete. Then there exists a non-trivial formal power series $f = \sum_{\nu=1}^{\infty} c_\nu \zeta^\nu \in K[\![\zeta]\!]$ such that:*

(i) *the coefficients $c_\nu \in K$ satisfy $|c_\nu| < 1$ and, hence, f converges on the open unit disk $\mathbb{B}^1_+ = \{x \in \overline{K} \, ; \, |x| < 1\}$,*

(ii) *f has infinitely many zeros on $\mathbb{B}^1_+$.*

Proof. We choose a sequence of coefficients $c_0, c_1, \ldots \in K$ such that the corresponding sequence of absolute values is strictly ascending and bounded by 1. Then $\lim_{\nu \to \infty} c_\nu \varepsilon^\nu = 0$ for all $\varepsilon \in \mathbb{R}$, $0 \leq \varepsilon < 1$, and the series $f = \sum_{\nu=1}^{\infty} c_\nu \zeta^\nu$ converges on $\mathbb{B}^1_+$.

For $\varepsilon \in |K|$, $0 < \varepsilon < 1$, let $\sigma(\varepsilon)$ be the largest index ν where the sequence $|c_\nu|\varepsilon^\nu$, $\nu = 0, 1, \ldots$, assumes its maximum. Note that $\sigma(\varepsilon)$ tends to infinity when ε approaches 1 from below. Now choose $c \in K$, $0 < |c| < 1$, and set $\varepsilon = |c|$. Using $\xi = c^{-1}\zeta$ as a new coordinate function, we can interpret the closed disk $\mathbb{B}_\varepsilon = \{x \in \mathbb{B}^1_+ \, ; \, |x| \leq \varepsilon\}$ as the affinoid unit disk $\operatorname{Sp} K\langle \xi \rangle$. Restricting f

to $\mathbb{B}_\varepsilon$ yields an element $f' \in K\langle\xi\rangle$ that is ξ-distinguished of order $\sigma(\varepsilon)$. But then f' is associated to a Weierstraß polynomial of degree $\sigma(\varepsilon)$ by 2.2/9, and it follows from 3.1/11 that f has $\sigma(\varepsilon)$ zeros in $\mathbb{B}_\varepsilon$. Thus, if ε approaches 1 and, hence, $\sigma(\varepsilon)$ approaches infinity, we see that f has infinitely many zeros on $\mathbb{B}_+^1 = \bigcup_{0<\varepsilon<1} \mathbb{B}_\varepsilon$.

$\square$

Example 22. Now, for K equipped with a non-discrete complete valuation as before, we can construct an affinoid K-space X with an affinoid subdomain $U \subset X$ where U admits a non-maximal prime ideal point that is not induced by a point of the same type on X. Let $X = \mathrm{Sp}\, T_2 = \mathrm{Sp}\, K\langle\zeta_1,\zeta_2\rangle$ be the two-dimensional unit ball and consider the Weierstraß domains

$$X_\varepsilon = \left\{x \in X \,;\, |\zeta_1(x)| \le \varepsilon\right\} \subset X, \qquad \varepsilon \in |K|, 0 < \varepsilon < 1.$$

Then the affinoid K-algebra corresponding to X_ε is $A_\varepsilon = T_2\langle c_\varepsilon^{-1}\zeta_1\rangle \simeq T_2$ where $c_\varepsilon \in K$ is a constant satisfying $|c_\varepsilon| = \varepsilon$. Furthermore, let $f \in K[\![\zeta_1]\!]$ be a formal power series as constructed in Example 21. So f is non-trivial, converges on the open unit disk $\mathbb{B}_+^1$, has infinitely many zeros on $\mathbb{B}_+^1$, and assumes values < 1. In particular, $\zeta_2 + f(\zeta_1)$ induces a well-defined element $h_\varepsilon \in A_\varepsilon$ for each ε as before. All elements h_ε are prime since the continuous morphism of K-algebras given by

$$A_\varepsilon \longrightarrow A_\varepsilon, \qquad \zeta_1 \longmapsto \zeta_1, \qquad \zeta_2 \longmapsto \zeta_2 + f(\zeta_1),$$

is an isomorphism and, hence, $A_\varepsilon/(h_\varepsilon) \simeq K\langle c_\varepsilon^{-1}\zeta_1\rangle$ is an integral domain. As all maximal ideals in A_ε are of height 2, cf. 2.2/17, we see clearly that the prime ideals $\mathfrak{p}_\varepsilon = h_\varepsilon A_\varepsilon \subset A_\varepsilon$ satisfy $\mathfrak{p}_\varepsilon \cap A_{\varepsilon'} = \mathfrak{p}_{\varepsilon'}$ as well as $\mathfrak{p}_{\varepsilon'} A_\varepsilon = \mathfrak{p}_\varepsilon$ for $\varepsilon < \varepsilon'$.

Next write $A = T_2$ for the affinoid K-algebra corresponding to X and look at some $\varepsilon \in |K|$, $0 < \varepsilon < 1$. Then the prime ideal $\mathfrak{p} = \mathfrak{p}_\varepsilon \cap A \subset A$ is independent of ε. We claim that, in fact, $\mathfrak{p} = 0$. First, $\mathfrak{p}$ cannot be maximal, since otherwise $\mathfrak{p}A_\varepsilon \subset \mathfrak{p}_\varepsilon$ would be maximal; use Lemma 10. Choosing an element $h \in \mathfrak{p}$, the inclusion $\mathfrak{p}A_\varepsilon \subset \mathfrak{p}_\varepsilon$ shows that the image of h in A_ε is a multiple of h_ε. Now let us restrict our situation to the Zariski closed subset $Y = V(\zeta_2) \subset X$, a process that on the level of affinoid K-algebras is realized by dividing out ideals generated by ζ_2. Then $Y = \mathrm{Sp}\, K\langle\zeta_1\rangle$ is the unit disk and the restriction $Y_\varepsilon = X_\varepsilon \cap Y$ gives rise to the closed subdisk $\mathrm{Sp}\, K\langle\zeta_1\rangle\langle c_\varepsilon^{-1}\zeta_1\rangle$ that is a Weierstraß domain in Y. Furthermore, h_ε induces on Y_ε the element given by the series f, as we have to divide out the ideal generated by ζ_2. Remembering that the image of h in A_ε is a multiple of h_ε and letting ε vary, we see that h restricts on each Y_ε to a multiple of f. Since f has an infinity of zeros on the open unit disk $\mathbb{B}_+^1$, the element h' induced by h on Y must have an infinity of zeros as well. However, due to Weierstraß theory, see 2.2/9, non-zero elements can only have finitely many zeros on Y. Hence, we must have $h' = 0$ and therefore $h \in \zeta_2 \cdot A$ so that if h varies over $\mathfrak{p}$ we get $\mathfrak{p} \subset \zeta_2 \cdot A$. As $\mathfrak{p}$ and $\zeta_2 \cdot A$ are prime ideals in A and $\zeta_2 \cdot A$ is of height 1 by Krull's Dimension Theorem, see [Bo], 2.4/6, we get $\mathfrak{p} = \zeta_2 \cdot A$ if $\mathfrak{p}$ is non-trivial. Then we would have

$\zeta_2 \in \mathfrak{p}_\varepsilon$ and, hence, $\mathfrak{p}_\varepsilon = \zeta_2 \cdot A_\varepsilon$ for all ε by an argument as before. However, h_ε is not divisible by ζ_2, which means that the only remaining possibility is $\mathfrak{p} = 0$.

To conclude our example, fix some $\varepsilon \in |K|$, $0 < \varepsilon < 1$, and consider the morphism of affinoid K-algebras $A \longrightarrow A_\varepsilon$ corresponding to the Weierstraß subdomain $X_\varepsilon \subset X$. Then it follows for the prime ideal $\mathfrak{p}_\varepsilon \subset A_\varepsilon$ that there cannot exist any ideal $\mathfrak{p} \subset A$ satisfying $\mathfrak{p} A_\varepsilon = \mathfrak{p}_\varepsilon$.

Chapter 4
Affinoid Functions

4.1 Germs of Affinoid Functions

Let X be an affinoid K-space. For any affinoid subdomain $U \subset X$ we denote by $\mathcal{O}_X(U)$ the affinoid K-algebra corresponding to U. Then, if $U \subset V$ is an inclusion of affinoid subdomains of X, there is a canonical morphism between the corresponding affinoid K-algebras $\mathcal{O}_X(V) \longrightarrow \mathcal{O}_X(U)$, which we might interpret as restriction of affinoid functions on V to affinoid functions on U. More precisely, $\mathcal{O}_X$ is a *presheaf* of affinoid K-algebras on the category of affinoid subdomains of X. This means that $\mathcal{O}_X$ associates to any affinoid subdomain $U \subset X$ an affinoid K-algebra $\mathcal{O}_X(U)$ and to any inclusion $U \subset V$ of affinoid subdomains in X a morphism of affinoid K-algebras $\rho_U^V \colon \mathcal{O}_X(V) \longrightarrow \mathcal{O}_X(U)$ (generally denoted by $f \longmapsto f|_U$) such that for subdomains $U \subset V \subset W$ of X the following conditions are fulfilled:

(i) $\rho_U^U = \mathrm{id}$,

(ii) $\rho_U^W = \rho_U^V \circ \rho_V^W$.

The presheaf $\mathcal{O}_X$ will be referred to as the *presheaf of affinoid functions* on X.

For any point $x \in X$ the ring

$$\mathcal{O}_{X,x} = \varinjlim_{x \in U} \mathcal{O}_X(U)$$

where the limit runs over all affinoid subdomains $U \subset X$ containing x, is called the *stalk* of $\mathcal{O}_X$ at x. Its elements are called *germs of affinoid functions at x*. To give a more explicit characterization of $\mathcal{O}_{X,x}$, we can say that any germ $f_x \in \mathcal{O}_{X,x}$ is represented by some function $f \in \mathcal{O}_X(U)$ for some affinoid subdomain $U \subset X$ containing x and that two functions $f_i \in \mathcal{O}_X(U_i)$, $i = 1, 2$, with $x \in U_1 \cap U_2$ represent the same germ $f_x \in \mathcal{O}_{X,x}$ if and only if there is an affinoid subdomain $U \subset X$ such that $x \in U \subset U_1 \cap U_2$ and $\rho_U^{U_1}(f_1) = \rho_U^{U_2}(f_2)$. It is clear that the construction of germs of affinoid functions is functorial in the sense that a morphism

S. Bosch, *Lectures on Formal and Rigid Geometry*, Lecture Notes
in Mathematics 2105, DOI 10.1007/978-3-319-04417-0_4,
© Springer International Publishing Switzerland 2014

of affinoid K-spaces $\varphi \colon Y \longrightarrow X$ induces a homomorphism

$$\varphi_y^* \colon \mathcal{O}_{X,\varphi(y)} \longrightarrow \mathcal{O}_{Y,y},$$

for any $y \in Y$. All such morphisms φ_y^* are local.

Proposition 1. *Let* X *be an affinoid* K-*space and* $x \in X$ *a point corresponding to the maximal ideal* $\mathfrak{m} \subset \mathcal{O}_X(X)$. *Then* $\mathcal{O}_{X,x}$ *is a local ring with maximal ideal* $\mathfrak{m}\mathcal{O}_{X,x}$.

Proof. For any affinoid subdomain $U \subset X$ containing x, we know from 3.3/10 that the morphism of affinoid K-algebras $\mathcal{O}_X(X) \longrightarrow \mathcal{O}_X(U)$ induces an isomorphism $\mathcal{O}_X(X)/\mathfrak{m} \overset{\sim}{\longrightarrow} \mathcal{O}_X(U)/\mathfrak{m}\mathcal{O}_X(U)$. Thus, passing to the direct limit, we get a surjective map $\mathcal{O}_{X,x} \longrightarrow K'$ where $K' \simeq \mathcal{O}_X(X)/\mathfrak{m}$ is a field that is finite over K. The map may be viewed as evaluation at x, and we will use the notation $f_x \longmapsto f_x(x)$ for it. Its kernel $\mathfrak{n}$ is a maximal ideal in $\mathcal{O}_{X,x}$, and we claim that $\mathfrak{n} = \mathfrak{m}\mathcal{O}_{X,x}$. Clearly we have $\mathfrak{m}\mathcal{O}_{X,x} \subset \mathfrak{n}$. To show the converse, consider an element $f_x \in \mathcal{O}_{X,x}$ represented by some $f \in \mathcal{O}_X(U)$ for some affinoid subdomain $U \subset X$. Then, if $f_x(x) = 0$ we must have $f(x) = 0$ and, hence, $f \in \mathfrak{m}\mathcal{O}_X(U)$, which implies $f_x \in \mathfrak{m}\mathcal{O}_{X,x}$. Alternatively we could have used the fact that $\varinjlim$ preserves exact sequences.

That $\mathfrak{n}$ is the only maximal ideal in $\mathcal{O}_{X,x}$ is easy to see. Consider an element $f_x \in \mathcal{O}_{X,x} - \mathfrak{n}$ represented by some $f \in \mathcal{O}_X(U)$ for some affinoid subdomain $U \subset X$. Then $f(x) \neq 0$ and, multiplying f by a suitable constant in K^*, we can even assume that $|f(x)| \geq 1$. But then $U(f^{-1})$ contains x and is an affinoid subdomain of X such that $f|_{U(f^{-1})}$ is a unit. Consequently, f_x is a unit in $\mathcal{O}_{X,x}$, and $\mathfrak{n}$ is the only maximal ideal in $\mathcal{O}_{X,x}$. $\square$

Proposition 2. *Let* $X = \mathrm{Sp}\, A$ *be an affinoid* K-*space and* $x \in X$ *a point corresponding to the maximal ideal* $\mathfrak{m} \subset A$. *Then the canonical map* $A \longrightarrow \mathcal{O}_{X,x}$ *decomposes into*

$$A \longrightarrow A_{\mathfrak{m}} \longrightarrow \mathcal{O}_{X,x}$$

where the first map is the canonical map of A *into its localization at* $\mathfrak{m}$ *and the second one is injective. Furthermore, these maps induce isomorphisms*

$$A/\mathfrak{m}^n \overset{\sim}{\longrightarrow} A_{\mathfrak{m}}/\mathfrak{m}^n A_{\mathfrak{m}} \overset{\sim}{\longrightarrow} \mathcal{O}_{X,x}/\mathfrak{m}^n \mathcal{O}_{X,x}, \qquad n \in \mathbb{N},$$

so that one obtains isomorphisms

$$\widehat{A} \overset{\sim}{\longrightarrow} \widehat{A_{\mathfrak{m}}} \overset{\sim}{\longrightarrow} \widehat{\mathcal{O}}_{X,x}$$

between the $\mathfrak{m}$-*adic completion of* A *and the maximal adic completions of* $A_{\mathfrak{m}}$ *and* $\mathcal{O}_{X,x}$.

Proof. For any affinoid subdomain $\operatorname{Sp} A' \subset \operatorname{Sp} A$ with $x \in \operatorname{Sp} A'$, the restriction maps $A \longrightarrow A' \longrightarrow \mathcal{O}_{X,x}$ induce maps

$$A/\mathfrak{m}^n \xrightarrow{\sigma_n} A'/\mathfrak{m}^n A' \xrightarrow{\tau_n} \mathcal{O}_{X,x}/\mathfrak{m}^n \mathcal{O}_{X,x}, \qquad n \in \mathbb{N},$$

and we claim that these are isomorphisms. For the σ_n this is clear from 3.3/10, and it is enough to show the same for the compositions $\tau_n \circ \sigma_n$. To do this, we may vary $\operatorname{Sp} A'$ as a neighborhood of x and take it as small as we want. As any element $f_x \in \mathcal{O}_{X,x}$ is represented by an element $f \in A'$ if $\operatorname{Sp} A'$ is small enough, we see that $\tau_n \circ \sigma_n$ is surjective. To show injectivity, consider an element $f \in A$ such that its image $f_x \in \mathcal{O}_{X,x}$ is contained in $\mathfrak{m}^n \mathcal{O}_{X,x}$. Writing $f_x = \sum_{i=1}^r g_{xi} \cdot m_i$ with germs $g_{xi} \in \mathcal{O}_{X,x}$ and elements $m_i \in \mathfrak{m}^n$, we can assume that the g_{xi} are represented by functions $g_i \in A'$. Choosing $\operatorname{Sp} A'$ small enough, we can even assume that $f|_{\operatorname{Sp} A'}$ coincides with $\sum_{i=1}^r g_i \cdot m_i$ on $\operatorname{Sp} A'$. But then we have $f|_{\operatorname{Sp} A'} \in \mathfrak{m}^n A'$ and, hence, by 3.3/10, even $f \in \mathfrak{m}^n$. This shows that $\tau_n \circ \sigma_n$ is injective and, hence, bijective. Alternatively, we could have used the fact that $\varinjlim$ is exact.

For $n = 1$, we see again that $\mathfrak{m}\mathcal{O}_{X,x}$ is a maximal ideal in $\mathcal{O}_{X,x}$ restricting to $\mathfrak{m}$ on A. As $\mathfrak{m}\mathcal{O}_{X,x}$ is the only maximal ideal in $\mathcal{O}_{X,x}$, it follows that the map $A \longrightarrow \mathcal{O}_{X,x}$ decomposes into the canonical map $A \longrightarrow A_\mathfrak{m}$ from A into its localization at $\mathfrak{m}$ and a map $A_\mathfrak{m} \longrightarrow \mathcal{O}_{X,x}$. As the canonical maps $A/\mathfrak{m}^n \longrightarrow A_\mathfrak{m}/\mathfrak{m}^n A_\mathfrak{m}$ are bijective, they induce a bijection

$$\widehat{A} = \varprojlim_n A/\mathfrak{m}^n \xrightarrow{\sim} \varprojlim_n A_\mathfrak{m}/\mathfrak{m}^n A_\mathfrak{m} = \widehat{A_\mathfrak{m}},$$

and we see that the map obtained from $A \longrightarrow A_\mathfrak{m}$ via $\mathfrak{m}$-adic completion is bijective. In the same way the bijective maps $\tau_n \circ \sigma_n$ give rise to a bijective map

$$\widehat{A} = \varprojlim_n A/\mathfrak{m}^n \xrightarrow{\sim} \varprojlim_n \mathcal{O}_{X,x}/\mathfrak{m}^n \mathcal{O}_{X,x} = \widehat{\mathcal{O}}_{X,x},$$

which is the $\mathfrak{m}$-adic completion of $A \longrightarrow \mathcal{O}_{X,x}$. As $\widehat{A} \longrightarrow \widehat{\mathcal{O}}_{X,x}$ is the composition of $\widehat{A} \longrightarrow \widehat{A}_\mathfrak{m}$ and $\widehat{A}_\mathfrak{m} \longrightarrow \widehat{\mathcal{O}}_{X,x}$, also the latter map is bijective. Finally, that $A_\mathfrak{m} \longrightarrow \mathcal{O}_{X,x}$ is injective, follows from the fact that, due to Krull's Intersection Theorem (see 7.1/2), the composition $A_\mathfrak{m} \longrightarrow \mathcal{O}_{X,x} \longrightarrow \widehat{\mathcal{O}}_{X,x} = \widehat{A}_\mathfrak{m}$ is injective. $\square$

We want to derive some direct consequences of the injectivity of the map $A_\mathfrak{m} \longrightarrow \mathcal{O}_{X,x}$ in Proposition 2.

Corollary 3. *An affinoid function f on some affinoid K-space X is trivial if and only if all its germs $f_x \in \mathcal{O}_{X,x}$ at points $x \in X$ are trivial.*

Proof. Writing $X = \operatorname{Sp} A$, the assertion is clear from the injections

$$A \lhook\joinrel\longrightarrow \prod_{\mathfrak{m} \in \operatorname{Max} A} A_{\mathfrak{m}} \lhook\joinrel\longrightarrow \prod_{x \in X} \mathcal{O}_{X,x}.$$

□

Corollary 4. *Let X be an affinoid K-space and $X = \bigcup_{i \in I} X_i$ a covering by affinoid subdomains. Then the restriction maps $\mathcal{O}_X(X) \longrightarrow \mathcal{O}_X(X_i)$ define an injection*

$$\mathcal{O}_X(X) \lhook\joinrel\longrightarrow \prod_{i \in I} \mathcal{O}_X(X_i).$$

Corollary 5. *For any affinoid subdomain $X' = \operatorname{Sp} A'$ of some affinoid K-space $X = \operatorname{Sp} A$, the restriction map $A \longrightarrow A'$ is flat.*

Proof. We use Bourbaki's criterion on flatness; see [AC], Chap. III, § 5, no. 2. For any maximal ideal $\mathfrak{m} \subset A$ corresponding to a point in X', we know from Proposition 2 that the map $A \longrightarrow A'_{\mathfrak{m}A'}$ gives rise to an isomorphism $\widehat{A} \overset{\sim}{\longrightarrow} \widehat{A'_{\mathfrak{m}A'}}$ between $\mathfrak{m}$-adic completions. It follows from loc. cit. § 5.4, Prop. 4, that $A \longrightarrow A'_{\mathfrak{m}A'}$ is flat. Varying $\mathfrak{m}$ over the points of X', we see that $A \longrightarrow A'$ is flat.   □

Proposition 6. *For any point x of an affinoid K-space X, the local ring $\mathcal{O}_{X,x}$ is Noetherian.*

Proof. Let $X = \operatorname{Sp} A$ and let $\mathfrak{m} \subset A$ be the maximal ideal corresponding to x. Then the local ring $\mathcal{O}_{X,x}$ is $\mathfrak{m}$-adically separated, i.e. $\bigcap_{n \in \mathbb{N}} \mathfrak{m}^n \mathcal{O}_{X,x} = 0$. In fact, consider an element $f_x \in \bigcap_{n \in \mathbb{N}} \mathfrak{m}^n \mathcal{O}_{X,x}$. There is an affinoid subdomain $U \subset X$ containing x such that f_x is represented by some element $f \in \mathcal{O}_X(U)$ and it follows $f \in \mathfrak{m}^n \mathcal{O}_X(U)$ for each $n \in \mathbb{N}$ by Proposition 2; we may write $X = \operatorname{Sp} A$ instead of U again. Then it follows from Krull's Intersection Theorem, see 7.1/2, that the image of f in $A_{\mathfrak{m}}$ is trivial. In particular, $f_x = 0$.

In the same way we can show for any finitely generated ideal $\mathfrak{a}_x \subset \mathcal{O}_{X,x}$ that the residue ring $\mathcal{O}_{X,x}/\mathfrak{a}_x$ is $\mathfrak{m}$-adically separated. Indeed, fixing a finite generating system of $\mathfrak{a}_x$, we may assume that these generators extend to functions in A and, hence, that $\mathfrak{a}_x$ is induced from an ideal $\mathfrak{a} \subset A$. Then we can interpret $\mathcal{O}_{X,x}/\mathfrak{a}_x$ as a stalk of the affinoid space $\operatorname{Sp} A/\mathfrak{a}$ and see that it is $\mathfrak{m}$-adically separated. The latter says that finitely generated ideals in $\mathcal{O}_{X,x}$ are closed with respect to the $\mathfrak{m}$-adic topology on $\mathcal{O}_{X,x}$.

Now consider an ascending sequence of finitely generated ideals

$$\mathfrak{a}_1 \subset \mathfrak{a}_2 \subset \ldots \subset \mathcal{O}_{X,x},$$

as well as the corresponding sequence of ideals

$$\hat{a}_1 \subset \hat{a}_2 \subset \ldots \subset \hat{\mathcal{O}}_{X,x}$$

where $\hat{a}_i$ is the closure of a_i in $\hat{\mathcal{O}}_{X,x}$. We use that $\hat{\mathcal{O}}_{X,x} = \hat{A}_{\mathfrak{m}}$ is Noetherian, as it is the maximal adic completion of a Noetherian local ring. So the chain in $\hat{\mathcal{O}}_{X,x}$ becomes stationary. As $\mathcal{O}_{X,x}$ is $\mathfrak{m}$-adically separated, the canonical map $\mathcal{O}_{X,x} \longrightarrow \hat{\mathcal{O}}_{X,x}$ is injective. But then the closedness of the ideals $a_i \subset \mathcal{O}_{X,x}$ implies that also the chain in $\mathcal{O}_{X,x}$ must become stationary. Thus, $\mathcal{O}_{X,x}$ is Noetherian. □

4.2   Locally Closed Immersions of Affinoid Spaces

In the present section we want to characterize affinoid subdomains of affinoid K-spaces in local terms and thereby provide a proof of the Theorem of Gerritzen–Grauert 3.3/20.

Definition 1. *A morphism of affinoid K-spaces $\varphi \colon X' \longrightarrow X$ is called a* closed immersion *if the morphism of affinoid K-algebras $\varphi^* \colon \mathcal{O}_X(X) \longrightarrow \mathcal{O}_{X'}(X')$ corresponding to φ is surjective. Furthermore, φ is called a* locally closed immersion (*resp. an* open immersion) *if it is injective and, for every $x \in X'$, the induced morphism $\varphi_x^* \colon \mathcal{O}_{X,\varphi(x)} \longrightarrow \mathcal{O}_{X',x}$ is surjective (resp. bijective).*

For example, any morphism of affinoid K-spaces $\varphi \colon X' \longrightarrow X$ defining X' as an affinoid subdomain of X is an open immersion, due to the transitivity of affinoid subdomains mentioned in 3.3/12. On the other hand, if φ is a closed immersion, one can see using Proposition 10 of Appendix B or, alternatively, with the help of 3.3/13 that φ is, in particular, a locally closed immersion. Furthermore, any composition of locally closed (resp. closed, resp. open) immersions is an immersion of the same type again.

At first sight it is not clear that the definition of a locally closed or open immersion $X' \hookrightarrow X$ will provide what is expected from such a terminology. However, we can conclude from 3.3/18 that there exists a family of special affinoid subdomains $U_i \subset X$, $i \in I$, such that $X' \subset \bigcup_{i \in I} U_i$ and the restrictions $X' \cap U_i \longrightarrow U_i$ are, indeed, closed immersions, respectively isomorphisms. The Theorem of Gerritzen–Grauert 3.3/20 will improve this fact and show that the U_i can be chosen large enough such that finitely many of them will suffice to cover X'.

Remark 2. *Let $\varphi \colon Y \longrightarrow X$ be a closed (resp. a locally closed, resp. an open) immersion of affinoid K-spaces. Then, for any affinoid subdomain $U \subset X$, the induced morphism $\varphi_U \colon \varphi^{-1}(U) \longrightarrow U$ is an immersion of the same type.*

Proof. The assertion is immediately clear for locally closed and open immersions φ, since these are characterized locally on X, and since affinoid subdomains $U \subset X$ are open, due to 3.3/19. Concerning closed immersions φ, the assertion is easily derived for Weierstraß, Laurent, or rational domains $U \subset X$, due to their explicit description. The remaining case of a closed immersion φ and a general affinoid subdomain $U \subset X$ is settled with the help of 3.3/13 from the fact that fiber products of affinoid K-spaces correspond to completed tensor products on the level of affinoid K-algebras and that, for two morphisms of affinoid K-algebras $\varphi^*: A \longrightarrow B$ and $A \longrightarrow A'$, the resulting morphism

$$\varphi \,\hat{\otimes}\, \mathrm{id}_{A'}: A' = A \,\hat{\otimes}_A\, A' \longrightarrow B \,\hat{\otimes}_A\, A'$$

is surjective when φ^* is surjective; use Proposition 10 of Appendix B. $\square$

More generally, one can show that closed (resp. locally closed, resp. open) immersions $Y \longrightarrow X$ are preserved under base change with *any* affinoid K-space Z over X, that is, the resulting morphism $Y \times_X Z \longrightarrow Z$ will be of the same type again.

Proposition 3. *Let $\varphi: X' \longrightarrow X$ be a locally closed immersion of affinoid K-spaces where the corresponding homomorphism of affinoid K-algebras is finite. Then φ is a closed immersion.*

Proof. Writing $X' = \mathrm{Sp}\, A'$ and $X = \mathrm{Sp}\, A$, the morphism φ induces for every $x \in X'$ a commutative diagram

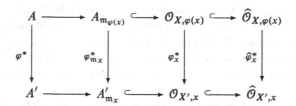

where $\mathfrak{m}_x \subset A'$ and $\mathfrak{m}_{\varphi(x)} \subset A$ denote the maximal ideals corresponding to x and $\varphi(x)$. Furthermore, $\varphi^*_{\mathfrak{m}_x}$, φ^*_x, and $\hat{\varphi}^*_x$ are the canonical extensions of φ^*. The injections in the middle of the first and second rows are due to 4.1/2, whereas the remaining ones on the right follow from the fact that $\mathcal{O}_{X,\varphi(x)}$ and $\mathcal{O}_{X',x}$, as Noetherian local rings (see 4.1/1 and 4.1/6) are maximal-adically separated. Since φ is injective, $\mathfrak{m}_x \subset A'$ is the only maximal ideal over $\mathfrak{m}_{\varphi(x)} \subset A$, and we therefore can view $A'_{\mathfrak{m}_x}$ as the localization of A' by the multiplicative system $\varphi^*(A - \mathfrak{m}_{\varphi(x)})$. Thus, since φ^* is finite, $\varphi^*_{\mathfrak{m}_x}$ will be finite, too.

The same argument shows that the $\mathfrak{m}_x$-adic topology of $A'_{\mathfrak{m}_x}$ coincides with the $\mathfrak{m}_{\varphi(x)}$-adic one, when $A'_{\mathfrak{m}_x}$ is viewed as an $A_{\mathfrak{m}_{\varphi(x)}}$-module via $\varphi^*_{\mathfrak{m}_x}$. Then, by Krull's Intersection Theorem, $\mathrm{im}\, \varphi^*_{\mathfrak{m}_x}$ is a closed submodule of $A'_{\mathfrak{m}_x}$. Now, since φ is a locally closed immersion, φ^*_x and, hence, $\hat{\varphi}^*_x$ are surjective. As a result, $\mathrm{im}\, \varphi^*_{\mathfrak{m}_x}$ is dense in $A'_{\mathfrak{m}_x}$ so that $\varphi^*_{\mathfrak{m}_x}$ must be surjective, too.

Now let $B = A/\ker\varphi^*$. Then φ^* gives rise to a homomorphism of Noetherian B-modules $B \longrightarrow A'$ that, as the above reasoning shows, reduces to an isomorphism when localized at any maximal ideal of B. Thus, by standard reasons, $B \longrightarrow A'$ is an isomorphism, and φ^* is surjective. Therefore φ is a closed immersion.   □

Proposition 4. *Let $\varphi\colon X' \longrightarrow X$ be a morphism of affinoid K-spaces that is an open and closed immersion. Then the image of X' is Zariski open and closed in X and, in particular, φ defines X' as a Weierstraß domain in X.*

Proof. Using notations as in the preceding proof, let $\varphi^*\colon A \longrightarrow A'$ be the morphism of affinoid K-algebras corresponding to φ, and consider the induced morphism $\varphi_{\mathfrak{m}_x}\colon A_{\mathfrak{m}_{\varphi(x)}} \longrightarrow A'_{\mathfrak{m}_x}$ at a point $x \in X'$. Then φ^* is surjective, since φ is a closed immersion, and $\varphi_{\mathfrak{m}_x}$ is, in fact, bijective. Indeed, $\varphi_{\mathfrak{m}_x}$ is surjective, since, as above, we may view $A'_{\mathfrak{m}_x}$ as the localization of A' by the multiplicative system $\varphi^*(A - \mathfrak{m}_{\varphi(x)})$. On the other hand, the commutative diagram

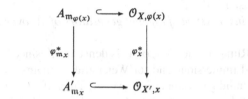

in conjunction with the injectivity of φ_x^* yields the injectivity of $\varphi_{\mathfrak{m}_x}^*$.

Now, since $\varphi_{\mathfrak{m}_x}^*$ is bijective, there is an element $f \in A$ such that $f(x) \neq 0$ and φ^* induces a bijection $A[f^{-1}] \overset{\sim}{\longrightarrow} A'[f^{-1}]$. Thus, letting x vary over X', we can conclude that $\varphi(X')$ is Zariski open in X. On the other hand, since $\varphi(X')$ is Zariski closed in X due to the fact that φ is a closed immersion, we see that A decomposes into a direct sum $A = A_1 \oplus A_2$ such that $\varphi^*\colon A \longrightarrow A'$ is the composition of the canonical projection $A \longrightarrow A_1$ and an isomorphism $A_1 \overset{\sim}{\longrightarrow} A'$. Choosing some unipotent element $e \in A$ that reduces to 0 on A_1 and to 1 on A_2, as well as a constant $c \in K$ with $|c| > 1$, it is easily seen that the projection $A \longrightarrow A_1$ corresponds to the Weierstraß subdomain $X(ce) \hookrightarrow X$, and we are done.   □

Next we introduce a particular class of locally closed immersions, so-called *Runge immersions*.

Definition 5. *A morphism of affinoid K-spaces $\varphi\colon X' \longrightarrow X$ is called a Runge immersion if it is the composition of a closed immersion $X' \longrightarrow W$ and an open immersion $W \longrightarrow X$ defining W as a Weierstraß domain in X.*

From Remark 2 we can immediately deduce:

Remark 6. *Let* $\varphi: X' \longrightarrow X$ *be a Runge immersion of affinoid K-spaces. Then, for any affinoid subdomain $U \subset X$, the induced morphism $\varphi_U: \varphi^{-1}(U) \longrightarrow U$ is a Runge immersion, too.*

If $\sigma: A \longrightarrow A'$ is a morphism of affinoid K-algebras, we call finitely many elements $h_1, \ldots, h_n \in A'$ a system of *affinoid generators* of A' over A (with respect to σ) if σ extends to an epimorphism

$$A\langle \zeta_1, \ldots, \zeta_n \rangle \longrightarrow A', \qquad \zeta_i \longmapsto h_i.$$

Of course, the $h_i \in A'$ are then necessarily power bounded.

Proposition 7. *For a morphism of affinoid K-algebras $\sigma: A \longrightarrow A'$ the following are equivalent:*

(i) *The morphism of affinoid K-spaces $\varphi: \operatorname{Sp} A' \longrightarrow \operatorname{Sp} A$ associated to σ is a Runge immersion.*
(ii) $\sigma(A)$ *is dense in A'.*
(iii) $\sigma(A)$ *contains a system of affinoid generators of A' over A.*

Proof. If φ is a Runge immersion, $\varphi(A)$ is dense in A', since the corresponding fact is true for closed immersions and for Weierstraß domains. Next, choose a system $h'_1, \ldots, h'_n$ of affinoid generators of A' over A. Then, if $\sigma(A)$ is dense in A', we can approximate each h'_i by some $h_i \in \sigma(A)$ in such a way that, using Lemma 8 below, $h_1, \ldots, h_n$ will be a system of affinoid generators of A' over A. Finally, assume that $h_1, \ldots, h_n \in \sigma(A)$ is a system of affinoid generators of A' over A. Then σ decomposes into the maps

$$A \longrightarrow A\langle h_1, \ldots, h_n \rangle \longrightarrow A'$$

where the first one corresponds to the inclusion of $X(h_1, \ldots, h_n)$ as a Weierstraß domain in $X = \operatorname{Sp} A$ and where the second is surjective and, hence, corresponds to a closed immersion $\operatorname{Sp} A' \longrightarrow X(h_1, \ldots, h_n)$. Thus, φ is a Runge immersion. $\square$

As a consequence we see that the composition of finitely many Runge immersions or, more specifically, closed immersions and inclusions of Weierstraß domains, yields a Runge immersion again.

Lemma 8. *Consider a morphism of affinoid K-algebras $\sigma: A \longrightarrow A'$ and a system $h' = (h'_1, \ldots, h'_r)$ of affinoid generators of A' over A. Fix a residue norm on A and consider on A' the residue norm via the epimorphism*

$$\pi': A\langle \zeta \rangle \longrightarrow A', \qquad \zeta \longmapsto h',$$

where we endow $A\langle\zeta\rangle = A\langle\zeta_1,\ldots,\zeta_n\rangle$ with the Gauß norm derived from the given residue norm on A. Then any system $h = (h_1,\ldots,h_n)$ in A' such that $|h'_i - h_i| < 1$ for all i, yields a system of affinoid generators of A' over A.

Proof. Since $|h'_i| \leq 1$, due to our assumption, we have $|h_i| \leq 1$ for all i and therefore can consider the morphism

$$\pi: A\langle\zeta\rangle \longrightarrow A', \qquad \zeta \longmapsto h.$$

Let $\varepsilon = \max_{i=1,\ldots,n} |h'_i - h_i|$ so that $\varepsilon < 1$. It is enough to show for any element $g \in A' = \mathrm{im}\,\pi'$ that there is some $f \in A\langle\zeta_1,\ldots,\zeta_n\rangle$ satisfying $|f| = |g|$ and $|\pi(f) - g| \leq \varepsilon|g|$. Then an iterative approximation argument shows that π is surjective.

Thus, start with an element $g \in A'$ and choose a π'-inverse $f = \sum_{\nu\in\mathbb{N}^n} a_\nu \zeta^\nu$ in $A\langle\zeta\rangle$ with coefficients $a_\nu \in A$; we may assume $|f| = |g|$ by 3.1/5. Then

$$|\pi(f) - g| = \left|\sum_{\nu\in\mathbb{N}^n} a_\nu h^\nu - \sum_{\nu\in\mathbb{N}^n} a_\nu h'^\nu\right|$$

$$= \left|\sum_{\nu\in\mathbb{N}^n} a_\nu (h^\nu - h'^\nu)\right| \leq \varepsilon \max_{\nu\in\mathbb{N}^n} |a_\nu| = \varepsilon|g|,$$

as required.   □

Next, we want to derive a certain extension lemma for Runge immersions. To do this, let K_a be an algebraic closure of K and write K_a^* for its multiplicative group, as well as $|K_a^*|$ for the corresponding value group. Then $|K_a^*|$ consists of all real numbers $\alpha > 0$ such that there is some integer $s > 0$ satisfying $\alpha^s \in K^*$. Furthermore, let $X = \mathrm{Sp}\,A$ be an affinoid K-space and consider functions $f_1,\ldots,f_r,g \in A$ generating the unit ideal. Then, for any $\varepsilon \in |K_a^*|$, we may consider the subset

$$X_\varepsilon = \{x \in X ; |f_j(x)| \leq \varepsilon|g(x)|, \ j = 1,\ldots,r\} \subset X.$$

If $\varepsilon^s = |c|$ for some $c \in K^*$, the set X_ε is characterized by the estimates

$$|f_j^s(x)| \leq |cg^s(x)|, \qquad j = 1,\ldots,r,$$

and therefore defines a *rational subdomain* in X. Given a morphism of affinoid K-spaces $\varphi: X' \longrightarrow X$, we set $X'_\varepsilon = \varphi^{-1}(X_\varepsilon)$ and consider the morphism $\varphi_\varepsilon: X'_\varepsilon \longrightarrow X_\varepsilon$ induced by φ.

Extension Lemma 9. *Assume that the morphism $\varphi_{\varepsilon_0}: X'_{\varepsilon_0} \longrightarrow X_{\varepsilon_0}$ defined as before is a Runge immersion for some $\varepsilon_0 \in |K_a^*|$. Then there is an $\varepsilon \in |K_a^*|, \varepsilon > \varepsilon_0$, such that $\varphi_\varepsilon: X'_\varepsilon \longrightarrow X_\varepsilon$ is a Runge immersion as well.*

Proof. Write $X = \operatorname{Sp} A$ and $X' = \operatorname{Sp} A'$, as well as $X_\varepsilon = \operatorname{Sp} A_\varepsilon$ and $X'_\varepsilon = \operatorname{Sp} A'_\varepsilon$ for $\varepsilon \in |K_a^*|$. Replacing X by $X_{\varepsilon'}$ and X' by $X'_{\varepsilon'}$ for some $\varepsilon' \in |K_a^*|$, $\varepsilon' > \varepsilon_0$, we may assume that all X_ε and X'_ε are Weierstraß domains in X and X', respectively. Then, for $\varepsilon \in |K_a^*|$, $\varepsilon \geq \varepsilon_0$, we have a canonical commutative diagram

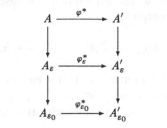

where the vertical maps all have dense images, since, on the level of affinoid spaces, they correspond to inclusions of Weierstraß domains. Now let $h' = (h'_1, \ldots, h'_n)$ be a system of affinoid generators of A' over A. Then h' gives rise to a system h'_ε of affinoid generators of A'_ε over A_ε, as well as to a system h'_{ε_0} of affinoid generators of A'_{ε_0} over A_{ε_0}.

Let us restrict ourselves for a moment to values $\varepsilon \in |K^*|$. In particular, we assume $\varepsilon_0 \in |K^*|$. Fixing a residue norm on A, we consider on A' the residue norm with respect to the epimorphism

$$\pi: A\langle \zeta_1, \ldots, \zeta_n \rangle \longrightarrow A', \qquad \zeta_i \longmapsto h'_i,$$

and on each A_ε the residue norm with respect to the epimorphism

$$p_\varepsilon: A\langle \varepsilon^{-1}\eta_1, \ldots, \varepsilon^{-1}\eta_r \rangle \longrightarrow A_\varepsilon, \qquad \eta_j \longmapsto \frac{f_j}{g},$$

where, strictly speaking, the element ε in the expression $\varepsilon^{-1}\eta_j$ has to be replaced by a constant $c \in K$ with $|c| = \varepsilon$ and where the elements $\varepsilon^{-1}\eta_j$ have to be viewed as variables. Then we can introduce on any A'_ε the residue norm via the epimorphism

$$\pi_\varepsilon: A_\varepsilon\langle \zeta_1, \ldots, \zeta_n \rangle \longrightarrow A'_\varepsilon, \qquad \zeta_i \longmapsto h'_i.$$

The latter equals the residue norm that is derived from the one of A via the epimorphism

$$\tau_\varepsilon: A\langle \zeta_1, \ldots, \zeta_n, \varepsilon^{-1}\eta_1, \ldots, \varepsilon^{-1}\eta_r \rangle \longrightarrow A'_\varepsilon, \qquad \zeta_i \longmapsto h'_i, \qquad \eta_j \longmapsto \frac{f_j}{g},$$

satisfying

$$\ker \tau_\varepsilon = (\ker \pi, g\eta_1 - f_1, \ldots, g\eta_r - f_r).$$

Now choose a system $h = (h_1, \ldots, h_n)$ of elements in A', having φ^*-inverses in A, and whose images in A'_{ε_0} satisfy

$$\left| h'_i \right|_{X'_{\varepsilon_0}} - h_i \left|_{X'_{\varepsilon_0}} \right| < 1, \qquad i = 1, \ldots, n.$$

The latter is possible, since the image of A is dense in A'_{ε_0}, due to the fact that X_{ε_0} is a Weierstraß domain in X and $\varphi_{\varepsilon_0} \colon X'_{\varepsilon_0} \longrightarrow X_{\varepsilon_0}$ is a Runge immersion. Then it follows from Lemma 8 that h gives rise to a system of affinoid generators of A'_{ε_0} over A_{ε_0}.

In order to settle the assertion of the Extension Lemma, it is enough to show that, in fact,

$$\left| h'_i \right|_{X'_{\varepsilon}} - h_i \left|_{X'_{\varepsilon}} \right| < 1, \qquad i = 1, \ldots, n, \qquad (*)$$

for some $\varepsilon > \varepsilon_0$. Then, using Lemma 8 again, $h|_{X'_{\varepsilon}}$ is a system of affinoid generators of A'_{ε} over A_{ε} belonging to the image of A_{ε}, and it follows from Proposition 7 that $\varphi_{\varepsilon} \colon X'_{\varepsilon} \longrightarrow X_{\varepsilon}$ is a Runge immersion in this case.

To abbreviate, let $d_{\varepsilon} = h'_i|_{X'_{\varepsilon}} - h_i|_{X'_{\varepsilon}} \in A'_{\varepsilon}$ for any $i \in \{1, \ldots, n\}$. Furthermore, fix $\varepsilon_1 \in |K^*|$ with $\varepsilon_1 > \varepsilon_0$ and choose an element $g_{\varepsilon_1} \in A\langle \zeta, \varepsilon_1^{-1}\eta \rangle$ with $\tau_{\varepsilon_1}(g_{\varepsilon_1}) = d_{\varepsilon_1}$ where $\zeta = (\zeta_1, \ldots, \zeta_n)$ and $\eta = (\eta_1, \ldots, \eta_r)$. For $\varepsilon \leq \varepsilon_1$, let g_{ε} be the image of g_{ε_1} in $A\langle \zeta, \varepsilon^{-1}\eta \rangle$ so that $\tau_{\varepsilon}(g_{\varepsilon}) = d_{\varepsilon}$ for all $\varepsilon \leq \varepsilon_1$. Now, by the choice of h_i, we have $|d_{\varepsilon_0}| < 1$. Thus, using 3.1/5, there is an element

$$g_0 \in \ker \tau_{\varepsilon_0} = (\ker \pi, g\eta_1 - f_1, \ldots, g\eta_r - f_r) A\langle \zeta, \varepsilon_0^{-1}\eta \rangle$$

such that $|g_{\varepsilon_0} + g_0| < 1$. Approximating functions in $A\langle \zeta, \varepsilon_0^{-1}\eta \rangle$ by polynomials in $A\langle \zeta \rangle [\varepsilon_0^{-1}\eta]$, we may assume that g_0 is induced by an element

$$g_1 \in \ker \tau_{\varepsilon_1} = (\ker \pi, g\eta_1 - f_1, \ldots, g\eta_r - f_r) A\langle \zeta, \varepsilon_1^{-1}\eta \rangle.$$

But then we may replace from the beginning g_{ε_1} by $g_{\varepsilon_1} - g_1$ and thereby assume $|g_{\varepsilon_0}| < 1$.

Now let

$$g_{\varepsilon_1} = \sum_{\mu \in \mathbb{N}^n, \nu \in \mathbb{N}^r} a_{\mu\nu} \zeta^{\mu} \eta^{\nu} \in A\langle \zeta, \varepsilon_1^{-1}\eta \rangle$$

with coefficients $a_{\mu\nu} \in A$. Since $|g_{\varepsilon_0} < 1|$, we get $\max_{\mu \in \mathbb{N}^n, \nu \in \mathbb{N}^r} |a_{\mu\nu}| \varepsilon_0^{|\nu|} < 1$. Passing from ε_0 to a slightly bigger ε (not necessarily contained in $|K^*|$), we still have

$$|g_{\varepsilon}| = \max_{\mu \in \mathbb{N}^n, \nu \in \mathbb{N}^r} |a_{\mu\nu}| \varepsilon^{|\nu|} < 1$$

for $\varepsilon > \varepsilon_0$ sufficiently close to ε_0. Thus, if such ε exist in $|K^*|$, the series g_ε is a well-defined element in $A\langle \zeta, \varepsilon^{-1}\eta \rangle$ satisfying $|d_\varepsilon| \leq |g_\varepsilon| < 1$ as required in (∗). This settles the assertion of the Extension Lemma in the case where $\varepsilon_0 \in |K^*|$ and the valuation on K is non-discrete.

In the general case, we can always enlarge the value group $|K^*|$ by passing to a suitable finite algebraic extension L/K. This way, we can assume $\varepsilon_0 \in |L^*|$ and, in addition, that the last step in the above argumentation works for some $\varepsilon > \varepsilon_0$ contained in $|L^*|$. In other words, the assertion of the Extension Lemma holds after replacing the base field K by a suitable finite algebraic extension L in the sense that we apply to our situation the base change functor

$$\operatorname{Sp} A \longmapsto \operatorname{Sp} A \,\hat{\otimes}_K L$$

where $\hat{\otimes}$ is the completed tensor product of Appendix B. Thus, it is enough to show that a morphism of affinoid K-spaces $X' \longrightarrow X$ is a Runge immersion if the corresponding morphism of affinoid L-spaces $X' \,\hat{\otimes}_K L \longrightarrow X \,\hat{\otimes}_K L$ has this property or, equivalently, that a morphism of affinoid K-algebras $A \longrightarrow A'$ has dense image if the corresponding morphism of affinoid L-algebras $A \,\hat{\otimes}_K L \longrightarrow A' \,\hat{\otimes}_K L$ has dense image. However, the latter is easy to see. Since the completed tensor product commutes with finite direct sums, see the discussion following Proposition 2 of Appendix B, it follows that the canonical morphism $A \otimes_K L \longrightarrow A \,\hat{\otimes}_K L$ is bijective for any affinoid K-algebra A and any *finite* extension L/K. Now consider a morphism of affinoid K-algebras $\sigma: A \longrightarrow A'$, and let $A'' \subset A'$ be the closure of $\sigma(A)$. Then the morphism $\sigma \otimes_K L: A \otimes_K L \longrightarrow A' \otimes_K L$ factors through the closed subalgebra $A'' \otimes_K L \subset A' \otimes_K L$. If $\sigma \otimes_K L$ has dense image, we see that $A'' \otimes_K L$ coincides with $A' \otimes_K L$ and, hence, by descent, that the same is true for A'' and A'. Thus, we are done.   □

Next, let us look more closely at the structure of locally closed immersions. We begin by stating the main structure theorem for such immersions and by deriving some of its consequences.

Theorem 10 (Gerritzen–Grauert). *Let* $\varphi: X' \longrightarrow X$ *be a locally closed immersion of affinoid K-spaces. Then there exists a covering* $X = \bigcup_{i=1}^r X_i$ *consisting of finitely many rational subdomains* $X_i \subset X$ *such that φ induces Runge immersions* $\varphi_i: \varphi^{-1}(X_i) \longrightarrow X_i$ *for* $i = 1, \ldots, r$.

Corollary 11. *If, in the situation of Theorem* 10, $\varphi: X' \longrightarrow X$ *is an open immersion, then the maps φ_i define* $\varphi^{-1}(X_i)$ *as a Weierstraß domain in X_i, for* $i = 1, \ldots, r$.

Proof. It is enough to show that a Runge immersion $\varphi: X' \longrightarrow X$ that at the same time is an open immersion, defines X' as a Weierstraß domain in X. Since φ is the composition of a closed immersion $X' \hookrightarrow W$ and of a Weierstraß domain

$W \hookrightarrow X$, we may assume $W = X$ and thereby are reduced to the case where φ is a closed immersion. But then the assertion follows from Proposition 4. $\qquad \square$

Corollary 12. *Let X be an affinoid K-space and $X' \subset X$ an affinoid subdomain. Then there exists a covering $X = \bigcup_{i=1}^{r} X_i$ consisting of finitely many rational subdomains $X_i \subset X$ such that $X_i \cap X'$ is a Weierstraß domain in X_i for every i. In particular, X' is a finite union of rational subdomains in X.*

Proof. The inclusion $X' \hookrightarrow X$ is an open immersion. Thus, we may apply Corollary 11 and use the fact that all intersections $X_i \cap X'$ are rational subdomains of X by 3.3/17. $\qquad \square$

To approach the proof of Theorem 10, we generalize the concept of Weierstraß division introduced in Sect. 2.2. In the following, let A be an affinoid K-algebra and $\zeta = (\zeta_1, \ldots, \zeta_n)$ a system of variables. For any point $x \in \mathrm{Sp}\, A$ denote by $\mathfrak{m}_x \subset A$ the corresponding maximal ideal. Furthermore, given any series $f \in A\langle \zeta \rangle$, let $|f|_x$ be the Gauß (or supremum) norm of the residue class of f in $(A/\mathfrak{m}_x)\langle \zeta \rangle$.

A series $f \in A\langle \zeta \rangle$ is called ζ_n-*distinguished of order s* at a point $x \in \mathrm{Sp}\, A$ if its residue class in $(A/\mathfrak{m}_x)\langle \zeta \rangle$ is ζ_n-distinguished of order s in the sense of Definition 2.2/6. Furthermore, if f is ζ_n-distinguished of some order $\leq s$ at each point $x \in \mathrm{Sp}\, A$, we say that f is ζ_n-*distinguished of order $\leq s$* on $\mathrm{Sp}\, A$. As a first step, we generalize 2.2/7.

Lemma 13. *Let $f = \sum_{\nu \in \mathbb{N}^n} a_\nu \zeta^\nu$ be a series in $A\langle \zeta \rangle$ such that its coefficients $a_\nu \in A$ have no common zero on $\mathrm{Sp}\, A$. Then there is an A-algebra automorphism $\sigma \colon A\langle \zeta \rangle \longrightarrow A\langle \zeta \rangle$ such that, for some $s \in \mathbb{N}$, the series $\sigma(f)$ is ζ_n-distinguished of order $\leq s$ on $\mathrm{Sp}\, A$.*

Proof. We may assume $A \neq 0$. For $x \in \mathrm{Sp}\, A$, let $t_x - 1$ be the least upper bound of all natural numbers that occur in multi-indices $\nu \in \mathbb{N}^n$ satisfying $|a_\nu(x)| = |f|_x$. We claim that

$$t = \sup_{x \in \mathrm{Sp}\, A} t_x$$

is finite. As the coefficients a_ν of f do not have a common zero on $\mathrm{Sp}\, A$, there are finitely many indices $\nu(1), \ldots, \nu(r) \in \mathbb{N}^n$ such that $a_{\nu(1)}, \ldots, a_{\nu(r)}$ generate the unit ideal in A; see 3.2/6. Fixing an equation $\sum_{i=1}^{r} c_i a_{\nu(i)} = 1$, the coefficients $c_i \in A$ have finite supremum norm by 2.2/5, and it follows that there is some $\gamma > 0$ such that

$$\max_{i=1,\ldots,r} |a_{\nu(i)}(x)| \geq \gamma$$

for all $x \in \operatorname{Sp} A$. However, since the a_ν form a zero sequence with respect to any residue norm on A and, hence by 3.1/9, also with respect to the supremum norm on A, we see that all t_x are bounded. Consequently, t is finite.

Now, proceeding as in 2.2/7, we set $\alpha_1 = t^{n-1}, \ldots, \alpha_{n-1} = t$ and consider the A-algebra automorphism

$$\sigma \colon A\langle \zeta \rangle \longrightarrow A\langle \zeta \rangle, \qquad \zeta_i \longmapsto \begin{cases} \zeta_i + \zeta_n^{\alpha_i} & \text{for } i < n \\ \zeta_n & \text{for } i = n \end{cases}.$$

Then, as in the proof of 2.2/7, $\sigma(f)$ is ζ_n-distinguished of order $\leq s = \sum_{i=1}^{n} t^i$ at each point $x \in \operatorname{Sp} A$.   □

Lemma 14. *Let $f \in A\langle \zeta \rangle$ be ζ_n-distinguished of order $\leq s$ on $\operatorname{Sp} A$. Then the set*

$$\{x \in \operatorname{Sp} A \; ; \; f \text{ is } \zeta_n\text{-distinguished of order } s \text{ at } x\}$$

is a rational subdomain in $\operatorname{Sp} A$.

Proof. We write $f = \sum_{\nu=0}^{\infty} f_\nu \zeta_n^\nu$ with coefficients $f_\nu \in A\langle \zeta_1, \ldots, \zeta_{n-1} \rangle$. Let $a_\nu \in A$ be the constant term of f_ν. That f is ζ_n-distinguished of some order $s_x \leq s$ at a point $x \in \operatorname{Sp} A$ means that

$$|f_\nu|_x \leq |f_{s_x}|_x \text{ for } \nu \leq s_x,$$

$$|f_\nu|_x < |f_{s_x}|_x \text{ for } \nu > s_x, \text{ and}$$

the residue class of f_{s_x} is a unit in $A/\mathfrak{m}_x \langle \zeta_1, \ldots, \zeta_{n-1} \rangle$.

Since $a_\nu(x)$ is the constant term of the residue class of f_ν in $A/\mathfrak{m}_x \langle \zeta_1, \ldots, \zeta_{n-1} \rangle$, we see that $|a_\nu(x)| \leq |f_\nu|_x$, which is, in fact, an equality for $\nu = s_x$ by 2.2/4. Thus we get

$$\left| a_\nu(x) \right| \leq \left| f_\nu \right|_x \leq \left| f_{s_x} \right|_x = \left| a_{s_x}(x) \right| \text{ for } \nu \leq s_x,$$

$$\left| a_\nu(x) \right| \leq \left| f_\nu \right|_x < \left| f_{s_x} \right|_x = \left| a_{s_x}(x) \right| \text{ for } \nu > s_x,$$

which shows, in particular, that $a_{s_x}(x) \neq 0$. Thus, since f is ζ_n-distinguished of some order $\leq s$ at each point $x \in \operatorname{Sp} A$, the elements $a_0, \ldots, a_s$ cannot have a common zero in $\operatorname{Sp} A$. Therefore

$$U = \{x \in \operatorname{Sp} A \; ; \; \left| a_\nu(x) \right| \leq \left| a_s(x) \right|, \nu = 0, \ldots, s-1\}$$

is a rational subdomain in $\operatorname{Sp} A$, and the above estimates show that f is ζ_n-distinguished of order s at a point $x \in \operatorname{Sp} A$ if and only if $x \in U$.   □

Proposition 15. *As before, let A be an affinoid K-algebra and $\zeta = (\zeta_1,\ldots,\zeta_n)$ a system of variables. Then, for any $f \in A\langle\zeta\rangle$ that is ζ_n-distinguished of order precisely s at each point $x \in \operatorname{Sp} A$, the canonical map*

$$A\langle\zeta_1,\ldots,\zeta_{n-1}\rangle \longrightarrow A\langle\zeta\rangle/(f)$$

is finite.

Proof. We write $f = \sum_{\nu=0}^{\infty} f_\nu \zeta_n^\nu$ with elements $f_\nu \in A\langle\zeta_1,\ldots,\zeta_{n-1}\rangle$. Then, for any $x \in \operatorname{Sp} A$, the residue class of f_s in $A/\mathfrak{m}_s\langle\zeta_1,\ldots,\zeta_{n-1}\rangle$ is a unit. Therefore f_s cannot have any zeros and, consequently, is a unit in $A\langle\zeta_1,\ldots,\zeta_{n-1}\rangle$ by 3.2/6. Replacing f by $f_s^{-1} f$, we may assume $f_s = 1$ and, furthermore,

$$|f_\nu|_{\sup} \leq 1 \quad \text{for} \quad \nu \leq s,$$

$$|f_\nu|_{\sup} < 1 \quad \text{for} \quad \nu > s,$$

where $|\cdot|_{\sup}$ denotes the supremum norm on $A\langle\zeta\rangle$. Thus $\overline{\zeta}_n$, the residue class of ζ_n in $A\langle\zeta\rangle/(f)$, satisfies the following estimate:

$$\left|\overline{\zeta}_n^{\,s} + f_{s-1}\overline{\zeta}_n^{\,s-1} + \ldots + f_0\right|_{\sup} < 1 \qquad (*)$$

Now choose a system of variables $\eta = (\eta',\eta_m) = (\eta_1,\ldots,\eta_m)$ with $m > s$ large enough such that there exists an epimorphism $\tau': K\langle\eta'\rangle \longrightarrow A\langle\zeta_1,\ldots,\zeta_{n-1}\rangle$ sending the first s variables $\eta_1,\ldots,\eta_s$ to $f_0,\ldots,f_{s-1}$. Since $|f_\nu|_{\sup} \leq 1$, the latter is possible due to 3.1/19. Furthermore, we can extend τ' to an epimorphism $\tau: K\langle\eta\rangle \longrightarrow A\langle\zeta\rangle/(f)$ by sending η_m to the residue class $\overline{\zeta}_n$. Then, due to $(*)$, the polynomial $\eta_m^s + \eta_{s-1}\eta_m^{s-1} + \ldots + \eta_0 \in K\langle\eta\rangle$ is a Weierstraß polynomial in η_m satisfying $|\tau(\omega)|_{\sup} < 1$ and, by 3.1/18, there is some $r \in \mathbb{N}$ such that the image $\tau(\omega^r)$ has residue norm < 1 with respect to the epimorphism τ. Hence, using 3.1/5, we can get an equation $\omega^r = g + h$ with an element $g \in K\langle\eta\rangle$ of Gauß norm $|g| < 1$ and some $h \in \ker\tau$. But then $h = \omega^r - g$ is a η_m-distinguished element of the kernel $\ker\tau$, and it follows from the Weierstraß division formula 2.2/8 that τ induces a finite morphism $K\langle\eta'\rangle \longrightarrow A\langle\zeta\rangle/(f)$. Consequently, $A\langle\zeta_1,\ldots,\zeta_{n-1}\rangle \longrightarrow A\langle\zeta\rangle/(f)$ is finite, as claimed. $\square$

After these preparations, we can start now with the *proof of Theorem 10*. Let $\varphi^*: A \longrightarrow A'$ be the morphism of affinoid K-algebras corresponding to the locally closed immersion $\varphi: X' \longrightarrow X$. Furthermore, let $\langle A' : A\rangle$ be the minimum of all integers n such that there exists a system of affinoid generators of A' over A of length n. We will proceed by induction on $\langle A' : A\rangle$, setting $n = \langle A' : A\rangle$. The case $n = 0$ is trivial. Then φ^* is an epimorphism, which means that φ a closed immersion and, hence, also a Runge immersion.

For $n \geq 1$ consider an epimorphism $\tau: A\langle\zeta\rangle \longrightarrow A'$ extending φ^* where ζ indicates a system of variables $(\zeta_1,\ldots,\zeta_n)$. We claim that there exists an element

$f = \sum_{v \in \mathbb{N}^n} a_v \zeta^v \in \ker \tau \subset A\langle \zeta \rangle$ whose coefficients $a_v \in A$ do not have a common zero on X. To justify this claim, consider for any $x \in X$ and its maximal ideal $\mathfrak{m}_x \subset A$ the morphism

$$\tau_x : A/\mathfrak{m}_x \langle \zeta \rangle \longrightarrow A'/\mathfrak{m}_x A',$$

obtained from τ by tensoring with $A/\mathfrak{m}_x$ over A. Then τ_x has a non-trivial kernel, due to the fact that φ is injective and, hence, $A'/\mathfrak{m}_x A'$ is a local ring, whereas $A/\mathfrak{m}_x \langle \zeta \rangle$ is not local. Since $\ker \tau$ is mapped surjectively onto $\ker \tau_x$, we see that, for each $x \in X$, there exists an element $g \in \ker \tau$ that is non-trivial modulo $\mathfrak{m}_x$. Thus, there are finitely many series $g_1, \ldots, g_r \in \ker \tau \subset A\langle \zeta \rangle$ whose coefficients generate the unit ideal in A. Since A is Noetherian, we can find an integer $d \in \mathbb{N}$ such that the zero set in X of the coefficients of any g_i is already defined by the coefficients of g_i with an index of total degree $< d$. Then, choosing some $v \in \mathbb{N}^n$ with $v_1 + \ldots + v_n = d$, the series

$$f = g_1 + \zeta^v g_2 + \ldots \zeta^{(r-1)v} g_r$$

belongs to $\ker \tau$, and its coefficients will have no common zero on X.

Thus, our claim is justified and, applying Lemma 13, we can assume that $\ker \tau$ contains a series $f \in A\langle \zeta \rangle$ that is ζ_n-distinguished of order $\leq s$ on X, for some $s \geq 0$. Now use Lemma 14 and let $X^{(s)} = \operatorname{Sp} A^{(s)} \subset X$ be the affinoid subdomain consisting of all points $x \in X$ where f is ζ_n-distinguished of order s. We want to show that we can apply the induction hypothesis to the restricted morphism $\varphi^{(s)} : \varphi^{-1}(X^{(s)}) \longrightarrow X^{(s)}$ that corresponds to the morphism $\tau_0^{(s)} : A^{(s)} \longrightarrow A' \widehat{\otimes}_A A^{(s)}$ obtained from φ^* by tensoring with $A^{(s)}$ over A. Tensoring τ in the same way, we get a morphism

$$\tau^{(s)} : A^{(s)} \langle \zeta \rangle \longrightarrow A' \widehat{\otimes}_A A^{(s)}$$

that is surjective by Remark 2 and extends $\tau_0^{(s)}$. Let $f^{(s)}$ be the image of f in $A^{(s)} \langle \zeta \rangle$. Then $f^{(s)}$ is ζ_n-distinguished of order s at all points $x \in X^{(s)}$ and, since $f^{(s)} \in \ker \tau^{(s)}$, we can conclude from Proposition 15 that $\tau^{(s)}$ gives rise to a finite morphism

$$\tau'^{(s)} : A^{(s)} \langle \zeta_1, \ldots, \zeta_{n-1} \rangle \longrightarrow A' \widehat{\otimes}_A A^{(s)}$$

extending $\tau_0^{(s)}$. Clearly, $\tau'^{(s)}$ corresponds to a locally closed immersion of affinoid K-spaces, as the same is true for $\tau_0^{(s)}$. But then $\tau'^{(s)}$ must be surjective by Proposition 3 so that $\langle A' \widehat{\otimes}_A A^{(s)} : A^{(s)} \rangle \leq n - 1$. Therefore we can apply the induction hypothesis to $\varphi^{(s)} : \varphi^{-1}(X^{(s)}) \longrightarrow X^{(s)}$, and it follows that there is a covering $X^{(s)} = \bigcup_{i=1}^r X_i$ consisting of rational subdomains $X_i \subset X^{(s)}$ such that the induced maps $\varphi_i : \varphi^{-1}(X_i) \longrightarrow X_i$ are Runge immersions.

Next, we want to apply the Extension Lemma 9 to the Runge immersions φ_i. Due to the transitivity property of rational subdomains, see 3.3/17, the X_i may be viewed as rational subdomains in X. Thus, choosing functions in A that describe X_i as a rational subdomain in X, we can introduce rational subdomains $X_{i,\varepsilon} \subset X$ for $\varepsilon \in |K_a^*|$ where $X_{i,1} = X_i$, as in the context of the Extension Lemma. Then we can fix an $\varepsilon \in |K_a^*|$, $\varepsilon > 1$, such that φ_i extends to a Runge immersion $\varphi_{i,\varepsilon} : \varphi^{-1}(X_{i,\varepsilon}) \longrightarrow X_{i,\varepsilon}$. The latter works for $i = 1, \ldots, r$, and we may even assume that ε is independent of i. Now it is enough to construct rational subdomains $V_1, \ldots, V_\ell \subset X - X^{(s)}$ with the property that

$$X = \bigcup_{i=1}^{r} X_{i,\varepsilon} \cup \bigcup_{\lambda=1}^{\ell} V_\lambda. \qquad (*)$$

Let us justify this claim. If $V_\lambda = \operatorname{Sp} B_\lambda$, the above epimorphism $\tau : A\langle \zeta \rangle \longrightarrow A'$ restricts to epimorphisms

$$\tau_\lambda : B_\lambda \langle \zeta \rangle \longrightarrow A' \hat{\otimes}_A B_\lambda, \qquad \lambda = 1, \ldots, \ell,$$

so that $\langle A' \hat{\otimes}_A B_\lambda : B_\lambda \rangle \leq n$. Furthermore, the above series $f \in A\langle \zeta \rangle$ induces series $f_\lambda \in \ker \tau_\lambda$, $\lambda = 1, \ldots, \ell$, that are ζ_n-distinguished of order $\leq s - 1$ since $V_\lambda \cap X^{(s)} = \emptyset$. Thus, proceeding with τ_λ in exactly the same way as we did with the epimorphism $\tau : A\langle \zeta \rangle \longrightarrow A'$, we may lower the order of distinguishedness of f until it is 0. But then f cannot have any zero. Thus, it must be a unit, and we can conclude $X' = \emptyset$ from $f \in \ker \tau$.

It remains to establish the covering $(*)$. Starting with the case $r = 1$, we drop the index i and write

$$X_\varepsilon = X\left(\varepsilon^{-1} \frac{g_1}{g_{\ell+1}}, \ldots, \varepsilon^{-1} \frac{g_\ell}{g_{\ell+1}}\right),$$

for $\varepsilon \in |K_a^*|$ and suitable functions $g_1, \ldots, g_{\ell+1} \in A$. Furthermore, let

$$V_\lambda = X\left(\varepsilon \frac{g_{\ell+1}}{g_\lambda}, \frac{g_j}{g_\lambda} ; j \neq \lambda, \ell + 1\right),$$

for $\lambda = 1, \ldots, \ell$. Then the V_λ are rational subdomains in X, disjoint from X_1, and the covering $X = X_\varepsilon \cup \bigcup_{\lambda=1}^{\ell} V_\lambda$ is as desired.

Finally, if $r > 1$, we construct as before rational subdomains $V_{i,\lambda} \subset X - X_{i,1}$ such that $X - X_{i,\varepsilon} \subset \bigcup_{\lambda=1}^{\ell} V_{i,\lambda}$ for $i = 1, \ldots, r$. Then

$$X - \bigcup_{i=1}^{r} X_{i,\varepsilon} \subset \bigcup_{\lambda_1, \ldots, \lambda_r} V_{1,\lambda_1} \cap \ldots \cap V_{r,\lambda_r},$$

and we can derive a covering of type $(*)$ as desired.   □

4.3   Tate's Acyclicity Theorem

Let X be an affinoid K-space and $\mathfrak{T} = \mathfrak{T}_X$ the category of affinoid subdomains in X, with inclusions as morphisms. A presheaf $\mathcal{F}$ (of groups, rings, ...) on $\mathfrak{T}$ is called a *sheaf* if for all objects $U \in \mathfrak{T}$ and all coverings $U = \bigcup_{i \in I} U_i$ by objects $U_i \in \mathfrak{T}$ the following hold:

(S₁)   If $f \in \mathcal{F}(U)$ satisfies $f|_{U_i} = 0$ for all $i \in I$, then $f = 0$.

(S₂)   Given elements $f_i \in \mathcal{F}(U_i)$ such that $f_i|_{U_i \cap U_j} = f_j|_{U_i \cap U_j}$ for all indices $i, j \in I$, there is an $f \in \mathcal{F}(U)$ (necessarily unique by (S₁)) such that $f|_{U_i} = f_i$ for all $i \in I$.

So if $\mathcal{F}$ is a sheaf, we can say that, in a certain sense, the elements of the groups (or rings etc.) $\mathcal{F}(U)$ with U varying over $\mathfrak{T}$ can be constructed locally. In the present section we are interested in the case where $\mathcal{F}$ equals the presheaf $\mathcal{O}_X$ of affinoid functions on X. We know from 4.1/4 that condition (S₁) holds for $\mathcal{O}_X$. However, due to the total disconnectedness of the canonical topology on X, condition (S₂) cannot be satisfied for $\mathcal{O}_X$, except for trivial cases. So, in strict terms, $\mathcal{O}_X$ cannot be called a sheaf. Nevertheless, we will see that $\mathcal{O}_X$ satisfies the sheaf condition (S₂) for *finite* coverings $U = \bigcup_{i \in I} U_i$. This fact, which is a special case of Tate's Acyclicity Theorem, is basic for rigid geometry, and we will give a direct proof for it.

Sheaf conditions (S₁) and (S₂) can conveniently be phrased by requiring that the sequence

$$\mathcal{O}_X(U) \longrightarrow \prod_{i \in I} \mathcal{O}_X(U_i) \rightrightarrows \prod_{i,j \in I} \mathcal{O}_X(U_i \cap U_j),$$

$$f \longmapsto \left(f|_{U_i}\right)_{i \in I}, \qquad \left(f_i\right)_{i \in I} \longmapsto \begin{cases} \left(f_i|_{U_i \cap U_j}\right)_{i,j \in I} \\ \left(f_j|_{U_i \cap U_j}\right)_{i,j \in I} \end{cases} \qquad (*)$$

be exact for every $U \in \mathfrak{T}$ and every covering $\mathfrak{U} = (U_i)_{i \in I}$ of U by sets $U_i \in \mathfrak{T}$. Note that a sequence of maps $A \longrightarrow B \rightrightarrows C$ is called *exact* if A is mapped bijectively onto the subset of B consisting of all elements having same image under the maps $B \rightrightarrows C$. For a presheaf $\mathcal{F}$ on X and a covering $\mathfrak{U} = (U_i)_{i \in I}$ of X by affinoid subdomains $U_i \subset X$, we will say that $\mathcal{F}$ is a $\mathfrak{U}$-*sheaf*, if for all affinoid subdomains $U \subset X$ the sequence $(*)$ applied to the covering $\mathfrak{U}|_U = (U_i \cap U)_{i \in I}$ is exact.

Theorem 1 (Tate).   *Let X be an affinoid K-space. The presheaf $\mathcal{O}_X$ of affinoid functions is a $\mathfrak{U}$-sheaf on X for all* finite *coverings $\mathfrak{U} = (U_i)_{i \in I}$ of X by affinoid subdomains $U_i \subset X$.*

The *proof* will be done by reducing to more simple coverings where, finally, a direct computation is possible. We begin by discussing the necessary reduction steps. Consider two coverings $\mathfrak{U} = (U_i)_{i \in I}$ and $\mathfrak{V} = (V_j)_{j \in J}$ of X. Then $\mathfrak{V}$ is called a *refinement* of $\mathfrak{U}$ if there exists a map $\tau: J \longrightarrow I$ such that $V_j \subset U_{\tau(j)}$ for all $j \in J$. In the following, $\mathcal{F}$ will be any presheaf on X.

Lemma 2. *Let $\mathfrak{U} = (U_i)_{i \in I}$ and $\mathfrak{V} = (V_j)_{j \in J}$ be coverings of X by affinoid subdomains where $\mathfrak{V}$ is a refinement of $\mathfrak{U}$. Then, if $\mathcal{F}$ is a $\mathfrak{V}$-sheaf, it is a $\mathfrak{U}$-sheaf as well.*

Proof. We will show the exactness of $(*)$ for the covering $\mathfrak{U}$; the proof for its restriction $\mathfrak{U}|_U$ on any affinoid subdomain $U \subset X$ works in the same way. So consider elements $f_i \in \mathcal{F}(U_i)$, $i \in I$, such that $f_i|_{U_i \cap U_{i'}} = f_{i'}|_{U_i \cap U_{i'}}$ for all $i, i' \in I$. Choosing a map $\tau: J \longrightarrow I$ such that $V_j \subset U_{\tau(j)}$, set $g_j = f_{\tau(j)}|_{V_j}$ for all $j \in J$. Then we have

$$
\begin{aligned}
g_j|_{V_j \cap V_{j'}} &= \left(f_{\tau(j)}|_{U_{\tau(j)} \cap U_{\tau(j')}} \right)|_{V_j \cap V_{j'}} \\
&= \left(f_{\tau(j')}|_{U_{\tau(j)} \cap U_{\tau(j')}} \right)|_{V_j \cap V_{j'}} = g_{j'}|_{V_j \cap V_{j'}}
\end{aligned}
$$

and, as $\mathcal{F}$ is a $\mathfrak{V}$-sheaf, there is a unique element $f \in \mathcal{F}(X)$ such that $f|_{V_j} = g_j$ for all $j \in J$. We claim that $f|_{U_i} = f_i$ for all $i \in I$. To check this, fix an index $i \in I$. Then

$$
\left(f|_{U_i} \right)|_{U_i \cap V_j} = f|_{U_i \cap V_j} = g_j|_{U_i \cap V_j}
$$

for $j \in J$. Furthermore,

$$
f_i|_{U_i \cap V_j} = f_i|_{U_i \cap U_{\tau(j)} \cap V_j} = f_{\tau(j)}|_{U_i \cap U_{\tau(j)} \cap V_j} = g_j|_{U_i \cap V_j},
$$

and, thus, we see that $f_i|_{U_i \cap V_j} = \left(f|_{U_i} \right)|_{U_i \cap V_j}$. Now, using the fact that $\mathcal{F}$ is a $\mathfrak{V}$-sheaf when restricted to U_i, we see that necessarily $f|_{U_i} = f_i$ for all $i \in I$. Clearly, f is uniquely determined by these conditions, and it follows that $\mathcal{F}$ is a $\mathfrak{U}$-sheaf. $\qquad\square$

Lemma 3. *Let $\mathfrak{U} = (U_i)_{i \in I}$ and $\mathfrak{V} = (V_j)_{j \in J}$ be coverings of X by affinoid subdomains. Assume that*

(i) *$\mathcal{F}$ is a $\mathfrak{V}$-sheaf, and*
(ii) *the restriction of $\mathcal{F}$ to V_j is a $\mathfrak{U}|_{V_j}$-sheaf for all $j \in J$.*

Then $\mathcal{F}$ is a $\mathfrak{U}$-sheaf as well.

Proof. Again, we will show the exactness of the sequence $(*)$ for the covering $\mathfrak{U}$; the proof for the restriction $\mathfrak{U}|_U$ on any affinoid subdomain $U \subset X$ works in the same way. Thus, consider elements $f_i \in \mathcal{F}(U_i)$ such that $f_i|_{U_i \cap U_{i'}} = f_{i'}|_{U_i \cap U_{i'}}$ for all $i, i' \in I$. Then, fixing $j \in J$, we have

$$
f_i|_{U_i \cap U_{i'} \cap V_j} = f_{i'}|_{U_i \cap U_{i'} \cap V_j},
$$

and condition (ii) implies that there exists a unique element $g_j \in \mathcal{F}(V_j)$ such that $g_j|_{U_i \cap V_j} = f_i|_{U_i \cap V_j}$ for all $i \in I$. Fixing $j, j' \in J$, we get

$$g_j|_{U_i \cap V_j \cap V_{j'}} = f_i|_{U_i \cap V_j \cap V_{j'}} = g_{j'}|_{U_i \cap V_j \cap V_{j'}}$$

for all $i \in I$. Hence, again by condition (ii), we have $g_j|_{V_j \cap V_{j'}} = g_{j'}|_{V_j \cap V_{j'}}$, and by condition (i) there exists a unique element $g \in \mathcal{F}(X)$ satisfying $g|_{V_j} = g_j$ for all $j \in J$. Now by construction, g coincides with f_i, when we restrict to $U_i \cap V_j$, for all $i \in I$, $j \in J$. But then, by condition (i), we must have $g|_{U_i} = f_i$ for all $i \in I$. As g is uniquely determined by these conditions, as is easily verified, we see that $\mathcal{F}$ is a $\mathfrak{U}$-sheaf. $\square$

Next we want to look at particular types of coverings of our affinoid K-space $X = \mathrm{Sp}\, A$ to which we want to apply Lemmata 2 and 3. We will call a *finite* covering of X by affinoid subdomains an *affinoid covering*. Furthermore, choosing elements $f_0, \ldots, f_r \in A$ without common zeros, we can write

$$U_i = X\left(\frac{f_0}{f_i}, \ldots, \frac{f_r}{f_i}\right), \qquad i = 0, \ldots, n,$$

thereby obtaining a finite covering $\mathfrak{U} = (U_i)_{i=0\ldots r}$ of X by rational subdomains. $\mathfrak{U}$ is called a *rational covering* or, more precisely, the *rational covering* associated to $f_0, \ldots, f_r$.

Lemma 4. *Every affinoid covering* $\mathfrak{U} = (U_i)_{i \in I}$ *of* X *admits a rational covering as a refinement.*

Proof. Using the Theorem of Gerritzen–Grauert in the version of 4.2/12, we can assume that $\mathfrak{U}$ consists of rational subdomains, say $\mathfrak{U} = (U_i)_{i=1\ldots n}$ with

$$U_i = X\left(\frac{f_1^{(i)}}{f_0^{(i)}}, \ldots, \frac{f_{r_i}^{(i)}}{f_0^{(i)}}\right).$$

Now, consider the set I of all tuples $(\nu_1, \ldots, \nu_n) \in \mathbb{N}^n$ with $0 \leq \nu_i \leq r_i$ and set

$$f_{\nu_1 \ldots \nu_n} = \prod_{i=1}^{n} f_{\nu_i}^{(i)}$$

for such tuples. Writing I' for the set of all $(\nu_1, \ldots, \nu_n) \in I$ such that $\nu_i = 0$ for at least one i, we claim that the functions

$$f_{\nu_1 \ldots \nu_n}, \qquad (\nu_1 \ldots \nu_n) \in I',$$

do not have a common zero on X and, thus, generate a rational covering $\mathfrak{V}$ on X. To verify this, look at a point $x \in X$ where all these functions might vanish. Then there is an index j such that $x \in U_j$ and, hence, $f_0^{(j)}(x) \neq 0$. It follows that all products

$$\prod_{i \neq j} f_{\nu_i}^{(i)}, \qquad 0 \leq \nu_i \leq r_i,$$

must vanish at x. But this is impossible since, for each i, the functions $f_0^{(i)}, \ldots, f_{r_i}^{(i)}$ generate the unit ideal in $A = \mathcal{O}_X(X)$. Thus, the rational covering $\mathfrak{V}$ is well-defined.

It remains to check that $\mathfrak{V}$ is a refinement of $\mathfrak{U}$. To do this, consider a tuple $(\nu_1, \ldots, \nu_n) \in I'$ and look at the set

$$X_{\nu_1, \ldots, \nu_n} = X\left(\frac{f_{\mu_1 \ldots \mu_n}}{f_{\nu_1, \ldots, \nu_n}} \,;\; (\mu_1, \ldots, \mu_n) \in I'\right) \in \mathfrak{V}$$

where, for example, $\nu_n = 0$. We want to show that $X_{\nu_1, \ldots, \nu_n} \subset U_n$. Thus, choosing a point $x \in X_{\nu_1, \ldots, \nu_n}$ and an index μ_n, $0 \leq \mu_n \leq r_n$, we have to show

$$\left| f_{\mu_n}^{(n)}(x) \right| \leq \left| f_0^{(n)}(x) \right| = \left| f_{\nu_n}^{(n)}(x) \right|.$$

There exists an index j such that $x \in U_j$. If $j = n$, nothing is to be proved. So assume that j is different from n, say $j = 1$. Then it follows $\left| f_{\mu_1}^{(1)}(x) \right| \leq \left| f_0^{(1)}(x) \right|$ for $0 \leq \mu_1 \leq r_1$ and

$$\left(\prod_{i=1}^{n-1} \left| f_{\nu_i}^{(i)}(x) \right|\right) \cdot \left| f_{\mu_n}^{(n)}(x) \right| \leq \left| f_0^{(1)}(x) \right| \cdot \left(\prod_{i=2}^{n-1} \left| f_{\nu_i}^{(i)}(x) \right|\right) \cdot \left| f_{\mu_n}^{(n)}(x) \right| \leq \prod_{i=1}^{n} \left| f_{\nu_i}^{(i)}(x) \right|,$$

as the tuple $(0, \nu_2, \ldots, \nu_{n-1}, \mu_n)$ belongs to I'. Now, since $\prod_{i=1}^{n} f_{\nu_i}^{(i)}(x)$ does not vanish, we can divide by $\prod_{i=1}^{n-1} f_{\nu_i}^{(i)}(x)$ to obtain the desired inequality showing $X_{\nu_1, \ldots, \nu_n} \subset U_n$. $\qquad \square$

It is necessary to consider another special class of coverings of affinoid K-spaces $X = \operatorname{Sp} A$. Choose elements $f_1, \ldots, f_r \in A$. Then the sets

$$X\left(f_1^{\alpha_1}, \ldots, f_r^{\alpha_r}\right), \qquad \alpha_i \in \{+1, -1\},$$

form a finite covering of X by Laurent domains; it is called a *Laurent covering* or, more precisely, the *Laurent covering* associated to $f_1, \ldots, f_r$.

Lemma 5. *Let $\mathfrak{U}$ be a rational covering of X. Then there exists a Laurent covering $\mathfrak{V}$ of X such that, for each $V \in \mathfrak{V}$, the covering $\mathfrak{U}|_V$ is a rational covering of V that is generated by units in $\mathcal{O}_X(V)$.*

Proof. Let $f_0, \ldots, f_r \in \mathcal{O}_X(X)$ be functions without common zeros on X generating the rational covering $\mathfrak{U}$. As f_i is invertible on $U_i = X(\frac{f_0}{f_i}, \ldots, \frac{f_r}{f_i})$ and since its inverse assumes its maximum on U_i, we can find an element $c \in K^*$ such that

$$|c|^{-1} < \inf_{x \in X} \left(\max_{i=0\ldots r} |f_i(x)| \right).$$

Let $\mathfrak{V}$ be the Laurent covering of X generated by the elements $cf_0, \ldots, cf_r$. We claim that $\mathfrak{V}$ is as desired. To justify this, consider a set

$$V = X\big((cf_0)^{\alpha_0}, \ldots, (cf_r)^{\alpha_r}\big) \in \mathfrak{V}$$

where $\alpha_0, \ldots, \alpha_r \in \{+1, -1\}$. We may assume that $\alpha_0 = \ldots = \alpha_s = +1$ and that $\alpha_{s+1} = \ldots = \alpha_r = -1$ for some $s \geq -1$. Then

$$X\left(\frac{f_0}{f_i}, \ldots, \frac{f_r}{f_i}\right) \cap V = \emptyset$$

for $i = 0, \ldots, s$, since

$$\max_{i=0\ldots s} |f_i(x)| \leq |c|^{-1} < \max_{i=0\ldots r} |f_i(x)|$$

for $x \in V$. In particular, we have

$$\max_{i=0\ldots r} |f_i(x)| = \max_{i=s+1\ldots r} |f_i(x)|$$

for all $x \in V$, and $\mathfrak{U}|_V$ is the rational covering generated by $f_{s+1}|_V, \ldots, f_r|_V$. By construction, these elements are units in $\mathcal{O}_X(V)$.   □

Lemma 6. *Let $\mathfrak{U}$ be a rational covering of $X = \mathrm{Sp}\, A$ that is generated by units $f_0, \ldots, f_r \in \mathcal{O}_X(X)$. Then there exists a Laurent covering $\mathfrak{V}$ of X that is a refinement of $\mathfrak{U}$.*

Proof. Let $\mathfrak{V}$ be the Laurent covering of X generated by all products

$$f_i f_j^{-1}, \qquad 0 \leq i < j \leq r.$$

We claim that $\mathfrak{V}$ refines $\mathfrak{U}$. To verify this, consider a set $V \in \mathfrak{V}$. Given elements $i, j \in S = \{0, \ldots, r\}$, we write $i \ll j$ if $|f_i(x)| \leq |f_j(x)|$ for all $x \in V$. The resulting relation $\ll$ on S is transitive and total in the sense that, for arbitrary $i, j \in S$, we have always $i \ll j$ or $j \ll i$. Thus, there is an element $i_s \in S$ that is maximal with respect to $\ll$, and it follows that

$$V \subset X\left(\frac{f_0}{f_{i_s}}, \dots, \frac{f_r}{f_{i_s}}\right).$$

Hence, $\mathfrak{V}$ is a refinement of $\mathfrak{U}$. □

We can now sum up the essence of Lemmata 2 to 6.

Proposition 7. *Let $\mathcal{F}$ be a presheaf on the affinoid K-space X. If $\mathcal{F}$ is a $\mathfrak{U}$-sheaf for all Laurent coverings $\mathfrak{U}$ of X, then it is a $\mathfrak{V}$-sheaf for all affinoid coverings $\mathfrak{V}$ of X.*

Proof. Start with a general affinoid covering $\mathfrak{V}$ of X. We have to show that $\mathcal{F}$ is a $\mathfrak{V}$-sheaf, provided it is a $\mathfrak{U}$-sheaf for every Laurent covering $\mathfrak{U}$ of X. By Lemma 4 there is a rational covering refining $\mathfrak{V}$ and we may assume that $\mathfrak{V}$ itself is a rational covering, due to Lemma 2. Furthermore, using Lemma 3 in conjunction with Lemma 5, we may even assume that $\mathfrak{V}$ is a rational covering that is generated by units in $\mathcal{O}_X(X)$. But then, Lemma 6 in conjunction with Lemma 2 again reduces everything to the case where $\mathfrak{V}$ is a Laurent covering of X and we are done. □

Thus, we have seen that it is only necessary to do the *proof of Theorem* 1 for Laurent coverings. In fact, combining Lemma 3 with an inductive argument it is only necessary, to consider a Laurent covering generated by one single function $f \in A = \mathcal{O}_X(X)$. Then we have to show that the sequence

$$0 \longrightarrow A \overset{\varepsilon}{\longrightarrow} A\langle f \rangle \times A\langle f^{-1} \rangle \overset{\delta}{\longrightarrow} A\langle f, f^{-1} \rangle \longrightarrow 0,$$

$$f \overset{\varepsilon}{\longmapsto} (f|_{X(f)}, f|_{X(f^{-1})}), \qquad (f, g) \overset{\delta}{\longmapsto} f|_{X(f,f^{-1})} - g|_{X(f,f^{-1})},$$

is left exact. The sequence is part of the following commutative diagram:

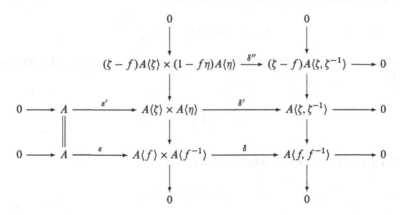

The symbols ζ, η denote indeterminates, ε' is the canonical injection, δ' is given by $(h_1(\zeta), h_2(\eta)) \longmapsto h_1(\zeta) - h_2(\zeta^{-1})$, and δ'' is induced by δ'. Furthermore, the vertical maps are characterized by $\zeta \longmapsto f$ and $\eta \longmapsto f^{-1}$, respectively. The first column of the diagram is exact due to the definition of $A\langle f \rangle$ and $A\langle f^{-1} \rangle$; cf. the proof of 3.3/11. Also the second column is exact since

$$A\langle f, f^{-1} \rangle = A\langle \zeta, \eta \rangle / (\zeta - f, 1 - f\eta)$$
$$= A\langle \zeta, \eta \rangle / (\zeta - f, 1 - \zeta\eta) = A\langle \zeta, \zeta^{-1} \rangle / (\zeta - f).$$

Clearly, δ' is surjective. Since

$$(\zeta - f)A\langle \zeta, \zeta^{-1} \rangle = (\zeta - f)A\langle \zeta \rangle + (1 - f\zeta^{-1})A\langle \zeta^{-1} \rangle,$$

the same is true for δ''. Thus, the first row is exact. Furthermore, also the second row is exact, since

$$0 = \delta'\left(\sum_{i=0}^{\infty} a_i \zeta^i, \sum_{i=0}^{\infty} b_i \eta^i\right) = \sum_{i=0}^{\infty} a_i \zeta^i - \sum_{i=0}^{\infty} b_i \zeta^{-i}$$

implies $a_i = b_i = 0$ for $i > 0$ and $a_0 - b_0 = 0$. Finally, looking at the third row, the injectivity of ε follows from 4.1/4, and the exactness of this row follows by diagram chase. This concludes the proof of Tate's Acyclicity Theorem in the version of Theorem 1. □

Next, without giving proofs, we want to discuss the general version of Tate's Acyclicity Theorem. For more details see [BGR], Chap. 8. We consider an affinoid K-space X and a finite covering $\mathfrak{U} = (U_i)_{i \in I}$ of it consisting of affinoid subdomains $U_i \subset X$. Furthermore, let us fix a presheaf $\mathcal{F}$, say of abelian groups, on the (category of) affinoid subdomains of X. Setting

$$U_{i_0 \ldots i_q} = U_{i_0} \cap \ldots \cap U_{i_q}$$

for indices $i_0, \ldots, i_q \in I$, we define the group of q-cochains on $\mathfrak{U}$ with values in $\mathcal{F}$ by

$$C^q(\mathfrak{U}, \mathcal{F}) = \prod_{i_0 \ldots i_q \in I} \mathcal{F}(U_{i_0 \ldots i_q}).$$

A cochain $f \in C^q(\mathfrak{U}, \mathcal{F})$ is called *alternating*, if

$$f_{i_{\pi(0)} \ldots i_{\pi(q)}} = \text{sgn}(\pi) f_{i_0 \ldots i_q}$$

for indices $i_0, \ldots, i_q \in I$ and any permutation $\pi \in \mathfrak{S}_{q+1}$ and if, furthermore, $f_{i_0 \ldots i_q} = 0$ for indices $i_0, \ldots, i_q$ that are not pairwise distinct. The alternating q-cochains form a subgroup $C_a^q(\mathfrak{U}, \mathcal{F})$ of $C^q(\mathfrak{U}, \mathcal{F})$.

There is a so-called *coboundary map*

$$d^q : C^q(\mathfrak{U}, \mathcal{F}) \longrightarrow C^{q+1}(\mathfrak{U}, \mathcal{F}),$$

given by

$$\left(d^q(f) \right)_{i_0 \ldots i_{q+1}} = \sum_{j=0}^{q+1} (-1)^j f_{i_0 \ldots \widehat{i_j} \ldots i_{q+1}} |_{U_{i_0 \ldots i_{q+1}}},$$

that satisfies $d^{q+1} \circ d^q = 0$ and maps alternating cochains into alternating ones, as is easily verified ($\widehat{i_j}$ means that the index i_j is to be omitted). Thus, we obtain a complex

$$0 \longrightarrow C^0(\mathfrak{U}, \mathcal{F}) \xrightarrow{d^0} C^1(\mathfrak{U}, \mathcal{F}) \xrightarrow{d^1} C^2(\mathfrak{U}, \mathcal{F}) \xrightarrow{d^2} \ldots,$$

which is called the *complex of Čech cochains on* $\mathfrak{U}$ with values in $\mathcal{F}$. In short, it is denoted by $C^\bullet(\mathfrak{U}, \mathcal{F})$. Similarly, there is the complex

$$0 \longrightarrow C_a^0(\mathfrak{U}, \mathcal{F}) \xrightarrow{d_a^0} C_a^1(\mathfrak{U}, \mathcal{F}) \xrightarrow{d_a^1} C_a^2(\mathfrak{U}, \mathcal{F}) \xrightarrow{d_a^2} \ldots,$$

of *alternating Čech cochains on* $\mathfrak{U}$ with values in $\mathcal{F}$, denoted by $C_a^\bullet(\mathfrak{U}, \mathcal{F})$. Associated to these complexes are the *Čech cohomology groups*

$$H^q(\mathfrak{U}, \mathcal{F}) = \ker d^q / \operatorname{im} d^{q-1},$$

$$H_a^q(\mathfrak{U}, \mathcal{F}) = \ker d_a^q / \operatorname{im} d_a^{q-1},$$

which are defined for $q \in \mathbb{N}$ (set $d^{-1} = 0$ and $d_a^{-1} = 0$). There is no difference, working with all cochains or merely with alternating ones, as is asserted by the following lemma:

Lemma 8. *The inclusion* $C_a^\bullet(\mathfrak{U}, \mathcal{F}) \lhook\joinrel\longrightarrow C^\bullet(\mathfrak{U}, \mathcal{F})$ *induces isomorphisms of cohomology groups*

$$H_a^q(\mathfrak{U}, \mathcal{F}) \xrightarrow{\sim} H^q(\mathfrak{U}, \mathcal{F}), \qquad q \in \mathbb{N}.$$

There is an immediate consequence:

Corollary 9. *If the covering $\mathfrak{U}$ consists of n elements, we have*

$$H^q(\mathfrak{U},\mathcal{F}) = H^q_a(\mathfrak{U},\mathcal{F}) = 0 \quad \text{for} \quad q \geq n.$$

The argument is, of course, that $C^q_a(\mathfrak{U}, \mathcal{F}) = 0$ for $q \geq n$ in the situation of Corollary 9. The covering $\mathfrak{U}$ of X is called $\mathcal{F}$-*acyclic* if the sequence

$$0 \longrightarrow \mathcal{F}(X) \overset{\varepsilon}{\longrightarrow} C^0(\mathfrak{U}, \mathcal{F}) \overset{d^0}{\longrightarrow} C^1(\mathfrak{U}, \mathcal{F}) \overset{d^1}{\longrightarrow} \cdots$$

is exact where $\varepsilon \colon \mathcal{F}(X) \longrightarrow C^0(\mathfrak{U}, \mathcal{F})$ is the so-called *augmentation* map given by $f \longmapsto (f|_{U_i})_{i \in I}$. Note that $\mathfrak{U}$ is $\mathcal{F}$-acyclic if and only if $\mathcal{F}$ satisfies the following conditions:

(i) the sequence $\mathcal{F}(X) \longrightarrow \prod_{i \in I} \mathcal{F}(U_i) \rightrightarrows \prod_{i,j \in I} \mathcal{F}(U_{ij})$ is exact, i.e. $\mathcal{F}$ satisfies the sheaf properties (S_1) and (S_2) for the covering $\mathfrak{U}$.

(ii) $H^q(\mathfrak{U}, \mathcal{F}) = 0$ for $q > 0$.

Now we can state Tate's Acyclicity Theorem in its general version:

Theorem 10 (Tate). *Let X be an affinoid K-space and $\mathfrak{U}$ a finite covering of X by affinoid subdomains. Then $\mathfrak{U}$ is acyclic with respect to the presheaf $\mathcal{O}_X$ of affinoid functions on X.*

The proof is the same as the one of Theorem 1; it is only necessary to establish Lemmata 2 and 3 in a more general cohomological context. Then, as exercised above, the assertion can be reduced to showing that for a Laurent covering of X generated by a single function $f \in \mathcal{O}_X(X)$, the augmented Čech complex of alternating cochains

$$0 \longrightarrow \mathcal{O}_X(X) \overset{\varepsilon}{\longrightarrow} C^0_a(\mathfrak{U}, \mathcal{O}_X) \overset{d^0}{\longrightarrow} C^1_a(\mathfrak{U}, \mathcal{O}_X) \overset{d^1}{\longrightarrow} 0$$

is exact.

Finally, if $X = \operatorname{Sp} A$ and M is an A-module, we can consider the presheaf $M \otimes_A \mathcal{O}_X$ on the affinoid subdomains of X given by

$$U \longmapsto M \otimes_A \mathcal{O}_X(U).$$

A simple argument shows that the assertion of Theorem 10 can be generalized to this presheaf in place of $\mathcal{O}_X$:

Corollary 11. *Let $X = \operatorname{Sp} A$ be an affinoid K-space, M an A-module, and $\mathfrak{U}$ a finite covering of X by affinoid subdomains. Then $\mathfrak{U}$ is acyclic with respect to the presheaf $M \otimes_A \mathcal{O}_X$.*

Proof. The assertion is a direct consequence of Theorem 10 if M is a free A-module, i.e. if $M = A^{(\Lambda)}$ for some index set Λ. Indeed, in this case the augmented Čech complex

$$0 \longrightarrow M \xrightarrow{\ \varepsilon\ } C^0(\mathfrak{U}, M \otimes_A \mathcal{O}_X) \xrightarrow{\ d^0\ } C^1(\mathfrak{U}, M \otimes_A \mathcal{O}_X) \xrightarrow{\ d^1\ } \cdots$$

is just the Λ-fold direct sum of the complex

$$0 \longrightarrow A \xrightarrow{\ \varepsilon\ } C^0(\mathfrak{U}, \mathcal{O}_X) \xrightarrow{\ d^0\ } C^1(\mathfrak{U}, \mathcal{O}_X) \xrightarrow{\ d^1\ } \cdots$$

If M is not free, we can choose a short exact sequence of A-modules

$$0 \longrightarrow M' \longrightarrow F \longrightarrow M \longrightarrow 0.$$

Associated to it is a sequence of augmented Čech complexes

$$0 \longrightarrow C^\bullet_{\mathrm{aug}}(\mathfrak{U}, M' \otimes_A \mathcal{O}_X) \longrightarrow C^\bullet_{\mathrm{aug}}(\mathfrak{U}, F \otimes_A \mathcal{O}_X) \longrightarrow C^\bullet_{\mathrm{aug}}(\mathfrak{U}, M \otimes_A \mathcal{O}_X) \longrightarrow 0,$$

which is exact, since for every affinoid subdomain $\operatorname{Sp} A' \hookrightarrow \operatorname{Sp} A$ the inherent morphism $A \longrightarrow A'$ is flat; see 4.1/5.

Now consider the long exact cohomology sequence induced from the preceding short exact sequence of Čech complexes. Since the complex $C^\bullet_{\mathrm{aug}}(\mathfrak{U}, F \otimes_A \mathcal{O}_X)$ has trivial cohomology, the long exact cohomology sequence contains isomorphisms of type

$$H^q_{\mathrm{aug}}(\mathfrak{U}, M \otimes_A \mathcal{O}_X) \xrightarrow{\ \sim\ } H^{q+1}_{\mathrm{aug}}(\mathfrak{U}, M' \otimes_A \mathcal{O}_X), \qquad q \geq 0,$$

where H^q_{aug} denotes the qth cohomology of augmented Čech complexes. If $\mathfrak{U}$ consists of n elements, we see from Corollary 9 that $H^q_{\mathrm{aug}}(\mathfrak{U}, N \otimes_A \mathcal{O}_X)$ is trivial for $q \geq n$ and all A-modules N. In particular, we have $H^q_{\mathrm{aug}}(\mathfrak{U}, M \otimes_A \mathcal{O}_X) = 0$ for $q \geq n$. Furthermore, the preceding isomorphism implies $H^{n-1}_{\mathrm{aug}}(\mathfrak{U}, M \otimes_A \mathcal{O}_X) = 0$. Replacing M by an arbitrary A-module N, it follows that $H^q_{\mathrm{aug}}(\mathfrak{U}, N \otimes_A \mathcal{O}_X)$ is trivial for $q \geq n - 1$. But then, by falling induction, we conclude that $\mathfrak{U}$ is acyclic for $M \otimes_A \mathcal{O}_X$. $\qquad\square$

Chapter 5
Towards the Notion of Rigid Spaces

5.1 Grothendieck Topologies

As we have already indicated at the end of Sect. 2.1, the presheaf of affinoid functions $\mathcal{O}_X$ on an affinoid K-space X cannot satisfy sheaf properties if we do not restrict the multitude of all possible open coverings. In fact, Tate's Acyclicity Theorem in the version of 4.3/1 or 4.3/10 is somehow the best result one can expect for general affinoid spaces, and we will base the construction of global rigid K-spaces by gluing local affinoid parts on this result. As a technical trick, we generalize the notion of a topology.

Definition 1. *A* Grothendieck topology $\mathfrak{T}$ *consists of a category* $\mathrm{Cat}\,\mathfrak{T}$ *and a set* $\mathrm{Cov}\,\mathfrak{T}$ *of families* $(U_i \longrightarrow U)_{i \in I}$ *of morphisms in* $\mathrm{Cat}\,\mathfrak{T}$, *called* coverings, *such that the following hold*:

(1) *If* $\Phi: U \longrightarrow V$ *is an isomorphism in* $\mathrm{Cat}\,\mathfrak{T}$, *then* $(\Phi) \in \mathrm{Cov}\,\mathfrak{T}$.

(2) *If* $(U_i \longrightarrow U)_{i \in I}$ *and* $(U_{ij} \longrightarrow U_i)_{j \in J_i}$ *for* $i \in I$ *belong to* $\mathrm{Cov}\,\mathfrak{T}$, *then the same is true for the composition* $(V_{ij} \longrightarrow U_i \longrightarrow U)_{i \in I, j \in J_i}$.

(3) *If* $(U_i \longrightarrow U)_{i \in I}$ *is in* $\mathrm{Cov}\,\mathfrak{T}$ *and if* $V \longrightarrow U$ *is a morphism in* $\mathrm{Cat}\,\mathfrak{T}$, *then the fiber products* $U_i \times_U V$ *exist in* $\mathrm{Cat}\,\mathfrak{T}$, *and* $(U_i \times_U V \longrightarrow V)_{i \in I}$ *belongs to* $\mathrm{Cov}\,\mathfrak{T}$.

We may think of the objects of $\mathrm{Cat}\,\mathfrak{T}$ as of the open sets of our topology and of the morphisms of $\mathrm{Cat}\,\mathfrak{T}$ as of the inclusions of open sets. Furthermore, a family $(U_i \longrightarrow U)_{i \in I}$ of $\mathrm{Cov}\,\mathfrak{T}$ has to be interpreted as a covering of U by the U_i and a fiber product of type $U_i \times_U V$ as the intersection of U_i with V. Thinking along these lines an ordinary topological space X is canonically equipped with a Grothendieck topology: $\mathrm{Cat}\,\mathfrak{T}$ is the category of open subsets of X, with inclusions as morphisms, and $\mathrm{Cov}\,\mathfrak{T}$ consists of all open covers of open subsets of X. However, there are more general examples where the morphisms of $\mathrm{Cat}\,\mathfrak{T}$ are far from being monomorphisms, like the étale topology, the *fppf*-topology, or the *fpqc*-topology

S. Bosch, *Lectures on Formal and Rigid Geometry*, Lecture Notes in Mathematics 2105, DOI 10.1007/978-3-319-04417-0_5,
© Springer International Publishing Switzerland 2014

in algebraic geometry. It should be pointed out that the "intersection" of "open" sets is dealt with in condition (3) of Definition 1, whereas we have refrained from giving any sense to the union of "open" sets. In fact, even in examples where the union of "open" sets does make sense, such a union will not necessarily yield an "open" set again.

The notion of a Grothendieck topology has been designed in such a way that the notion of presheaf or sheaf can easily be adapted to such a situation:

Definition 2. *Let $\mathfrak{T}$ be a Grothendieck topology and $\mathfrak{C}$ a category admitting cartesian products. A* presheaf *on $\mathfrak{T}$ with values in $\mathfrak{C}$ is defined as a contravariant functor $\mathcal{F}: \mathrm{Cat}\,\mathfrak{T} \longrightarrow \mathfrak{C}$. We call $\mathcal{F}$ a* sheaf *if the diagram*

$$\mathcal{F}(U) \longrightarrow \prod_{i \in I} \mathcal{F}(U_i) \rightrightarrows \prod_{i,j \in I} \mathcal{F}(U_i \times_U U_j)$$

is exact for any covering $(U_i \longrightarrow U)_{i \in I}$ in $\mathrm{Cov}\,\mathfrak{T}$.

From now on we will exclusively consider Grothendieck topologies $\mathfrak{T}$ of a special type. More specifically, the category $\mathrm{Cat}\,\mathfrak{T}$ will always be a category of certain subsets of a given set X, with inclusions as morphisms. The objects of $\mathrm{Cat}\,\mathfrak{T}$ will be referred to as the *admissible open subsets* of X. Likewise, the elements of $\mathrm{Cov}\,\mathfrak{T}$ are called the *admissible coverings*, and we will only consider those cases where admissible coverings $(U_i \longrightarrow U)_{i \in I}$ are, indeed, true coverings of U by admissible open sets U_i. To let the set X intervene, we will talk about a Grothendieck topology $\mathfrak{T}$ on X and call X a *G-topological space*. Of course, we are interested in the case where X is an affinoid K-space, and in Grothendieck topologies on X with respect to which the presheaf $\mathcal{O}_X$ of affinoid functions is actually a sheaf. A straightforward possibility to define such a Grothendieck topology is as follows:

Definition 3. *For any affinoid K-space X, let $\mathrm{Cat}\,\mathfrak{T}$ be the category of affinoid subdomains of X with inclusions as morphisms. Furthermore, let $\mathrm{Cov}\,\mathfrak{T}$ be the set of all finite families $(U_i \longrightarrow U)_{i \in I}$ of inclusions of affinoid subdomains in X such that $U = \bigcup_{i \in I} U_i$. Then $\mathfrak{T}$ is called the* weak Grothendieck topology *on X.*

That we really get a Grothendieck topology on X is easily verified. It follows from 3.3/8 and 3.3/19 that all admissible open subsets of X (in the sense of the weak Grothendieck topology) are open with respect to the canonical topology, as introduced in 3.3/1. Furthermore, if $\varphi: Z \longrightarrow X$ is a morphism of affinoid K-spaces, the inverse image $\varphi^{-1}(U)$ of any admissible open subset $U \subset X$ is admissible open in Z by 3.3/13, and the inverse image of any admissible covering in X yields an admissible covering in Z. To characterize such a behavior we will say that the map φ is *continuous* with respect to the relevant Grothendieck topologies, in this case the weak Grothendieck topology on Z and on X.

It has to be pointed out that the presheaf $\mathcal{O}_X$ of affinoid functions on any affinoid K-space X really *is* a presheaf in the sense of Definition 2, if X is equipped with the weak Grothendieck topology. Even better, Tate's Acyclicity Theorem 4.3/1 says that $\mathcal{O}_X$ is a sheaf in this context.

There is a canonical way to enlarge the weak Grothendieck topology on affinoid K-spaces by adding more admissible open sets and more admissible coverings in such a way that morphisms of affinoid K-spaces remain continuous and sheaves extend to sheaves with respect to this new topology. The resulting Grothendieck topology is the *strong Grothendieck topology* on affinoid K-spaces, which we will define now.

Definition 4. *Let X be an affinoid K-space. The* strong Grothendieck topology *on X is given as follows.*

(i) *A subset $U \subset X$ is called* admissible open *if there is a (not necessarily finite) covering $U = \bigcup_{i \in I} U_i$ of U by affinoid subdomains $U_i \subset X$ such that for all morphisms of affinoid K-spaces $\varphi: Z \longrightarrow X$ satisfying $\varphi(Z) \subset U$ the covering $(\varphi^{-1}(U_i))_{i \in I}$ of Z admits a refinement that is a finite covering of Z by affinoid subdomains.*

(ii) *A covering $V = \bigcup_{j \in J} V_j$ of some admissible open subset $V \subset X$ by means of admissible open sets V_j is called* admissible *if for each morphism of affinoid K-spaces $\varphi: Z \longrightarrow X$ satisfying $\varphi(Z) \subset V$ the covering $(\varphi^{-1}(V_j))_{j \in J}$ of Z admits a refinement that is a finite covering of Z by affinoid subdomains.*

Note that any covering $(U_i)_{i \in I}$ as in (i) is admissible by (ii). It is easily checked that the strong Grothendieck topology on X really is a Grothendieck topology such that any *finite* union of affinoid subdomains of X is admissible open. Furthermore, a direct verification shows that it satisfies certain completeness conditions. These allow, as we will see, to construct Grothendieck topologies on global spaces from local ones.

Proposition 5. *Let X be an affinoid K-space. The strong Grothendieck topology is a Grothendieck topology on X satisfying the following completeness conditions:*

(G_0) *$\emptyset$ and X are admissible open.*

(G_1) *Let $(U_i)_{i \in I}$ be an admissible covering of an admissible open subset $U \subset X$. Furthermore, let $V \subset U$ be a subset such that $V \cap U_i$ is admissible open for all $i \in I$. Then V is admissible open in X.*

(G_2) *Let $(U_i)_{i \in I}$ be a covering of an admissible open set $U \subset X$ by admissible open subsets $U_i \subset X$ such that $(U_i)_{i \in I}$ admits an admissible covering of U as refinement. Then $(U_i)_{i \in I}$ itself is admissible.*

Again, let X be an affinoid K-space. Then, if $U \subset X$ is an affinoid subdomain, the strong Grothendieck topology on X restricts to the strong Grothendieck topology on U, viewed as an affinoid K-space of its own. More generally, we show:

Proposition 6. *Let* $\varphi: Y \longrightarrow X$ *be a morphism of affinoid K-spaces. Then φ is continuous with respect to strong Grothendieck topologies on X and Y.*

Proof. Consider an admissible open set $U \subset X$ and, furthermore, an admissible covering $\mathfrak{U} = (U_i)_{i \in I}$ of it where all U_i are affinoid subdomains of X; such a covering exists by Definition 4 (i). To show that $V = \varphi^{-1}(U)$ is admissible open in Y, consider a morphism of affinoid K-spaces $\tau: Z \longrightarrow Y$ such that $\tau(Z) \subset V$. Then $\varphi \circ \tau$ maps Z into U and we see that the covering $(\tau^{-1}\varphi^{-1}(U_i))_{i \in I}$ of Z is refined by a (finite) affinoid covering. But then, as the sets $\varphi^{-1}(U_i)$ are affinoid subdomains of Y covering V, it follows that V is admissible open in Y.

More generally, if $(U_i)_{i \in I}$ is an arbitrary admissible covering of an admissible open set $U \subset X$, the same argument shows that $(\varphi^{-1}(U_i))_{i \in I}$ is an admissible covering of $\varphi^{-1}(U)$. Thus, φ is continuous with respect to strong Grothendieck topologies. $\square$

Next we want to relate the strong Grothendieck topology of an affinoid K-space to the Zariski topology.

Proposition 7. *Let X be an affinoid K-space. For $f \in \mathcal{O}_X(X)$ consider the following sets:*

$$U = \{x \in X ; |f(x)| < 1\}$$

$$U' = \{x \in X ; |f(x)| > 1\}$$

$$U'' = \{x \in X ; |f(x)| > 0\}$$

Any finite union of sets of this type is admissible open. Any finite covering by finite unions of sets of this type is admissible.

Proof. We write $\sqrt{|K^*|}$ for the group of nth roots of elements in $|K^*|$ where n varies over $\mathbb{N}$. Choosing a sequence $\varepsilon_\nu \in \sqrt{|K^*|}$ satisfying $\varepsilon_\nu < 1$ and $\lim_{\nu \to \infty} \varepsilon_n = 1$, we have

$$U = \bigcup_{\nu=0}^{\infty} X(\varepsilon_\nu^{-1} f)$$

where we have used the notation

$$X(\varepsilon_\nu^{-1} f) = \{x \in X ; |f(x)| \le \varepsilon_\nu\} = X(c^{-1} f^r)$$

for $c \in K^*$ being chosen in such a way that $|c| = \varepsilon_\nu^r$ for some integer $r > 0$. To see that U is admissible open in X, consider a morphism of affinoid spaces $\varphi: Z \longrightarrow X$ satisfying $\varphi(Z) \subset U$. If φ^* is the associated homomorphism of affinoid K-algebras, we have $|\varphi^*(f)(z)| = |f(\varphi(z))| < 1$ for all $z \in Z$ and, thus, by the Maximum Principle 3.1/15, $|\varphi^*(f)|_{\sup} < 1$. But then the covering

$$Z = \bigcup_{\nu=0}^{\infty} \varphi^{-1}\big(X(\varepsilon_\nu^{-1} f)\big) = \bigcup_{\nu=0}^{\infty} Z\big(\varepsilon_\nu^{-1} \varphi^*(f)\big)$$

admits a finite subcover, since $Z(\varepsilon_\nu^{-1}\varphi^*(f)) = Z$ for almost all ν. This shows that U is admissible open and that $(X(\varepsilon_\nu^{-1} f))_{\nu \in \mathbb{N}}$ is an admissible covering of U.

That U' and U'' are admissible open is shown similarly. However, that finite unions of sets of type U, U', U'' are admissible open requires a more sophisticated application of the maximum principle, which we give below in Lemma 8. Along the same lines one proves the assertion on admissible coverings.   □

Lemma 8. *Let A be an affinoid K-algebra and*

$$f = (f_1, \ldots, f_r), \qquad g = (g_1, \ldots, g_s), \qquad h = (h_1, \ldots, h_t)$$

systems of functions in A such that each $x \in \operatorname{Sp} A$ satisfies at least one of the equations

$$|f_\rho(x)| < 1, \qquad |g_\sigma(x)| > 1, \qquad |h_\tau(x)| > 0.$$

Then there exist constants $\alpha, \beta, \gamma \in \sqrt{|K^*|}$ where $\alpha < 1 < \beta$, such that each $x \in \operatorname{Sp} A$ satisfies, in fact, one of the equations

$$|f_\rho(x)| \le \alpha, \qquad |g_\sigma(x)| \ge \beta, \qquad |h_\tau(x)| \ge \gamma.$$

Proof. The problem is local on $X = \operatorname{Sp} A$ in the sense that we may choose a (finite) affinoid cover $(U_i)_{i \in I}$ of X and verify the assertion for the restrictions of f, g, h onto each U_i. In particular, we may choose an $\alpha \in \sqrt{|K^*|}, \alpha < 1$, and consider the covering

$$X = X(\alpha f_1^{-1}, \ldots, \alpha f_r^{-1}) \cup \bigcup_{\rho=1}^{r} X(\alpha^{-1} f_\rho).$$

As the assertion is clear on all affinoid subdomains $X(\alpha^{-1} f_\rho)$, we may replace X by $X(\alpha f_1^{-1}, \ldots, \alpha f_r^{-1})$. Thereby we can assume that all f_ρ are units in A, and we can look at the inequalities $|f_\rho^{-1}(x)| > 1$ instead of $|f_\rho(x)| < 1$. Thus, replacing the system g by $(f_1^{-1}, \ldots, f_r^{-1}, g_1, \ldots, g_s)$, we have transferred our problem to the case where the system f is not present and only g and h are of interest.

In this situation, $h_1, \ldots, h_t$ cannot have a common zero on $X(g_1, \ldots, g_s)$. Thus they generate the unit ideal in $\mathcal{O}_X(X(g_1, \ldots, g_s))$, and there is a $\gamma \in \sqrt{|K^*|}$ such that $\max_{\tau=1 \ldots t} |h_\tau(x)| > \gamma$ for all $x \in X(g_1, \ldots, g_s)$. Equivalently, there is for any $x \in X(\gamma^{-1} h_1, \ldots, \gamma^{-1} h_t)$ an index $\sigma \in \{1, \ldots, s\}$ such that $|g_\sigma(x)| > 1$. Hence, considering the covering

$$X = X(\gamma^{-1}h_1, \ldots, \gamma^{-1}h_t) \cup \bigcup_{\tau=1}^{t} X(\gamma h_\tau^{-1}),$$

we may replace X by $X(\gamma^{-1}h_1, \ldots, \gamma^{-1}h_t)$. Thereby h can be dropped and we might assume that only the system g is present.

In this special case, the functions $g_1, \ldots, g_s$ do not have a common zero on X, and

$$X = \bigcup_{\sigma=1}^{s} X\left(\frac{g_1}{g_\sigma}, \ldots, \frac{g_r}{g_\sigma}\right)$$

is a well-defined rational covering of X such that

$$\max_{\sigma'=1\ldots s} |g_{\sigma'}(x)| = |g_\sigma(x)| > 1$$

for all $x \in X(\frac{g_1}{g_\sigma}, \ldots, \frac{g_r}{g_\sigma})$. But then we are done, since g_σ^{-1} assumes its maximum on $X(\frac{g_1}{g_\sigma}, \ldots, \frac{g_r}{g_\sigma})$ by the Maximum Principle 3.1/15.   □

Since any Zariski open subset of an affinoid K-space X is a finite union of sets of type U'' as mentioned in Proposition 7, we can conclude from this result:

Corollary 9. *Let X be an affinoid K-space. Then the strong Grothendieck topology on X is finer than the Zariski topology, i.e. every Zariski open subset $U \subset X$ is admissible open and every Zariski covering is admissible.*

We end this section by some remarks on how to construct global Grothendieck topologies from local data.

Proposition 10. *Let $\mathfrak{T}$ be a Grothendieck topology on a set X such that conditions (G_0), (G_1), and (G_2) of Proposition 5 are satisfied. Let $(X_i)_{i \in I}$ be an admissible covering of X. Then:*

(i) *A subset $U \subset X$ is admissible open if and only if all intersections $U \cap X_i$, $i \in I$, are admissible open.*

(ii) *A covering $(U_j)_{j \in J}$ of some admissible open subset $U \subset X$ is admissible if and only if $(X_i \cap U_j)_{j \in J}$ is an admissible covering of $X_i \cap U$, for all $i \in I$.*

The *proof* is straightforward. Assertion (i) is a direct consequence of condition (G_1), whereas (ii) follows from (G_2), since $(X_i \cap U)_{i \in I}$ and $(X_i \cap U_j)_{i \in I, j \in J}$ are admissible coverings of U.   □

Proposition 11. *Let X be a set and $(X_i)_{i \in I}$ a covering of X. Furthermore, let $\mathfrak{T}_i$ be a Grothendieck topology on X_i, $i \in I$, such that conditions (G_0), (G_1), and (G_2) of Proposition 5 are satisfied. For all $i, j \in I$, assume that $X_i \cap X_j$ is $\mathfrak{T}_i$-open (i.e. admissible open with respect to $\mathfrak{T}_i$) in X_i and that $\mathfrak{T}_i$ and $\mathfrak{T}_j$ restrict to the same*

Grothendieck topology on $X_i \cap X_j$. Then there is a unique Grothendieck topology $\mathfrak{T}$ on X such that the following hold:

(i) X_i *is* $\mathfrak{T}$*-open in* X*, and* $\mathfrak{T}$ *induces* $\mathfrak{T}_i$ *on* X_i*.*

(ii) $\mathfrak{T}$ *satisfies conditions* (G_0)*,* (G_1)*, and* (G_2) *of Proposition 5.*

(iii) $(X_i)_{i \in I}$ *is a* $\mathfrak{T}$*-covering of* X *(i.e. admissible with respect to* $\mathfrak{T}$*).*

Proof. Due to Proposition 10, there is at most one possibility to define $\mathfrak{T}$. Call a subset $U \subset X$ $\mathfrak{T}$-open if each intersection $X_i \cap U$, $i \in I$, is $\mathfrak{T}_i$-open. Similarly, we call a covering $\mathfrak{U} = (U_j)_{j \in J}$ consisting of $\mathfrak{T}$-open sets $U_j \subset X$ a $\mathfrak{T}$-covering if, for each $i \in I$, the covering $\mathfrak{U}|_{X_i} = (X_i \cap U_j)_{j \in J}$ is a $\mathfrak{T}_i$-covering of $X_i \cap U$. That $\mathfrak{T}$ is a Grothendieck topology as required is easily checked. $\qquad\square$

5.2 Sheaves

In the following, let X be a G-topological space, i.e. a set with a Grothendieck topology $\mathfrak{T}$ on it. As in 5.1/2 we define a presheaf of groups, rings, etc. on X as a contravariant functor $\mathcal{F}$ from Cat $\mathfrak{T}$ to the category of groups, rings, etc. Furthermore, $\mathcal{F}$ is called a sheaf if for each admissible covering $(U_i)_{i \in I}$ of an admissible open set $U \subset X$ the diagram

$$\mathcal{F}(U) \longrightarrow \prod_{i \in I} \mathcal{F}(U_i) \rightrightarrows \prod_{i,j \in I} \mathcal{F}(U_i \times_U U_j)$$

is exact.

We are, of course, interested in the case where X is an affinoid K-space. Considering the weak Grothendieck topology on X, we have introduced the presheaf $\mathcal{O}_X$ of affinoid functions on X, and we have seen from Tate's Acyclicity Theorem 4.3/1 that $\mathcal{O}_X$ even is a sheaf. One of the objectives of this section is to pass on to the strong Grothendieck topology on X and to show that sheaves extend canonically from the weak to the strong Grothendieck topology on X.

Let X be an arbitrary G-topological space again. For any presheaf $\mathcal{F}$ on X and a point $x \in X$, we define

$$\mathcal{F}_x = \varinjlim \mathcal{F}(U)$$

as the *stalk* of $\mathcal{F}$ at the point x where the limit runs over all admissible open $U \subset X$ containing x. Next, let $\sigma \colon \mathcal{F} \longrightarrow \mathcal{F}'$ be a morphism of presheaves on X. Thereby we mean a system of morphisms $\sigma_U \colon \mathcal{F}(U) \longrightarrow \mathcal{F}'(U)$ for U varying over all admissible open subsets of X such that the σ_U are compatible with the restriction morphisms of $\mathcal{F}$ and $\mathcal{F}'$. Such a morphism induces for any $x \in X$ a morphism $\sigma_x \colon \mathcal{F}_x \longrightarrow \mathcal{F}'_x$ on the level of stalks.

Definition 1. *Let $\mathcal{F}$ be a presheaf on a G-topological space X. A sheafification of $\mathcal{F}$ is a morphism $\mathcal{F} \longrightarrow \mathcal{F}'$, where $\mathcal{F}'$ is a sheaf such that the following universal property is satisfied:*

Each morphism $\mathcal{F} \longrightarrow \mathcal{G}$ where $\mathcal{G}$ is a sheaf, factors through $\mathcal{F} \longrightarrow \mathcal{F}'$ via a unique morphism $\mathcal{F}' \longrightarrow \mathcal{G}$.

In the situation of Definition 1, $\mathcal{F}'$ is called the *sheaf associated to* $\mathcal{F}$. Such a sheaf can always be constructed, as we will see below. The classical construction of associated sheaves on a topological space X is to consider the disjoint union $E = \coprod_{x \in X} \mathcal{F}_x$, which comes equipped with a canonical projection $\pi \colon E \longrightarrow X$. For any open subset $U \subset X$, a map $f \colon U \longrightarrow E$ is called a *section* of π if $\pi \circ f = \mathrm{id}$; so f associates to each point $x \in U$ an element $f_x \in \mathcal{F}_x$. Now let $\mathcal{F}'(U)$ be the set of those sections $f \colon U \longrightarrow E$ of π such that, for all $x \in U$, there are an open neighborhood $U(x) \subset U$ of x and an element $g \in \mathcal{F}(U(x))$ with the property that $g_y = f_y$ for all $y \in U(x)$. So $\mathcal{F}'(U)$ consists of all sections over U that, locally, are induced from elements of the presheaf $\mathcal{F}$, and it is easily checked that $\mathcal{F} \longrightarrow \mathcal{F}'$ is a sheafification of $\mathcal{F}$.

For G-topological spaces X the classical construction cannot work properly, since there can exist non-zero sheaves on X having zero stalks at all points $x \in X$. We give an example.

Example 2. Consider the unit disk $X = \mathrm{Sp}\, T_1$ over a field K that, for simplicity, is supposed to be algebraically closed. Then, pointwise, we can identify X with the closed unit disk around 0 in K. A subset $U \subset X$ is called a *standard set* if it is empty or of type

$$U = D^+(a,r) - \bigcup_{i=1}^{s} D^-(a_i,r_i)$$

for points $a \in X$, $a_1,\ldots,a_s \in D^+(a,r)$ and radii $r_1,\ldots,r_s \le r$ in $|K^*|$. We set $d(U) = r$ as well as $d(\emptyset) = 0$ and call this the *diameter* of U. Of course, every standard set in X gives rise to an affinoid subdomain of X. Conversely, one can show that every non-empty affinoid subdomain $U \subset X$ is a finite and, in fact, unique *disjoint* union of standard sets; see [BGR], 9.7.2/2. Let us write $d(U)$ for the *maximum* of all diameters of the occurring standard sets.

Now we can define a sheaf of abelian groups $\mathcal{F}$ on X by setting

$$\mathcal{F}(U) = \begin{cases} \mathbb{Z} & \text{if } d(U) = 1 \\ 0 & \text{if } d(U) < 1 \end{cases}$$

with the obvious restriction morphisms. Then it is easily checked that $\mathcal{F}$ is a sheaf with respect to the weak Grothendieck topology on X. The reason is that for any affinoid subdomain $U \subset X$ and a finite covering $U = \bigcup_{i=1}^{n} U_i$ by affinoid subdomains $U_i \subset U$ one has

$$d(U) = \max_{i=1,\dots,n} d(U_i).$$

In particular, due to the restrictiveness of allowed coverings on X, there is no admissible affinoid covering $(U_i)_{i \in I}$ of X itself satisfying $d(U_i) < 1$ for all i that would force all global sections of $\mathcal{F}$ to vanish. The same is true for affinoid subdomains $U \subset X$ satisfying $d(U) = 1$. Therefore, indeed, $\mathcal{F}$ is a non-zero sheaf having zero stalks at all points of X.

Returning to the construction of sheafifications on arbitrary G-topological spaces X, we will assume that $\mathcal{F}$ is at least a presheaf of *abelian* groups so that techniques from Čech cohomology can be used, for example as presented in [BGR], Chaps. 8 and 9. For any admissible open subset $U \subset X$ we set

$$\check{H}^q(U, \mathcal{F}) = \varinjlim H^q(\mathfrak{U}, \mathcal{F}), \qquad q \in \mathbb{N},$$

where the direct limit runs over all admissible coverings $\mathfrak{U}$ of U, using the relation of being finer as a partial ordering. Clearly, the ordering is directed since any two such coverings $(U_i)_{i \in I}$, $(V_j)_{j \in J}$ admit a common admissible refinement, for example $(U_i \cap V_j)_{i \in I, j \in J}$. To execute the direct limit, we use, of course, the fact that, for a refinement $\mathfrak{V}$ of some admissible covering $\mathfrak{U}$ of U, there is always a canonical morphism $H^q(\mathfrak{U}, \mathcal{F}) \longrightarrow H^q(\mathfrak{V}, \mathcal{F})$ (which, for the purpose of sheafifications, will only be needed for $q = 0$). Furthermore, varying U, we get the presheaf $\check{\mathcal{H}}^q(X, \mathcal{F})$ that associates to an admissible open subset $U \subset X$ the cohomology group $\check{H}^q(U, \mathcal{F}|_U)$. Note that, for any admissible covering $(U_i)_{i \in I}$ of some admissible open subset $U \subset X$, we have a canonical morphism $\mathcal{F}(U) \longrightarrow H^0(\mathfrak{U}, \mathcal{F})$ and, hence, varying $\mathfrak{U}$, a canonical morphism $\mathcal{F}(U) \longrightarrow \check{H}^0(U, \mathcal{F})$. The morphisms of the latter type give rise to a canonical morphism $\mathcal{F} \longrightarrow \check{\mathcal{H}}^0(X, \mathcal{F})$.

Proposition 3. *Let $\mathcal{F}$ be a presheaf (of abelian groups, rings, etc.) on a G-topological space X.*

(i) *The presheaf $\mathcal{F}^+ = \check{\mathcal{H}}^0(X, \mathcal{F})$ satisfies sheaf property (S_1) of Sect. 4.3, i.e. the canonical map $\mathcal{F}^+(U) \longrightarrow \prod_{i \in I} \mathcal{F}^+(U_i)$ is injective for any admissible covering $(U_i)_{i \in I}$ of an admissible open subset $U \subset X$.*

(ii) *If $\mathcal{F}$ satisfies sheaf property (S_1) of Sect. 4.3, then $\mathcal{F}^+$ satisfies (S_1) and (S_2) and, thus, is a sheaf.*

(iii) *$\mathcal{F}^{++} = \check{\mathcal{H}}^0(X, \check{\mathcal{H}}^0(X, \mathcal{F}))$ is a sheaf, and the composition of canonical morphisms $\mathcal{F} \longrightarrow \mathcal{F}^+ \longrightarrow \mathcal{F}^{++}$ is a sheafification of $\mathcal{F}$.*

The proof will be omitted; it is straightforward, although a little bit technical; cf. [BGR], 9.2.2/3 and 9.2.2/4. As an application of the existence of associated sheaves, we can define as usual the sheaf image of a morphism $\sigma: \mathcal{F} \longrightarrow \mathcal{G}$ of abelian sheaves, i.e. of sheaves of abelian groups. It is the sheaf associated to the presheaf $U \longmapsto \sigma_U(\mathcal{F}(U))$ where U varies over all admissible open subsets of X. Similarly, the quotient $\mathcal{F}/\mathcal{F}_0$ of an abelian sheaf $\mathcal{F}$ by a subsheaf $\mathcal{F}_0$ is defined as the sheaf associated to the presheaf $U \longmapsto \mathcal{F}(U)/\mathcal{F}_0(U)$.

Finally, we want to attack the problem of extending sheaves from the weak to the strong Grothendieck topology on affinoid K-spaces.

Proposition 4. *Let X be a set with Grothendieck topologies $\mathfrak{T}$ and $\mathfrak{T}'$ such that*:

(i) *$\mathfrak{T}'$ is finer than $\mathfrak{T}$.*

(ii) *Each $\mathfrak{T}'$-open set $U \subset X$ admits a $\mathfrak{T}'$-covering $(U_i)_{i \in I}$ where all U_i are $\mathfrak{T}$-open in X.*

(iii) *Each $\mathfrak{T}'$-covering of a $\mathfrak{T}$-open subset $U \subset X$ admits a $\mathfrak{T}$-covering as a refinement.*

Then each $\mathfrak{T}$-sheaf $\mathcal{F}$ on X admits an extension $\mathcal{F}'$ as a $\mathfrak{T}'$-sheaf on X. The latter is unique up to canonical isomorphism.

We give only some indications on how to construct $\mathcal{F}'$. Consider the presheaf $\mathcal{F}'$ with respect to $\mathfrak{T}'$ on X that is given by

$$U \longmapsto \varinjlim_{\mathfrak{U}} H^0(\mathfrak{U}, \mathcal{F})$$

where the limit runs over all $\mathfrak{T}'$-coverings $\mathfrak{U} = (U_i)_{i \in I}$ of U consisting of $\mathfrak{T}$-open sets U_i. Due to condition (iii), $\mathcal{F}'$ is an extension of $\mathcal{F}$. Using the fact that $\mathcal{F}$ is a sheaf, it is easily checked that $\mathcal{F}'$ is a sheaf as well. In fact, we may interpret $\mathcal{F}'$ as the sheaf $\check{\mathcal{H}}^0(X_{\mathfrak{T}'}, \mathcal{F})$, just observing that, in order to construct the latter object, we need only to know $\mathcal{F}$ on the $\mathfrak{T}$-open subsets of X.

As a direct consequence, we can state:

Corollary 5. *Let X be an affinoid K-space. Then any sheaf $\mathcal{F}$ on X with respect to the weak Grothendieck topology admits a unique extension with respect to the strong Grothendieck topology. The latter applies, in particular, to the presheaf of affinoid functions $\mathcal{F} = \mathcal{O}_X$, which is a sheaf with respect to the weak Grothendieck topology by 4.3/1.*

Extending $\mathcal{O}_X$ with respect to the strong Grothendieck topology on X, we will call the resulting sheaf *the sheaf of rigid analytic functions on X* and use the notation $\mathcal{O}_X$ for it again. Anyway, from now on we will always consider, unless stated otherwise, the strong Grothendieck topology on affinoid K-spaces.

5.3 Rigid Spaces

A *ringed K-space* is a pair $(X, \mathcal{O}_X)$ where X is a topological space and $\mathcal{O}_X$ a sheaf of K-algebras on it. This concept can be adapted in a natural way to G-topological spaces.

Definition 1. *A G-ringed K-space is a pair $(X, \mathcal{O}_X)$ where X is a G-topological space and $\mathcal{O}_X$ a sheaf of K-algebras on it. $(X, \mathcal{O}_X)$ is called a* locally *G-ringed K-space if, in addition, all stalks $\mathcal{O}_{X,x}$, $x \in X$, are local rings.*

A morphism of G-ringed K-spaces $(X, \mathcal{O}_X) \longrightarrow (Y, \mathcal{O}_Y)$ is a pair (φ, φ^*) where $\varphi: X \longrightarrow Y$ is a map, continuous with respect to Grothendieck topologies, and where φ^* is a system of K-homomorphisms $\varphi_V^*: \mathcal{O}_Y(V) \longrightarrow \mathcal{O}_X(\varphi^{-1}(V))$ with V varying over the admissible open subsets of Y. It is required that the φ_V^* are compatible with restriction homomorphisms, i.e. for $V' \subset V$ the diagram

$$
\begin{array}{ccc}
\mathcal{O}_Y(V) & \xrightarrow{\varphi_V^*} & \mathcal{O}_X(\varphi^{-1}(V)) \\
\downarrow & & \downarrow \\
\mathcal{O}_Y(V') & \xrightarrow{\varphi_{V'}^*} & \mathcal{O}_X(\varphi^{-1}(V'))
\end{array}
$$

must be commutative.

Furthermore, assuming that $(X, \mathcal{O}_X)$ and $(Y, \mathcal{O}_Y)$ are locally G-ringed K-spaces, a morphism $(\varphi, \varphi^*): (X, \mathcal{O}_X) \longrightarrow (Y, \mathcal{O}_Y)$ is called a morphism of locally *G-ringed K-spaces if the ring homomorphisms*

$$
\varphi_x^*: \mathcal{O}_{Y, \varphi(x)} \longrightarrow \mathcal{O}_{X,x}, \qquad x \in X,
$$

induced from the φ_V^* are local in the sense that the maximal ideal of $\mathcal{O}_{Y, \varphi(x)}$ is mapped into the one of $\mathcal{O}_{X,x}$.

For example, if X is an affinoid K-space, we can consider the associated locally G-ringed K-space $(X, \mathcal{O}_X)$ where X, as a G-topological space, is endowed with the strong G-topology and $\mathcal{O}_X$ is the structure sheaf on X, as introduced in 5.2/5. As all stalks of $\mathcal{O}_X$ are local rings by 4.1/1, $(X, \mathcal{O}_X)$ is even a locally G-ringed K-space. Furthermore, it is more or less clear that each morphism of affinoid K-spaces $\varphi: X \longrightarrow Y$ induces a morphism $(\varphi, \varphi^*): (X, \mathcal{O}_X) \longrightarrow (Y, \mathcal{O}_Y)$ between associated locally G-ringed K-spaces. To justify this claim, note first that φ defines a continuous morphism of G-topological spaces if X and Y are endowed with the strong G-topology; cf. 5.1/6. Next, consider an affinoid subdomain $V \subset Y$. Then $\varphi^{-1}(V)$ is an affinoid subdomain in X by 3.3/13. Therefore φ induces a morphism of affinoid K-algebras $\varphi_V^*: \mathcal{O}_Y(V) \longrightarrow \mathcal{O}_X(\varphi^{-1}(V))$ that, varying V, clearly is compatible with restriction of V. If, more generally, V is just an admissible open subset in Y, we can choose an admissible affinoid covering $(V_i)_{i \in I}$ of V and obtain a well-defined morphism $\varphi_V^*: \mathcal{O}_Y(V) \longrightarrow \mathcal{O}_X(\varphi^{-1}(V))$ in a similar way by using the exact diagrams

$$\mathcal{O}_Y(V) \longrightarrow \prod_{i \in I} \mathcal{O}_Y(V_i) \rightrightarrows \prod_{i,j \in I} \mathcal{O}_Y(V_i \cap V_j),$$

$$\mathcal{O}_X(\varphi^{-1}(V)) \longrightarrow \prod_{i \in I} \mathcal{O}_X(\varphi^{-1}(V_i)) \rightrightarrows \prod_{i,j \in I} \mathcal{O}_Y(\varphi^{-1}(V_i) \cap \varphi^{-1}(V_j)),$$

in conjunction with the maps

$$\varphi_{V_i}^*: \mathcal{O}_Y(V_i) \longrightarrow \mathcal{O}_X(\varphi^{-1}(V_i)),$$

$$\varphi_{V_i \cap V_j}^*: \mathcal{O}_Y(V_i \cap V_j) \longrightarrow \mathcal{O}_X(\varphi^{-1}(V_i \cap V_j));$$

note that, just as the V_i, all intersections $V_i \cap V_j$ are affinoid subdomains of Y by 3.3/14. Now, writing φ^* for the system of all maps φ_V^*, it is easily seen that the pair (φ, φ^*) constitutes a morphism of ringed K-spaces $(X, \mathcal{O}_X) \longrightarrow (Y, \mathcal{O}_Y)$. That this morphism is, in fact, a morphism of locally G-ringed K-spaces, is seen as follows. Consider a point $x \in X$ with maximal ideal $\mathfrak{m}_x \subset \mathcal{O}_X(X)$ and image $\varphi(x) \in Y$ corresponding to the maximal ideal $\mathfrak{m}_{\varphi(x)} \subset \mathcal{O}_Y(Y)$. Then, $\varphi_Y^*: \mathcal{O}_Y(Y) \longrightarrow \mathcal{O}_X(X)$, the map between affinoid K-algebras associated to φ, maps $\mathfrak{m}_{\varphi(x)}$ into $\mathfrak{m}_x$, as we have $\mathfrak{m}_{\varphi(x)} = (\varphi_Y^*)^{-1}(\mathfrak{m}_x)$ by definition. Consequently, the morphism $\varphi_x^*: \mathcal{O}_{Y,\varphi(x)} \longrightarrow \mathcal{O}_{X,x}$ must map the maximal ideal of $\mathcal{O}_{Y,\varphi(x)}$, which is generated by $\mathfrak{m}_{\varphi(x)}$ due to 4.1/1, into the maximal ideal of $\mathcal{O}_{X,x}$, which again due to 4.1/1, is generated by $\mathfrak{m}_x$. Hence, we have constructed a map from the set of morphisms $X \longrightarrow Y$ between affinoid K-spaces X and Y to the set of morphisms of locally G-ringed K-spaces $(X, \mathcal{O}_X) \longrightarrow (Y, \mathcal{O}_Y)$. We want to show that this map is actually a bijection.

Proposition 2. *Let X and Y be affinoid K-spaces. Then the map from morphisms of affinoid K-spaces $X \longrightarrow Y$ to morphisms of locally G-ringed K-spaces $(X, \mathcal{O}_X) \longrightarrow (Y, \mathcal{O}_Y)$, as constructed above, is bijective.*

Proof. To exhibit an inverse of the above constructed map, associate to any morphism of locally G-ringed K-spaces $(\varphi, \varphi^*): (X, \mathcal{O}_X) \longrightarrow (Y, \mathcal{O}_Y)$ the morphism of affinoid K-spaces $X \longrightarrow Y$ corresponding to the morphism of affinoid K-algebras $\varphi_Y^*: \mathcal{O}_Y(Y) \longrightarrow \mathcal{O}_X(X)$. To see that it really is an inverse, it is enough to establish the following auxiliary result:

Lemma 3. *Let X and Y be affinoid K-spaces. For every morphism of affinoid K-algebras $\sigma^*: \mathcal{O}_Y(Y) \longrightarrow \mathcal{O}_X(X)$ there exists a unique morphism of locally G-ringed K-spaces $(\varphi, \varphi^*): (X, \mathcal{O}_X) \longrightarrow (Y, \mathcal{O}_Y)$ satisfying $\varphi_Y^* = \sigma^*$.*

Proof. Only the uniqueness assertion has to be verified. So consider a morphism of locally G-ringed K-spaces $(\varphi, \varphi^*) \colon (X, \mathcal{O}_X) \longrightarrow (Y, \mathcal{O}_Y)$ satisfying $\varphi_Y^* = \sigma^*$ for a given morphism of affinoid K-algebras $\sigma^* \colon \mathcal{O}_Y(Y) \longrightarrow \mathcal{O}_X(X)$. Then for each $x \subset X$, there is a commutative diagram

$$
\begin{CD}
\mathcal{O}_Y(Y) @>{\sigma^* = \varphi_Y^*}>> \mathcal{O}_X(X) \\
@VVV @VVV \\
\mathcal{O}_{Y,\varphi(x)} @>{\varphi_x^*}>> \mathcal{O}_{X,x} .
\end{CD}
$$

Let $\mathfrak{m}_x \subset \mathcal{O}_X(X)$ be the maximal ideal corresponding to x and $\mathfrak{m}_{\varphi(x)} \subset \mathcal{O}_Y(Y)$ the maximal ideal corresponding to $\varphi(x)$. Since φ_x^* is local, it maps the maximal ideal $\mathfrak{m}_{\varphi(x)} \mathcal{O}_{Y,\varphi(x)}$ of the local ring $\mathcal{O}_{Y,\varphi(x)}$ into the maximal ideal $\mathfrak{m}_x \mathcal{O}_{X,x}$ of the local ring $\mathcal{O}_{X,x}$. Hence, using the isomorphism $\mathcal{O}_X(X)/\mathfrak{m}_x \overset{\sim}{\longrightarrow} \mathcal{O}_{X,x}/\mathfrak{m}_x \mathcal{O}_{X,x}$ of 4.1/2, we see that σ^* maps $\mathfrak{m}_{\varphi(x)}$ into $\mathfrak{m}_x$, so that we have $\mathfrak{m}_{\varphi(x)} = (\sigma^*)^{-1}(\mathfrak{m}_x)$. From this it follows that φ, as a map of sets, coincides with the morphism of affinoid K-spaces $X \longrightarrow Y$ given by σ^*. Thus, at least φ is uniquely determined by σ^*.

To show that all maps φ_V^* are unique, we may restrict ourselves to affinoid subdomains $V \subset Y$. Then $\varphi^{-1}(V)$ is an affinoid subdomain in X, and there is a commutative diagram

$$
\begin{CD}
\mathcal{O}_Y(Y) @>{\sigma^* = \varphi_Y^*}>> \mathcal{O}_X(X) \\
@VVV @VVV \\
\mathcal{O}_Y(V) @>{\varphi_V^*}>> \mathcal{O}_X(\varphi^{-1}(V)),
\end{CD}
$$

the vertical maps being restriction homomorphisms. Since the first one corresponds to the inclusion of the affinoid subdomain $V \hookrightarrow Y$, it follows from the defining properties of affinoid subdomains that φ_V^* is uniquely determined by $\sigma^* = \varphi_Y^*$. $\square$

The assertion of Proposition 2 enables us to identify morphisms of affinoid K-spaces with morphisms of their associated locally G-ringed K-spaces. In other words, the functor from the category of affinoid K-spaces to the category of locally G-ringed K-spaces that we have constructed is fully faithful. Therefore, in general, we will make no notational difference between an affinoid K-space and its associated locally G-ringed K-space, writing simply X instead of $(X, \mathcal{O}_X)$. Also note that, due to our construction, an inclusion $U \hookrightarrow X$ of an affinoid subdomain U into X gives rise to an open immersion of locally G-ringed K-spaces $(U, \mathcal{O}_U) \longrightarrow (X, \mathcal{O}_X)$. The latter means that U is an admissible open subset of X, that $\mathcal{O}_U$ is the restriction of $\mathcal{O}_X$ to U, and that $(U, \mathcal{O}_U) \longrightarrow (X, \mathcal{O}_X)$ is the canonical morphism. Now it is easy to define global objects that look locally like affinoid K-spaces.

Definition 4. *A* rigid (analytic) *K-space is a locally G-ringed K-space $(X, \mathcal{O}_X)$ such that*

(i) *the G-topology of X satisfies conditions (G_0), (G_1), and (G_2) of 5.1/5, and*
(ii) *X admits an admissible covering $(X_i)_{i \in I}$ where $(X_i, \mathcal{O}_X|_{X_i})$ is an affinoid K-space for all $i \in I$.*

A morphism of rigid K-spaces $(X, \mathcal{O}_X) \longrightarrow (Y, \mathcal{O}_Y)$ is a morphism in the sense of locally G-ringed K-spaces.

It follows for an admissible open subset $U \subset X$ that the induced locally G-ringed K-space $(U, \mathcal{O}_X|_U)$ is a rigid K-space again; we will call $(U, \mathcal{O}_X|_U)$ an *open subspace* of $(X, \mathcal{O}_X)$. In most cases, however, rigid K-spaces will simply be denoted by a single symbol, say X, instead of $(X, \mathcal{O}_X)$. As usual, global rigid K-spaces can be constructed by gluing local ones.

Proposition 5. *Consider the following data:*

(i) *rigid K-spaces X_i, $i \in I$, and*
(ii) *open subspaces $X_{ij} \subset X_i$ and isomorphisms $\varphi_{ij} \colon X_{ij} \xrightarrow{\sim} X_{ji}$, for all $i, j \in I$,*

and assume that these are subject to the following conditions:

(a) *$\varphi_{ij} \circ \varphi_{ji} = \mathrm{id}$, $X_{ii} = X_i$, and $\varphi_{ii} = \mathrm{id}$, for all $i, j \in I$,*
(b) *φ_{ij} induces isomorphisms $\varphi_{ijk} \colon X_{ij} \cap X_{ik} \xrightarrow{\sim} X_{ji} \cap X_{jk}$ that satisfy $\varphi_{ijk} = \varphi_{kji} \circ \varphi_{ikj}$ for all $i, j, k \in I$.*

Then the X_i can be glued by identifying X_{ij} with X_{ji} via φ_{ij} to yield a rigid K-space X admitting $(X_i)_{i \in I}$ as an admissible covering.

More precisely, there exists a rigid K-space X together with an admissible covering $(X'_i)_{i \in I}$ and isomorphisms $\psi_i \colon X_i \xrightarrow{\sim} X'_i$ restricting to isomorphisms $\psi_{ij} \colon X_{ij} \xrightarrow{\sim} X'_i \cap X'_j$ such that the diagram

$$
\begin{array}{ccc}
X_{ij} & \xrightarrow{\psi_{ij}} & X'_i \cap X'_j \\
{\scriptstyle \varphi_{ij}} \downarrow & & \| \\
X_{ji} & \xrightarrow{\psi_{ji}} & X'_j \cap X'_i
\end{array}
$$

is commutative. Furthermore, X is unique up to canonical isomorphism.

The *proof* is straightforward. To construct X as a set, we glue the X_i, using the isomorphisms φ_{ij} as identifications. In more precise terms, we start out from the disjoint union $X' = \coprod_{i \in I} X_i$ and call $x, y \in X'$ equivalent, say $x \in X_i$ and $y \in X_j$, if $\varphi_{ij}(x) = y$. The relation we get is symmetric and reflexive by the conditions in (a) and transitive by (b). Thus, we really get an equivalence relation $\sim$ and can define X as the quotient $X' / \sim$. Then we may view X as being covered by the X_i and, applying 5.1/11, we get a unique Grothendieck topology on it such that $(X_i)_{i \in I}$ is an admissible covering of X. Next, one constructs the structure sheaf $\mathcal{O}_X$ by gluing the sheaves $\mathcal{O}_{X_i}$. In a first step one identifies rings of type $\mathcal{O}_{X_i}(U)$ and $\mathcal{O}_{X_j}(U)$, in case U is contained in both, X_i and X_j, just by using the isomorphism φ_{ij}. This way one obtains a sheaf $\mathcal{O}_X$ on X with respect to some Grothendieck topology that is weaker than the one we have to consider. In a second step one applies 5.2/4, thereby extending $\mathcal{O}_X$ with respect to the Grothendieck topology we are considering on X. ☐

More easy is the gluing of morphisms:

Proposition 6. *Let X, Y be rigid K-spaces and let $(X_i)_{i \in I}$ be an admissible covering of X. Furthermore, let $\varphi_i : X_i \longrightarrow Y$ be morphisms of rigid K-spaces such that $\varphi_i|_{X_i \cap X_j} : X_i \cap X_j \longrightarrow Y$ coincides with $\varphi_j|_{X_i \cap X_j} : X_i \cap X_j \longrightarrow Y$ for all $i, j \in I$. Then there is a unique morphism of rigid K-spaces $\varphi : X \longrightarrow Y$ satisfying $\varphi|_{X_i} = \varphi_i$ for all $i \in I$.*

The *proof* is straightforward by using the sheaf property of $\mathcal{O}_X$.

Corollary 7. *Let X be a rigid K-space and Y an affinoid K-space. Then the canonical map*

$$\operatorname{Hom}(X, Y) \longrightarrow \operatorname{Hom}(\mathcal{O}_Y(Y), \mathcal{O}_X(X)), \qquad \varphi \longmapsto \varphi_Y^*,$$

is bijective.

Proof. The assertion follows from Proposition 2 if X is affinoid. In the general case it is only necessary to consider a homomorphism $\sigma^* : \mathcal{O}_Y(Y) \longrightarrow \mathcal{O}_X(X)$ and to show that there is a unique morphism of rigid K-spaces $\varphi : X \longrightarrow Y$ satisfying $\varphi_Y^* = \sigma^*$. To do this, choose an admissible affinoid covering $(X_i)_{i \in I}$ of X and write σ_i^* for the composition of σ^* with the canonical map $\mathcal{O}_X(X) \longrightarrow \mathcal{O}_X(X_i)$. Again by Proposition 2, each σ_i^* corresponds to a morphism of affinoid K-spaces $\varphi_i : X_i \longrightarrow Y$, and one concludes with the help of Proposition 6 that the φ_i can be glued to yield a unique morphism $\varphi : X \longrightarrow Y$, corresponding to σ^*. ☐

Corollary 8. *For two rigid K-spaces X, Y over a third one Z, the fiber product $X \times_Z Y$ can be constructed.*

Proof. The category of affinoid K-spaces admits fiber products, since, dually, the category of affinoid K-algebras admits amalgamated sums; see Appendix B. Thus, we have

$$\operatorname{Sp} A \times_{\operatorname{Sp} C} \operatorname{Sp} B = \operatorname{Sp}(A \,\hat{\otimes}_C\, B)$$

for morphisms of affinoid K-algebras $C \longrightarrow A$ and $C \longrightarrow B$. But then one can construct fiber products of global rigid K-spaces as usual by gluing local affinoid ones. ☐

In the next section, dealing with the GAGA-functor, we will give another application of Propositions 5 and 6.

Finally, let us briefly touch the subject of *connectedness* and of *connected components* for rigid spaces, as we will need these concepts later (see the proof of 8.4/4 (e)).

Definition 9. *A rigid K-space X is called* connected *if there do not exist non-empty admissible open subspaces $X_1, X_2 \subset X$ such that $X_1 \cap X_2 = \emptyset$ and (X_1, X_2) is an admissible covering of X.*

It follows from Tate's Acyclicity Theorem 4.3/10 that an affinoid K-space $\mathrm{Sp}\, A$ is connected if and only if A cannot be written as a non-trivial cartesian product of two K-algebras. The latter amounts to the fact that $\mathrm{Sp}\, A$ is connected with respect to the Zariski topology. In general, an affinoid K-space $\mathrm{Sp}\, A$ can be decomposed into its Zariski-connected components. These are affinoid subdomains of $\mathrm{Sp}\, A$ and define an admissible affinoid covering as they are of finite number, due to the fact that affinoid algebras are Noetherian by 3.1/3 (i). To check whether or not a global rigid K-space X is connected, one can consider an admissible covering $(U_i)_{i \in J}$ of X by non-empty connected admissible open subspaces $U_i \subset X$. For example, one may take the U_i to be affinoid and connected. If there is no partition of J into non-empty subsets $J_1, J_2 \subset J$ such that

$$\bigcup_{i \in J_1} U_i \cap \bigcup_{i \in J_2} U_i = \emptyset, \qquad\qquad (*)$$

then X is connected, otherwise it is not. Indeed, if there is a partition $J = J_1 \amalg J_2$ satisfying $(*)$, then $X_1 = \bigcup_{i \in J_1} U_i$ and $X_2 = \bigcup_{i \in J_2} U_i$ are admissible open in X by condition (G_1) of 5.1/5 and $X = X_1 \cup X_2$ is an admissible covering of X by condition (G_2) of 5.1/5. Conversely, assume there are admissible open subspaces $X_1, X_2 \subset X$ satisfying $X_1 \cap X_2 = \emptyset$, which define an admissible covering of X. Then consider an admissible covering $(U_i)_{i \in J}$ of X consisting of connected admissible open subsets. The admissible covering (X_1, X_2) of X restricts to an admissible covering $(X_1 \cap U_i, X_2 \cap U_i)$ on each U_i. Since U_i is supposed to be connected, we get $X_1 \cap U_i = U_i$ or $X_2 \cap U_i = U_i$ and, thus, $U_i \subset X_1$ or $U_i \subset X_2$. This leads to a partition $J = J_1 \amalg J_2$ such that $(*)$ is satisfied.

To define the connected components of a rigid K-space X, write $x \sim y$ for two points $x, y \in X$ if there exist finitely many connected admissible open subsets $U_0, \ldots, U_n \subset X$ such that $x \in U_0$, $y \in U_n$, and $U_{i-1} \cap U_i \neq \emptyset$ for $i = 1, \ldots, n$.

Proposition 10. *Let X be a rigid K-space and consider the just defined relation " $\sim$ " on it.*

(i) *" $\sim$ " is an equivalence relation.*

(ii) *For any $x \in X$ the corresponding equivalence class $Z(x)$ is admissible open in X. It is called the* connected component *of X that contains x.*

(iii) *The connected components of X form an admissible covering of X.*

Proof. First, that " $\sim$ " is an equivalence relation is clear from the definition of the relation. Next, consider an admissible open subset $U \subset X$ that is connected and assume $U \cap Z(x) \neq \emptyset$. Then we must have $U \subset Z(x)$ by the definition of $Z(x)$. In particular, consider an admissible covering $(U_i)_{i \in J}$ of X where all U_i are connected. It follows $U_i \subset Z(x)$ or $U_i \cap Z(x) = \emptyset$, depending on $i \in J$. Thus, we can conclude from condition (G_1) of 5.1/5 that $Z(x)$ is admissible open in X. By a similar reasoning one concludes from condition (G_2) of 5.1/5 that $(Z(x))_{x \in X}$ is an admissible covering of X. $\square$

5.4 The GAGA-Functor

We want to construct a functor that associates to any K-scheme Z of locally finite type a rigid K-space Z^{rig}, called the rigid analytification of Z. The corresponding functor in the classical complex case was first investigated by Serre in his fundamental paper [S]. Taking initials of the main words in the title, the functor has been referred to as the GAGA-functor since.

Let us start by constructing the rigid version of the affine n-space $\mathbb{A}_K^n$. To do this, we denote by $T_n(r)$ for $r > 0$ the K-algebra of all power series $\sum_\nu a_\nu \zeta^\nu$ in n variables $\zeta = (\zeta_1, \dots, \zeta_n)$ and with coefficients in K satisfying $\lim_\nu a_\nu r^{|\nu|} = 0$. Thus, $T_n(r)$ consists of all power series converging on a closed n-dimensional ball of radius r. Now choose $c \in K$, $|c| > 1$. Then we may identify $T_n^{(i)} = T_n(|c|^i)$ with the Tate algebra $K \langle c^{-i} \zeta_1, \dots, c^{-i} \zeta_n \rangle$. The inclusions

$$T_n = T_n^{(0)} \longleftarrow T_n^{(1)} \longleftarrow T_n^{(2)} \longleftarrow \dots \longleftarrow K[\zeta]$$

give rise to inclusions of affinoid subdomains

$$\mathbb{B}^n = \operatorname{Sp} T_n^{(0)} \longhookrightarrow \operatorname{Sp} T_n^{(1)} \longhookrightarrow \operatorname{Sp} T_n^{(2)} \longhookrightarrow \dots$$

where $\operatorname{Sp} T_n^{(i)}$ can be interpreted as the n-dimensional ball of radius $|c^i|$. Using 5.3/5, the "union" of all these balls can be constructed. The resulting rigid K-space $\mathbb{A}_K^{n,\text{rig}}$ comes equipped with the admissible covering $\mathbb{A}_K^{n,\text{rig}} = \bigcup_{i=0}^\infty \operatorname{Sp} T_n^{(i)}$, and we refer to it as the *rigid analytification* of the affine n-space $\mathbb{A}_K^n$. In particular, we will see that $\mathbb{A}_K^{n,\text{rig}}$ is independent of the choice of c and that it satisfies the universal property of an affine n-space in the category of rigid K-spaces. In a first step, we want to show that, pointwise, $\mathbb{A}_K^{n,\text{rig}}$ coincides with the set of closed points in $\mathbb{A}_K^n$.

Lemma 1. *The inclusions*

$$T_n^{(0)} \supset T_n^{(1)} \supset T_n^{(2)} \supset \dots \supset K[\zeta]$$

induce inclusions of spectra of maximal ideals

$$\text{Max } T_n^{(0)} \subset \text{Max } T_n^{(1)} \subset \text{Max } T_n^{(2)} \subset \ldots \subset \text{Max } K[\zeta]$$

such that $\text{Max } K[\zeta] = \bigcup_{i=0}^{\infty} \text{Max } T_n^{(i)}$.

Proof. As we have inclusions of affinoid subdomains $\text{Sp } T_n^{(i)} \hookrightarrow \text{Sp } T_n^{(i+1)}$ the inclusions between maximal spectra of the above affinoid K-algebras are clear. Next we show the following assertions:

(a) *Let* $\mathfrak{m} \subset K\langle\zeta\rangle$ *be a maximal ideal. Then* $\mathfrak{m}' = \mathfrak{m} \cap K[\zeta]$ *is a maximal ideal in* $K[\zeta]$ *satisfying* $\mathfrak{m} = \mathfrak{m}' K\langle\zeta\rangle$.
(b) *Given a maximal ideal* $\mathfrak{m}' \subset K[\zeta]$, *there is an index* $i_0 \in \mathbb{N}$ *such that* $\mathfrak{m}' K\langle c^{-i}\zeta\rangle$ *is maximal in* $K\langle c^{-i}\zeta\rangle = T_n^{(i)}$ *for all* $i \geq i_0$.

Let us start with assertion (a). There is a commutative diagram

$$
\begin{array}{ccc}
K[\zeta] & \longrightarrow & K\langle\zeta\rangle \\
\downarrow & & \downarrow \\
K[\zeta]/\mathfrak{m}' & \longrightarrow & K\langle\zeta\rangle/\mathfrak{m}
\end{array}
$$

with horizontal maps being injections. As $K\langle\zeta\rangle/\mathfrak{m}$ is a field that is finite over K by 2.2/12, the same must be true for $K[\zeta]/\mathfrak{m}'$, and it follows that $\mathfrak{m}'$ is maximal in $K[\zeta]$.

To see $\mathfrak{m} = \mathfrak{m}' K\langle\zeta\rangle$, look at the following commutative diagram:

$$
\begin{array}{ccc}
K[\zeta]/\mathfrak{m}' & \longrightarrow & K\langle\zeta\rangle/\mathfrak{m}' K\langle\zeta\rangle \\
\| & & \downarrow \\
K[\zeta]/\mathfrak{m}' & \longrightarrow & K\langle\zeta\rangle/\mathfrak{m}
\end{array}
$$

As $K[\zeta]$ is dense in $K\langle\zeta\rangle$, and as finite-dimensional K-vector spaces are complete (and, hence, closed if they are subspaces, see Theorem 1 of Appendix A), it follows that the horizontal maps are surjective. As the lower horizontal map is injective by definition of $\mathfrak{m}'$, it is, in fact, bijective. Then the upper horizontal map is injective and, hence, bijective as well. Consequently, the right vertical map is bijective, and assertion (a) is clear. Thereby we see that the canonical map $\text{Max } T_n^{(i)} \longrightarrow \text{Max } K[\zeta]$ is a well-defined injection for $i = 0$ and, in a similar way, for all i.

To verify (b), consider a maximal ideal $\mathfrak{m}' \subset K[\zeta]$. Then, by the analog of 2.2/12 for polynomial rings, $K[\zeta]/\mathfrak{m}'$ is a finite extension of K and, as such, carries a well-defined absolute value extending the one of K. Choosing an integer $i_0 \in \mathbb{N}$ such that the absolute values of the residue classes $\overline{\zeta}_j \in K[\zeta]/\mathfrak{m}'$ of all components of

ζ satisfy $|\bar{\zeta}_j| \leq |c|^{i_0}$, it follows that the projection $K[\zeta] \longrightarrow K[\zeta]/\mathfrak{m}'$ factors for $i \geq i_0$ through $T_n^{(i)} = K\langle c^{-i}\zeta \rangle$ via a unique K-morphism $T_n^{(i)} \longrightarrow K[\zeta]/\mathfrak{m}'$ sending ζ_j to $\bar{\zeta}_j$. The kernel $\mathfrak{m}$ of the latter map is a maximal ideal in $T_n^{(i)}$ satisfying $\mathfrak{m} \cap K[\zeta] = \mathfrak{m}'$. Consequently, (a) and (b) together imply that $\operatorname{Max} K[\zeta]$ is the union of the $\operatorname{Max} T_n^{(i)}$. $\qquad\square$

To construct the rigid analytification of an affine K-scheme of finite type, say of $\operatorname{Spec} K[\zeta]/\mathfrak{a}$ with an ideal $\mathfrak{a} \subset K[\zeta]$ and a system ζ of n variables $\zeta_1, \ldots, \zeta_n$, we proceed similarly by looking at the maps

$$T_n^{(0)}/(\mathfrak{a}) \longleftarrow T_n^{(1)}/(\mathfrak{a}) \longleftarrow T_n^{(2)}/(\mathfrak{a}) \longleftarrow \cdots \longleftarrow K[\zeta]/\mathfrak{a}$$

and the associated sequence of inclusions

$$\operatorname{Max} T_n^{(0)}/(\mathfrak{a}) \hookrightarrow \operatorname{Max} T_n^{(1)}/(\mathfrak{a}) \hookrightarrow \operatorname{Max} T_n^{(2)}/(\mathfrak{a}) \hookrightarrow \cdots$$
$$\hookrightarrow \operatorname{Max} K[\zeta]/\mathfrak{a}$$

where, again, we may interpret the first maps as inclusions of affinoid subdomains $\operatorname{Sp} T_n^{(i)}/(\mathfrak{a}) \hookrightarrow \operatorname{Sp} T_n^{(i+1)}/(\mathfrak{a})$, for all i. Furthermore, we see from Lemma 1 that all maps into $\operatorname{Max} K[\zeta]/\mathfrak{a}$ are injective and that $\operatorname{Max} K[\zeta]/\mathfrak{a}$ equals the union of all $\operatorname{Max} T_n^{(i)}/(\mathfrak{a})$. Thus, the union $\bigcup_{i=0}^{\infty} \operatorname{Sp} T_n^{(i)}/(\mathfrak{a})$ can be constructed as a rigid K-space using 5.3/5, and we call it the *rigid analytification* of $\operatorname{Spec} K[\zeta]/\mathfrak{a}$.

We want to show that, for any K-scheme of locally finite type Z and its analytification Z^{rig}, there is a canonical morphism of locally G-ringed K-spaces $(\iota, \iota^*) \colon (Z^{\text{rig}}, \mathcal{O}_{Z^{\text{rig}}}) \longrightarrow (Z, \mathcal{O}_Z)$ where, of course, Z is provided with the Zariski topology. Adapting 5.3/6 to our situation, the existence of such a morphism is a consequence of the following auxiliary result.

Lemma 2. *Let Z be an affine K-scheme of finite type and Y a rigid K-space. Then the set of morphisms of locally G-ringed K-spaces $(Y, \mathcal{O}_Y) \longrightarrow (Z, \mathcal{O}_Z)$ corresponds bijectively to the set of K-algebra homomorphisms $\mathcal{O}_Z(Z) \longrightarrow \mathcal{O}_Y(Y)$.*

Proof. We can conclude similarly as in 5.3/2 and 5.3/7. Let us first consider the case where Y is affinoid. Set $B = \mathcal{O}_Y(Y)$ and $C = \mathcal{O}_Z(Z)$ and consider a K-morphism $\sigma \colon C \longrightarrow B$. By the usual reasoning involving 2.2/12, taking inverse images of maximal ideals yields a map $\operatorname{Max} B \to \operatorname{Max} C \hookrightarrow \operatorname{Spec} C$ and, thus, a well-defined map $\varphi \colon Y \longrightarrow Z$ that is easily seen to be continuous with respect to Grothendieck topologies. For $f \in C$ and $\varepsilon \in K^*$ there is a commutative diagram

$$
\begin{array}{ccc}
C & \xrightarrow{\sigma} & B \\
\downarrow & & \downarrow \\
C[f^{-1}] & \longrightarrow & B\langle \varepsilon \cdot \sigma(f)^{-1} \rangle
\end{array}
$$

with a unique lower map, due to the fact that $\sigma(f)$ is invertible in $B\langle\varepsilon\cdot\sigma(f)^{-1}\rangle$. Thus, varying ε yields a commutative diagram

$$
\begin{array}{ccc}
\mathcal{O}_Z(Z) & \xrightarrow{\ \sigma\ } & \mathcal{O}_Y(Y) \\
\downarrow & & \downarrow \\
\mathcal{O}_Z(Z_f) & \longrightarrow & \mathcal{O}_Y\big(\varphi^{-1}(Z_f)\big)
\end{array}
$$

with a unique lower map; Z_f is the part of Z where f does not vanish. From this and the standard globalization argument one concludes that there is a morphism of ringed K-spaces $(\varphi,\varphi^*)\colon (Y,\mathcal{O}_Y) \longrightarrow (Z,\mathcal{O}_Z)$ satisfying $\varphi_Z^* = \sigma$. The morphism is a morphism of locally ringed K-spaces, as for any point $z \in Z$ and its corresponding prime ideal $\mathfrak{p} \in C$, the maximal ideal of the local ring $\mathcal{O}_{Z,z} = C_{\mathfrak{p}}$ is generated by $\mathfrak{p}$.

Just as in 5.3/2, it remains to show that there is at most one morphism of locally G-ringed K-spaces $(\varphi,\varphi^*)\colon (Y,\mathcal{O}_Y) \longrightarrow (Z,\mathcal{O}_Z)$ satisfying $\varphi_Z^* = \sigma$. The proof is the same as the one of 5.3/3. Finally, the generalization to the case where $(Y,\mathcal{O}_Y)$ is not necessarily affinoid is done as in 5.3/7.   □

To show that rigid analytifications are independent of the choice of the constant $c \in K$ and of the representation of K-algebras of finite type as quotients $K[\zeta]/\mathfrak{a}$, we want to characterize them by a universal property.

Definition and Proposition 3. *Let $(Z,\mathcal{O}_Z)$ be a K-scheme of locally finite type. A rigid analytification of $(Z,\mathcal{O}_Z)$ is a rigid K-space $(Z^{\mathrm{rig}},\mathcal{O}_{Z^{\mathrm{rig}}})$ together with a morphism of locally G-ringed K-spaces $(\iota,\iota^*)\colon (Z^{\mathrm{rig}},\mathcal{O}_{Z^{\mathrm{rig}}}) \longrightarrow (Z,\mathcal{O}_Z)$ satisfying the following universal property:*

Given a rigid K-space $(Y,\mathcal{O}_Y)$ and a morphism of locally G-ringed K-spaces $(Y,\mathcal{O}_Y) \longrightarrow (Z,\mathcal{O}_Z)$, the latter factors through (ι,ι^*) via a unique morphism of rigid K-spaces $(Y,\mathcal{O}_Y) \longrightarrow (Z^{\mathrm{rig}},\mathcal{O}_{Z^{\mathrm{rig}}})$.

For example, the analytifications Z^{rig} constructed above for affine K-schemes of finite type, give rise to analytifications in the sense of this definition.

Proof. To show that the rigid analytifications as constructed in the beginning are analytifications in the sense of the definition, we look at an affine K-scheme of finite type $Z = \operatorname{Spec} K[\zeta]/\mathfrak{a}$ and consider its associated rigid K-space that is given by $Z^{\mathrm{rig}} = \bigcup_{i=0}^{\infty} \operatorname{Sp} T_n^{(i)}/(\mathfrak{a})$. The canonical morphisms $K[\zeta]/\mathfrak{a} \longrightarrow T_n^{(i)}/(\mathfrak{a})$ constitute a morphism $\mathcal{O}_Z(Z) \longrightarrow \mathcal{O}_{Z^{\mathrm{rig}}}(Z^{\mathrm{rig}})$ and, using Lemma 2, the latter gives rise to a well-defined morphism of locally G-ringed K-spaces

$$
(\iota,\iota^*)\colon (Z^{\mathrm{rig}},\mathcal{O}_{Z^{\mathrm{rig}}}) \longrightarrow (Z,\mathcal{O}_Z).
$$

We claim that (ι, ι^*) satisfies the universal property of rigid analytifications. To justify this, look at a morphism of locally G-ringed K-spaces $(Y, \mathcal{O}_Y) \longrightarrow (Z, \mathcal{O}_Z)$ where $(Y, \mathcal{O}_Y)$ is a rigid K-space that we may assume to be affinoid.

Using Lemma 2, the morphism $(Y, \mathcal{O}_Y) \longrightarrow (Z, \mathcal{O}_Z)$ corresponds to a K-morphism $\sigma \colon K[\zeta]/\mathfrak{a} \longrightarrow B$ where $B = \mathcal{O}_Y(Y)$, and it is enough to show that, for all $i \in \mathbb{N}$ sufficiently large, there is a factorization $K[\zeta]/\mathfrak{a} \longrightarrow T_n^{(i)}/(\mathfrak{a}) \longrightarrow B$ of σ with a unique map $T_n^{(i)}/(\mathfrak{a}) \longrightarrow B$. Choose $i \in \mathbb{N}$ such that the residue classes $\bar{\zeta}_j \in K[\zeta]/\mathfrak{a}$ satisfy $|\sigma(\bar{\zeta}_j)|_{\sup} \leq |c|^i$ in B. Then the K-morphism $K[\zeta] \longrightarrow B$ obtained from σ extends uniquely to $T_n^{(i)}$, and we see that σ admits a unique factorization through $T_n^{(i)}/(\mathfrak{a})$, as claimed. $\qquad\square$

Proposition 4. *Every K-scheme Z of locally finite type admits an analytification $Z^{\mathrm{rig}} \longrightarrow Z$. Furthermore, the underlying map of sets identifies the points of Z^{rig} with the closed points of Z.*

Proof. We know the assertion already if Z is affine. In the general case we choose a covering of Z by affine open subschemes Z_i, $i \in J$. The latter admit analytifications $\iota_i \colon Z_i^{\mathrm{rig}} \longrightarrow Z_i$. It follows then from the definition of analytifications that $\iota_i^{-1}(Z_i \cap Z_j) \longrightarrow Z_i \cap Z_j$ is an analytification of $Z_i \cap Z_j$ for all $i, j \in J$. Thus, we can transport the gluing data we have on the Z_i to the analytifications Z_i^{rig} and thereby construct a rigid K-space Z^{rig} using 5.3/5. By 5.3/6 we get a morphism of locally G-ringed K-spaces $Z^{\mathrm{rig}} \longrightarrow Z$ that is easily seen to be an analytification of Z. Finally, the assertion on the underlying map of point sets follows from the construction of Z^{rig}, since the assertion is known over the affine open parts of Z. $\qquad\square$

The characterizing universal property of rigid analytifications shows that morphisms between K-schemes of locally finite type admit analytifications as well. Thus we can state:

Corollary 5. *Rigid analytification defines a functor from the category of K-schemes of locally finite type to the category of rigid K-spaces, the so-called GAGA-functor.*

Relying on the relevant universal properties, one can even show that rigid analytification respects fiber products, see Köpf [Kö], Satz 1.8. Furthermore, for a K-scheme Z of locally finite type and its rigid analytification Z^{rig}, the maximal adic completion of the stalk $\mathcal{O}_{Z^{\mathrm{rig}}, z}$ at a point $z \in Z^{\mathrm{rig}}$ coincides canonically with the maximal adic completion of the stalk $\mathcal{O}_{Z, z}$ at the corresponding closed point in Z; see [Kö], Satz 2.1. One can conclude from this that the GAGA-functor is faithful. However, it is not fully faithful since there exist K-schemes of locally finite type Y and Z, for example take $Y = Z$ as the affine line $\mathbb{A}_K^1$, such that there are morphisms of rigid K-spaces $Y^{\mathrm{rig}} \longrightarrow Z^{\mathrm{rig}}$ that cannot be viewed as analytifications of morphisms $Y \longrightarrow Z$.

Let us conclude the section by looking at some examples. First we want to show that the analytification $\mathbb{A}_K^{n,\mathrm{rig}}$ of the affine n-space $\mathbb{A}_K^n$ satisfies the universal property of an n-dimensional affine space, namely that for any rigid K-space Y the set of morphisms of rigid K-spaces $Y \longrightarrow \mathbb{A}_K^{n,\mathrm{rig}}$ is in one-to-one correspondence with $\mathcal{O}_Y(Y)^n$, the n-fold cartesian product of the set of global sections on Y. Indeed, composition with the canonical morphism of locally G-ringed spaces $\mathbb{A}_K^{n,\mathrm{rig}} \longrightarrow \mathbb{A}_K^n$ of Definition 3 yields a bijection

$$\mathrm{Hom}_K(Y, \mathbb{A}_K^{n,\mathrm{rig}}) \xrightarrow{\sim} \mathrm{Hom}_K(Y, \mathbb{A}_K^n)$$

between the set of rigid morphisms $Y \longrightarrow \mathbb{A}_K^{n,\mathrm{rig}}$ and the set of morphisms $Y \longrightarrow \mathbb{A}_K^n$ of locally G-ringed spaces over K. Furthermore, by Lemma 2, we get bijections

$$\mathrm{Hom}_K(Y, \mathbb{A}_K^n) \xrightarrow{\sim} \mathrm{Hom}_K\big(K[\zeta_1,\ldots,\zeta_n], \mathcal{O}_Y(Y)\big) \xrightarrow{\sim} \mathcal{O}_Y(Y)^n$$

so that the desired property $\mathrm{Hom}_K(Y, \mathbb{A}_K^n) \xrightarrow{\sim} \mathcal{O}_Y(Y)^n$ follows.

Let us have a particular look at the analytification $\mathbb{A}_K^{1,\mathrm{rig}}$ of the affine 1-space $\mathbb{A}_K^1$ that is constructed by gluing the ascending sequence of affinoid spaces

$$\mathrm{Sp}\, T_1^{(0)} \lhook\joinrel\longrightarrow \mathrm{Sp}\, T_1^{(1)} \lhook\joinrel\longrightarrow \mathrm{Sp}\, T_1^{(2)} \lhook\joinrel\longrightarrow \cdots$$

where, for some $c \in K$ with $|c| > 1$, we may interpret $\mathrm{Sp}\, T_1^{(i)} = K\langle c^{-i}\zeta \rangle$ as the disk with radius $|c|^i$ centered at the origin. Writing $R^{(i)} = \mathrm{Sp}\, T_1^{(i+1)}\langle c^i \zeta^{-1}\rangle$ for the annulus with radii $|c|^i$ and $|c|^{i+1}$, we obtain for each $i \in \mathbb{N}$

$$\mathrm{Sp}\, T_1^{(i+1)} = \mathrm{Sp}\, T_1^{(i)} \cup R^{(i)}$$

as an admissible affinoid covering of $\mathrm{Sp}\, T_1^{(i+1)}$ and, hence,

$$\mathbb{A}_K^{1,\mathrm{rig}} = \mathrm{Sp}\, T_1^{(0)} \cup \bigcup_{i \in \mathbb{N}} R^{(i)}$$

as an admissible affinoid covering of the analytification $\mathbb{A}_K^{1,\mathrm{rig}}$. Thus, we could just as well define the rigid analytification of $\mathbb{A}_K^1$ by relying on such a covering consisting of a disk and an infinite sequence of annuli. Removing the origin from $\mathbb{A}_K^1$, we get $\mathbb{A}_K^{1,\mathrm{rig}} - \{0\}$ as its analytification, and the latter admits

$$\mathbb{A}_K^{1,\mathrm{rig}} - \{0\} = \bigcup_{i \in \mathbb{Z}} R^{(i)}$$

as a convenient admissible affinoid covering by annuli.

Let us assume for a moment that K is algebraically closed, but not spherically complete. Not spherically complete means that there exists a descending sequence $D_0 \supset D_1 \supset D_2 \supset \ldots$ of disks of type $D^-(a, r)$ in K, centered at points $a \in K$ and with radii $r \in |K|$, such that the intersection $\bigcap_{i \in \mathbb{N}} D_i$ is *empty*. For example, it can be shown that the field $\mathbb{C}_p$ for any prime p is not spherically complete. Interpreting $\mathbb{B}^1 = \operatorname{Sp} T_1$ as the unit disk, we may assume that all disks D_i are contained in $\mathbb{B}^1$. Then $\mathbb{B}^1$ is covered by the ascending sequence of annuli $\mathbb{B}^1 - D_i$ where all these annuli may be interpreted as affinoid subdomains of $\mathbb{B}^1$. However, as this covering does not admit a finite refinement, it is not admissible. On the other hand, we are free to use the covering

$$D^+(0, 1) = \bigcup_{i \in \mathbb{N}} (\mathbb{B}^1 - D_i)$$

in order to define an "exotic" structure of rigid K-space on the points of the unit disk. In the same way the covering

$$K = \bigcup_{i \in \mathbb{N}} (\mathbb{B}^1 - D_i) \cup \bigcup_{i > 1} R^{(i)}$$

leads to an "exotic" structure of rigid K-space on the affine line over K. But let us point out that in more refined theories allowing additional points like Berkovich or Huber theory, these "exotic" structures become quite natural as they give rise to subspace structures on suitable subspaces of the unit disk or the affine line.

Finally let us look at the projective n-space

$$\mathbb{P}_K^n = \operatorname{Proj} K[\zeta_0, \ldots, \zeta_n]$$

where $\zeta_0, \ldots, \zeta_n$ denote variables and K is not necessarily algebraically closed any more. Writing

$$A_i = K\left[\frac{\zeta_0}{\zeta_i}, \ldots, \frac{\zeta_n}{\zeta_i}\right]$$

for the homogeneous localization of $K[\zeta_0, \ldots, \zeta_n]$ by ζ_i, the projective n-space $\mathbb{P}_K^n$ is covered by the open affine subschemes $U_i = \operatorname{Spec} A_i \simeq \mathbb{A}_K^n$. Accordingly, the rigid analytification $\mathbb{P}_K^{n,\mathrm{rig}}$ admits an admissible covering consisting of the rigid analytifications

$$U_i^{\mathrm{rig}} = \bigcup_{j \in \mathbb{N}} \operatorname{Sp} K\left\langle c^{-j} \frac{\zeta_0}{\zeta_i}, \ldots, c^{-j} \frac{\zeta_n}{\zeta_i} \right\rangle \simeq \mathbb{A}_K^{n,\mathrm{rig}}, \qquad i = 0, \ldots, n,$$

for some $c \in K$, $|c| > 1$. We claim that, in fact, $\mathbb{P}_K^{n,\mathrm{rig}}$ is already covered by the unit balls

$$\operatorname{Sp} K\left\langle \frac{\zeta_0}{\zeta_i}, \ldots, \frac{\zeta_n}{\zeta_i} \right\rangle \subset U_i^{\mathrm{rig}}, \qquad i = 0, \ldots, n.$$

As a consequence, the latter covering is admissible, since it is a refinement of the previous one. To justify that $\mathbb{P}_K^{n,\mathrm{rig}}$ is a union of the $n + 1$ unit balls in $U_0^{\mathrm{rig}}, \ldots, U_n^{\mathrm{rig}}$, consider a closed point $x \in \mathbb{P}_K^n$, say with residue field $L = K(x)$, and view it as an L-valued point in $\mathbb{P}_K^n(L)$. As the latter set can be interpreted as the ordinary projective n-space $\mathbb{P}^n(L) = (L^{n+1} - \{0\})/L^*$, we may represent x in terms of homogeneous coordinates, say $x = (x_0 : \ldots : x_n)$ with components $x_i \in L$. Extending the absolute value of K to L, which is finite over K, choose an index i such that

$$|x_i| = \max\{|x_0|, \ldots, |x_n|\}.$$

Then x factors through $\operatorname{Sp} K\langle \frac{\zeta_0}{\zeta_i}, \ldots, \frac{\zeta_n}{\zeta_i} \rangle$ and, consequently, $\mathbb{P}_K^{n,\mathrm{rig}}$ is covered by unit balls as claimed.

Chapter 6
Coherent Sheaves on Rigid Spaces

6.1 Coherent Modules

Consider an affinoid K-space $X = \operatorname{Sp} A$ and an A-module M. We can look at the functor $\mathcal{F}$ from affinoid subdomains in X to abelian groups that associates to any affinoid subdomain $\operatorname{Sp} A' \subset X$ the tensor product $M \otimes_A A'$. The latter is, of course, an abelian group, but we can also view it as an A-module or even as an A'-module. $\mathcal{F}$ is a presheaf on X with respect to the weak G-topology, and this presheaf is, in fact, a sheaf, as we have already remarked within the context of Tate's Acyclicity Theorem in 4.3/11. In particular, using 5.2/4, we see that $\mathcal{F}$ extends to a sheaf with respect to the strong G-topology, again denoted by $\mathcal{F}$.

It follows from the construction that $\mathcal{F}$ is a so-called $\mathcal{O}_X$-*module*. This means that, for any admissible open $U \subset X$, the abelian group $\mathcal{F}(U)$ is equipped with an $\mathcal{O}_X(U)$-module structure, in a way that all these module structures are compatible with restriction homomorphisms. We call $\mathcal{F}$ the $\mathcal{O}_X$-*module associated to the A-module M*, writing $\mathcal{F} = M \otimes_A \mathcal{O}_X$. Note that we have

$$\mathcal{F}|_{X'} = (M \otimes_A A') \otimes_{A'} \mathcal{O}_X|_{X'}$$

for the restriction of $\mathcal{F}$ to any affinoid subdomain $X' = \operatorname{Sp} A'$ in X.

Proposition 1. *Let $X = \operatorname{Sp} A$ be an affinoid K-space.*

(i) *The functor*

$$\cdot \otimes_A \mathcal{O}_X \colon \quad M \longmapsto M \otimes_A \mathcal{O}_X$$

from A-modules to $\mathcal{O}_X$-modules is fully faithful.

(ii) *It commutes with the formation of kernels, images, cokernels, and tensor products.*

S. Bosch, *Lectures on Formal and Rigid Geometry*, Lecture Notes in Mathematics 2105, DOI 10.1007/978-3-319-04417-0_6, © Springer International Publishing Switzerland 2014

(iii) *A sequence of A-modules* $0 \longrightarrow M' \longrightarrow M \longrightarrow M'' \longrightarrow 0$ *is exact if and only if the associated sequence of $\mathcal{O}_X$-modules is exact:*

$$0 \longrightarrow M' \otimes_A \mathcal{O}_X \longrightarrow M \otimes_A \mathcal{O}_X \longrightarrow M'' \otimes_A \mathcal{O}_X \longrightarrow 0$$

Proof. It is clear that the canonical map

$$\operatorname{Hom}_A(M, M') \longrightarrow \operatorname{Hom}_{\mathcal{O}_X}(M \otimes_A \mathcal{O}_X, M' \otimes_A \mathcal{O}_X)$$

is bijective, since an $\mathcal{O}_X$-morphism $M \otimes_A \mathcal{O}_X \longrightarrow M' \otimes_A \mathcal{O}_X$ is uniquely determined by its inherent A-morphism

$$M = M \otimes_A \mathcal{O}_X(X) \longrightarrow M' \otimes_A \mathcal{O}_X(X) = M'.$$

Thus, the functor $\cdot \otimes_A \mathcal{O}_X$ is fully faithful, which settles assertion (i). Furthermore, by its construction, it commutes with tensor products.
    Next, if

$$0 \longrightarrow M' \longrightarrow M \longrightarrow M'' \longrightarrow 0$$

is an exact sequence of A-modules, the induced sequence

$$0 \longrightarrow M' \otimes_A A' \longrightarrow M \otimes_A A' \longrightarrow M'' \otimes_A A' \longrightarrow 0$$

is exact for any affinoid subdomain $\operatorname{Sp} A' \subset X$, since the corresponding map $A \longrightarrow A'$ is flat by 4.1/5. From this one easily concludes that the functor of taking associated $\mathcal{O}_X$-modules is exact, i.e. carries short exact sequences over to short exact sequences. Then assertion (ii) becomes clear and, furthermore, also (iii), using the fact that an A-module M is trivial if and only if $M \otimes_A \mathcal{O}_X$ is trivial. $\square$

Definition 2. *Let X be a rigid K-space and $\mathcal{F}$ an $\mathcal{O}_X$-module.*

(i) *$\mathcal{F}$ is called of* finite type *if there exists an admissible covering $(X_i)_{i \in I}$ of X together with exact sequences of type*

$$\mathcal{O}_X^{s_i}|_{X_i} \longrightarrow \mathcal{F}|_{X_i} \longrightarrow 0, \qquad i \in I.$$

(ii) *$\mathcal{F}$ is called of* finite presentation, *if there exists an admissible covering $(X_i)_{i \in I}$ of X together with exact sequences of type*

$$\mathcal{O}_X^{r_i}|_{X_i} \longrightarrow \mathcal{O}_X^{s_i}|_{X_i} \longrightarrow \mathcal{F}|_{X_i} \longrightarrow 0, \qquad i \in I.$$

(iii) *$\mathcal{F}$ is called* coherent *if $\mathcal{F}$ is of finite type and if for every admissible open subspace $U \subset X$ the kernel of a morphism $\mathcal{O}_X^s|_U \longrightarrow \mathcal{F}|_U$ is of finite type.*

For affinoid K-spaces $X = \operatorname{Sp} A$, we have $\mathcal{O}_X^r = A^r \otimes_A \mathcal{O}_X$. Furthermore, as A is Noetherian, we conclude from Proposition 1 that kernels and cokernels of morphisms of type $\mathcal{O}_X^r \longrightarrow \mathcal{O}_X^s$ are associated to A-modules of finite type. Therefore we can state:

Remark 3. *An $\mathcal{O}_X$-module $\mathcal{F}$ on a rigid K-space X is coherent if and only if there exists an admissible affinoid covering $\mathfrak{U} = (X_i)_{i \in I}$ of X such that $\mathcal{F}|_{X_i}$ is associated to a finite $\mathcal{O}_{X_i}(X_i)$-module for all $i \in I$. More precisely, we will say that $\mathcal{F}$ is $\mathfrak{U}$-coherent in this case.*

There is a basic result that fully clarifies the structure of coherent modules on affinoid K-spaces, see Kiehl [K1]:

Theorem 4 (Kiehl). *Let $X = \operatorname{Sp} A$ be an affinoid K-space and $\mathcal{F}$ an $\mathcal{O}_X$-module. Then $\mathcal{F}$ is coherent if and only if $\mathcal{F}$ is associated to a finite A-module.*

Before we give the proof, let us observe that this result allows a characterization of coherent $\mathcal{O}_X$-modules as follows:

Corollary 5. *Let X be a rigid K-space and $\mathcal{F}$ an $\mathcal{O}_X$-module on it. The following are equivalent:*

(i) *$\mathcal{F}$ is coherent, i.e. $\mathcal{F}$ is $\mathfrak{U}$-coherent for some admissible affinoid covering $\mathfrak{U}$ of X.*
(ii) *$\mathcal{F}$ is $\mathfrak{U}$-coherent for all admissible affinoid coverings $\mathfrak{U}$ of X.*

Proof. We have only to show that (i) implies (ii). So assume that $\mathcal{F}$ is coherent. In order to derive assertion (ii), we may assume that X is affinoid, say $X = \operatorname{Sp} A$. But then, applying Theorem 4, $\mathcal{F}$ is associated to a finite A-module and (ii) is obvious. $\square$

To start the *proof of Theorem* 4, observe that the if-part of the assertion is trivial. So assume that $\mathcal{F}$ is $\mathfrak{U}$-coherent for some admissible affinoid covering $\mathfrak{U}$ of X. To show that $\mathcal{F}$ is associated to a finite A-module, we may apply Lemmata 4.3/4, 4.3/5, and 4.3/6, and thereby restrict ourselves to the case where $\mathfrak{U}$ is a Laurent covering of X. Furthermore, using an inductive argument, it is only necessary to treat the case where $\mathfrak{U}$ is a Laurent covering generated by a single function $f \in A$. Then it is enough to establish the following facts:

Lemma 6. *Let $\mathcal{F}$ be $\mathfrak{U}$-coherent. Then $H^1(\mathfrak{U},\mathcal{F}) = 0$.*

Lemma 7. *Assume $H^1(\mathfrak{U},\mathcal{F}) = 0$ for all $\mathfrak{U}$-coherent $\mathcal{O}_X$-modules $\mathcal{F}$. Then any such module is associated to a finite A-module.*

Proof of Lemma 6. Let $\mathfrak{U} = (U_1, U_2)$ with $U_1 = X(f)$ and $U_2 = X(f^{-1})$. Due to our assumption,

$$M_1 = \mathcal{F}(U_1), \qquad M_2 = \mathcal{F}(U_2), \qquad M_{12} = \mathcal{F}(U_1 \cap U_2)$$

are finite modules over $A\langle f\rangle$, $A\langle f^{-1}\rangle$, and $A\langle f, f^{-1}\rangle$, respectively, and the Čech complex of alternating cochains $C_a^{\bullet}(\mathfrak{U}, \mathcal{F})$ degenerates to

$$0 \longrightarrow M_1 \times M_2 \overset{d^0}{\longrightarrow} M_{12} \longrightarrow 0.$$

Since $H^1(\mathfrak{U}, \mathcal{F})$ can be computed using alternating cochains, see 4.3/8, it is only necessary to show that $d^0 \colon M_1 \times M_2 \longrightarrow M_{12}$ is surjective.

To do so, we fix an arbitrary residue norm on A and consider on $A\langle\zeta\rangle$, $A\langle\eta\rangle$, as well as $A\langle\zeta, \eta\rangle$ the Gauß norm, and on $A\langle f\rangle$, $A\langle f^{-1}\rangle$, and $A\langle f, f^{-1}\rangle$ the residue norms induced from the canonical epimorphisms

$$
\begin{aligned}
A\langle\zeta\rangle &\longrightarrow & A\langle\zeta\rangle/(\zeta - f) &= A\langle f\rangle, \\
A\langle\eta\rangle &\longrightarrow & A\langle\eta\rangle/(f\eta - 1) &= A\langle f^{-1}\rangle, \\
A\langle\zeta, \eta\rangle &\longrightarrow & A\langle\zeta, \eta\rangle/(\zeta - f, f\eta - 1) &= A\langle f, f^{-1}\rangle.
\end{aligned}
$$

Then all restriction morphisms of the commutative diagram

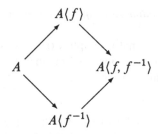

are contractive. Choosing a constant $\beta > 1$, any $g \in A\langle f, f^{-1}\rangle$ can be represented by a power series

$$g' = \sum c_{\mu\nu} \zeta^{\mu} \eta^{\nu} \in A\langle\zeta, \eta\rangle$$

where the coefficients $c_{\mu\nu} \in A$ form a zero sequence satisfying $|c_{\mu\nu}| \le \beta|g|$. Thereby we see:

(∗) *Let $\beta > 1$. For any $g \in A\langle f, f^{-1}\rangle$, there exist elements $g^+ \in A\langle f\rangle$ and $g^- \in A\langle f^{-1}\rangle$ such that*

$$|g^+| \le \beta|g|, \qquad |g^-| \le \beta|g|, \qquad g = g^+|_{U_1\cap U_2} + g^-|_{U_1\cap U_2}.$$

Next choose elements $v'_1, \ldots, v'_m \in M_1$ and $w'_1, \ldots, w'_n \in M_2$ generating M_1 as an $A\langle f \rangle$-module and M_2 as an $A\langle f^{-1} \rangle$-module. Using the fact that $\mathscr{F}$ is $\mathfrak{U}$-coherent, the restrictions $v_1, \ldots, v_m$ of the v'_i to $U_1 \cap U_2$, as well as the restrictions $w_1, \ldots, w_n$ of the w'_j to $U_1 \cap U_2$, will generate M_{12} as $A\langle f, f^{-1} \rangle$-module. Now look at the epimorphisms

$$\left(A\langle f \rangle \right)^m \longrightarrow M_1, \qquad \left(A\langle f^{-1} \rangle \right)^n \longrightarrow M_2, \qquad \left(A\langle f, f^{-1} \rangle \right)^m \longrightarrow M_{12},$$

given by mapping unit vectors to the $v'_i \in M_1$, to the $w'_j \in M_2$, and to the $v_i \in M_{12}$, respectively. Just as in the case of affinoid algebras, we can consider the attached residue norms on M_1, M_2, and M_{12}, starting out from the maximum norms on $(A\langle f \rangle)^m$, $(A\langle f^{-1} \rangle)^n$, and $(A\langle f, f^{-1} \rangle)^m$. These residue norms will be complete, as any Cauchy sequence, for example in M_1, can be lifted to a Cauchy sequence in $(A\langle f \rangle)^m$. Furthermore, M_1 will be a normed $A\langle f \rangle$-module in the sense that we have $|av| \le |a||v|$ for $a \in A\langle f \rangle$ and $v \in M_1$; likewise for M_2 and M_{12}. Thus, using a standard approximation procedure, the surjectivity of the coboundary morphism $d^0 \colon M_1 \times M_2 \longrightarrow M_{12}$ will be a consequence of the following assertion:

(∗∗) *Let $\varepsilon > 0$. Then there is a constant $\alpha > 1$ such that for each $u \in M_{12}$, there exist elements $u^+ \in M_1$ and $u^- \in M_2$ with*

$$|u^+| \le \alpha |u|, \qquad |u^-| \le \alpha |u|, \qquad \left| u - (u^+|_{U_1 \cap U_2}) - (u^-|_{U_1 \cap U_2}) \right| \le \varepsilon |u|.$$

To justify the assertion, recall that the elements v_i as well as the w_j generate M_{12} as an $A\langle f, f^{-1} \rangle$-module. Hence, there are equations

$$v_i = \sum_{j=1}^{n} c_{ij} w_j, \qquad i = 1, \ldots, m,$$

$$w_j = \sum_{l=1}^{m} d_{jl} v_l, \qquad j = 1, \ldots, n,$$

with coefficients $c_{ij}, d_{jl} \in A\langle f, f^{-1} \rangle$. Using the fact that the image of $A\langle f^{-1} \rangle$ is dense in $A\langle f, f^{-1} \rangle$, there are elements $c'_{ij} \in A\langle f^{-1} \rangle$ such that

$$\max_{ijl} |c_{ij} - c'_{ij}||d_{jl}| \le \beta^{-2} \varepsilon,$$

where $\beta > 1$ is a constant as in (∗) and where, in more precise terms, we should have used the restriction $c'_{ij}|_{U_1 \cap U_2}$ in place of c'_{ij}. We claim that assertion (∗∗) holds for

$$\alpha = \beta^2 \max(|c'_{ij}| + 1).$$

Indeed, consider any element $u \in M_{12}$, and write it as $u = \sum_{i=1}^{m} a_i v_i$ with coefficients $a_i \in A\langle f, f^{-1}\rangle$. Due to the choice of the norm on M_{12}, we may assume $|a_i| \leq \beta |u|$ for all i. Furthermore, using (∗), we can write

$$a_i = a_i^+ |_{U_1 \cap U_2} + a_i^- |_{U_1 \cap U_2}$$

with elements $a_i^+ \in A\langle f\rangle, a_i^- \in A\langle f^{-1}\rangle$ satisfying $|a_i^+| \leq \beta |a_i|$ and $|a_i^-| \leq \beta |a_i|$. Now consider the elements

$$u^+ = \sum_{i=1}^{m} a_i^+ v_i' \in M_1,$$

$$u^- = \sum_{i=1}^{m} \sum_{j=1}^{n} a_i^- c_{ij}' w_j' \in M_2.$$

We have

$$|u^+| \leq \max_i |a_i^+| \leq \max_i \beta |a_i| \leq \beta^2 |u| \leq \alpha |u|,$$

$$|u^-| \leq \max_{ij} |a_i^-| |c_{ij}'| \leq \max_i \beta |a_i| \max_{ij} |c_{ij}'| \leq \beta^2 |u| \max_{ij} |c_{ij}'| \leq \alpha |u|,$$

and, omitting restrictions to $U_1 \cap U_2$,

$$u = \sum_{i=1}^{m} (a_i^+ + a_i^-) v_i = u^+ + \sum_{i=1}^{m} \sum_{j=1}^{n} a_i^- c_{ij} w_j$$

$$= u^+ + u^- + \sum_{i=1}^{m} \sum_{j=1}^{n} a_i^- (c_{ij} - c_{ij}') w_j.$$

Hence,

$$|u - u^+ - u^-| = \left| \sum_{i=1}^{m} \sum_{j=1}^{n} \sum_{l=1}^{m} a_i^- (c_{ij} - c_{ij}') d_{jl} v_l \right|$$

$$\leq \max_{ijl} |a_i^-| |c_{ij} - c_{ij}'| |d_{jl}| \leq \beta^2 |u| \beta^{-2} \varepsilon = \varepsilon |u|,$$

which justifies assertion (∗∗) and thereby the assertion of the lemma.   □

Proof of Lemma 7. Here it is not necessary to make a difference between Laurent and general affinoid coverings. Therefore, consider a covering $\mathfrak{U} = (U_i)_{i=1,\ldots,n}$ of $X = \mathrm{Sp}\, A$ consisting of affinoid subdomains $U_i = \mathrm{Sp}\, A_i \subset X$. Since $\mathcal{F}$ is $\mathfrak{U}$-coherent, $\mathcal{F}|_{U_i}$ is associated to a finite A_i-module M_i, $i = 1, \ldots, n$. For x a point in X, we denote by $\mathfrak{m}_x \subset A$ its corresponding maximal ideal and by $\mathfrak{m}_x \mathcal{O}_X$

the associated coherent ideal of the structure sheaf $\mathcal{O}_X$. Its product with $\mathcal{F}$ yields a submodule $\mathfrak{m}_x \mathcal{F} \subset \mathcal{F}$ that is $\mathfrak{U}$-coherent, since its restriction to each U_i is associated to the submodule $\mathfrak{m}_x M_i \subset M_i$; the latter is finite, since M_i is a finite module over a Noetherian ring. Then $\mathcal{F}/\mathfrak{m}_x\mathcal{F}$ is $\mathfrak{U}$-coherent by Proposition 1 and

$$0 \longrightarrow \mathfrak{m}_x\mathcal{F} \longrightarrow \mathcal{F} \longrightarrow \mathcal{F}/\mathfrak{m}_x\mathcal{F} \longrightarrow 0$$

is a short exact sequence of $\mathfrak{U}$-coherent $\mathcal{O}_X$-modules.

If $U' = \operatorname{Sp} A'$ is an affinoid subdomain of X, which is contained in U_i for some index i, then the above short exact sequence restricts to a short exact sequence of coherent modules on U'. More precisely, as the modules $\mathfrak{m}_x\mathcal{F}$, $\mathcal{F}$, and $\mathcal{F}/\mathfrak{m}_x\mathcal{F}$ are $\mathfrak{U}$-coherent, their restrictions to U_i are associated to finite A_i-modules and the same is true for restrictions to U' in terms of A'-modules. Thus, by Proposition 1, the above short exact sequence leads to a short exact sequence of A'-modules

$$0 \longrightarrow \mathfrak{m}_x\mathcal{F}(U') \longrightarrow \mathcal{F}(U') \longrightarrow \mathcal{F}/\mathfrak{m}_x\mathcal{F}(U') \longrightarrow 0.$$

In particular, U' can be any intersection of sets in $\mathfrak{U}$, and we thereby see that the canonical sequence of Čech complexes

$$0 \longrightarrow C^\bullet(\mathfrak{U}, \mathfrak{m}_x\mathcal{F}) \longrightarrow C^\bullet(\mathfrak{U}, \mathcal{F}) \longrightarrow C^\bullet(\mathfrak{U}, \mathcal{F}/\mathfrak{m}_x\mathcal{F}) \longrightarrow 0$$

is exact. As $H^1(\mathfrak{U}, \mathfrak{m}_x\mathcal{F}) = 0$ by our assumption, the associated long cohomology sequence yields an exact sequence

$$0 \longrightarrow \mathfrak{m}_x\mathcal{F}(X) \longrightarrow \mathcal{F}(X) \longrightarrow \mathcal{F}/\mathfrak{m}_x\mathcal{F}(X) \longrightarrow 0. \qquad (*)$$

Next we claim:

$(**)$ *The restriction homomorphism $\mathcal{F}/\mathfrak{m}_x\mathcal{F}(X) \longrightarrow \mathcal{F}/\mathfrak{m}_x\mathcal{F}(U_j)$ is bijective for any index j such that $x \in U_j$.*

To justify the claim, consider an affinoid subdomain $U' = \operatorname{Sp} A' \subset X$ such that $\mathcal{F}|_{U'}$ is associated to a finite A'-module M' and write $U' \cap U_j = \operatorname{Sp} A'_j$. Then $\mathcal{F}/\mathfrak{m}_x|_{U'}$ is associated to the quotient $M'/\mathfrak{m}_x M'$, and the canonical map

$$M'/\mathfrak{m}_x M' \longrightarrow M'/\mathfrak{m}_x M' \otimes_{A'} A'_j \xrightarrow{\sim} M'/\mathfrak{m}_x M' \otimes_{A'/\mathfrak{m}_x A'} A'_j/\mathfrak{m}_x A'_j$$

is bijective for $x \in U_j$. This follows from 3.3/10 if $x \in U' \cap U_j$, since the restriction map $A'/\mathfrak{m}_x A' \longrightarrow A'_j/\mathfrak{m}_x A'_j$ is bijective then. However, the latter is also true for $x \notin U'$ since in this case the quotients $A'/\mathfrak{m}_x A'$ and $A'_j/\mathfrak{m}_x A'_j$ are trivial.

Now if $\mathcal{F}$ is known to be $\mathfrak{U}$-coherent, we look at the canonical diagram

$$\mathcal{F}/\mathfrak{m}_x\mathcal{F}(X) \longrightarrow \prod_{i=1}^{n} \mathcal{F}/\mathfrak{m}_x\mathcal{F}(U_i) \rightrightarrows \prod_{i,i'=1}^{n} \mathcal{F}/\mathfrak{m}_x\mathcal{F}(U_i \cap U_{i'})$$

$$\mathcal{F}/\mathfrak{m}_x\mathcal{F}(U_j) \longrightarrow \prod_{i=1}^{n} \mathcal{F}/\mathfrak{m}_x\mathcal{F}(U_i \cap U_j) \rightrightarrows \prod_{i,i'=1}^{n} \mathcal{F}/\mathfrak{m}_x\mathcal{F}(U_i \cap U_{i'} \cap U_j)$$

with exact rows. By the consideration above, the middle and right restriction morphisms are bijective. Thus, the same will hold for the left one, which settles assertion $(**)$.

Looking at the commutative diagram

$$
\begin{array}{ccc}
\mathcal{F}(X) & \longrightarrow & \mathcal{F}/\mathfrak{m}_x(X) \\
\downarrow & & \downarrow \\
M_j = \mathcal{F}(U_j) & \longrightarrow & \mathcal{F}/\mathfrak{m}_x(U_j) = M_j/\mathfrak{m}_x M_j
\end{array}
$$

for $x \in U_j$, the exact sequence $(*)$ shows in conjunction with $(**)$ that $M_j/\mathfrak{m}_x M_j$, as an A_j-module, is generated by the image of $\mathcal{F}(X)$. Hence, by the classical Lemma of Nakayama, $\mathcal{F}(X)$ generates M_j locally at each point $x \in U_j$. But then the submodule of M_j generated by the image of $\mathcal{F}(X)$ must coincide with M_j. Therefore we can choose elements $f_1, \ldots, f_s \in \mathcal{F}(X)$ such that their images generate all modules $M_i = \mathcal{F}(U_i)$ simultaneously for $i = 1, \ldots, n$. As a consequence, the morphism of $\mathcal{O}_X$-modules $\varphi: \mathcal{O}_X^s \longrightarrow \mathcal{F}$ given by $f_1, \ldots, f_s$ is an epimorphism of $\mathfrak{U}$-coherent $\mathcal{O}_X$-modules, and its kernel $\ker \varphi$ is a $\mathfrak{U}$-coherent submodule of $\mathcal{O}_X^s$ by Proposition 1.

We can work now in the same way as before with $\ker \varphi$ in place of $\mathcal{F}$ and construct an epimorphism $\psi: \mathcal{O}_X^r \longrightarrow \ker \varphi$, thus obtaining an exact sequence

$$\mathcal{O}_X^r \xrightarrow{\psi} \mathcal{O}_X^s \xrightarrow{\varphi} \mathcal{F} \longrightarrow 0$$

of $\mathcal{O}_X$-modules. Thereby we see that $\mathcal{F}$ is isomorphic to the cokernel of ψ, and so $\mathcal{F}$ is associated to the cokernel of the A-module morphism $\psi(X): A^r \longrightarrow A^s$ by Proposition 1. The latter is finite and, hence, $\mathcal{F}$ is associated to a finite A-module. This finishes the proof of Lemma 7 and thereby also the proof of Theorem 4. $\square$

If $\varphi: X \longrightarrow Y$ is a morphism of rigid K-spaces and $\mathcal{F}$ an $\mathcal{O}_X$-module, we can construct its *direct image* $\varphi_*\mathcal{F}$. In terms of abelian groups, the latter sheaf associates to any admissible open subspace $V \subset Y$ the abelian group $\mathcal{F}(\varphi^{-1}(V))$. Clearly, $\mathcal{F}(\varphi^{-1}(V))$ is an $\mathcal{O}_X(\varphi^{-1}(V))$-module and, via the morphism $\varphi_V^*: \mathcal{O}_Y(V) \longrightarrow \mathcal{O}_X(\varphi^{-1}(V))$, also an $\mathcal{O}_Y(V)$-module. Thereby the sheaf $\varphi_*\mathcal{F}$ inherits the structure of an $\mathcal{O}_Y$-module. The picture is quite simple for associated modules on affinoid K-spaces. So assume that $\varphi\colon X \longrightarrow Y$ is a morphism of affinoid K-spaces, say $X = \operatorname{Sp} A$ and $Y = \operatorname{Sp} B$, and let $\mathcal{F} = M \otimes_A \mathcal{O}_X$ for some A-module M. Then the definition shows that $\varphi_* \mathcal{F}$ coincides with the $\mathcal{O}_Y$-module associated to M viewed as a B-module via the morphism $\varphi_Y^*\colon B \longrightarrow A$. In particular, if φ is finite in the sense that A is a finite B-module via φ_Y^*, it follows that the direct image $\varphi_* \mathcal{F}$ is coherent if the same is true for $\mathcal{F}$. The latter statement is more generally true for so-called *proper* morphisms of rigid K-spaces, as we will explain later.

Considering a morphism $\varphi\colon X \longrightarrow Y$ of rigid K-spaces again, we may view φ_* as a functor from $\mathcal{O}_X$-modules to $\mathcal{O}_Y$-modules, a functor that is easily seen to be *left-exact*. There is a so-called *left-adjoint* φ^* of φ_*, which is *right-exact*. Given an $\mathcal{O}_Y$-module $\mathcal{E}$, the $\mathcal{O}_X$-module $\varphi^* \mathcal{E}$ is uniquely characterized (up to canonical isomorphism) by the equation

$$\operatorname{Hom}_{\mathcal{O}_X}(\varphi^* \mathcal{E}, \mathcal{F}) = \operatorname{Hom}_{\mathcal{O}_Y}(\mathcal{E}, \varphi_* \mathcal{F}),$$

which is supposed to be functorial in $\mathcal{F}$ varying over all $\mathcal{O}_X$-modules. $\varphi^* \mathcal{E}$ is called the *inverse image* of $\mathcal{E}$. Of course, one has to show that an $\mathcal{O}_X$-module $\varphi^* \mathcal{E}$ satisfying these equations really exists. There is a general procedure for showing the existence, which we will not explain at this place. We just look at the special case where X and Y are affinoid, say $X = \operatorname{Sp} A$ and $Y = \operatorname{Sp} B$, and where $\mathcal{E}$ is associated to a B-module N. In this situation, it is easy to see that the $\mathcal{O}_X$-module associated to $N \otimes_B A$ satisfies the above equations and, hence, must coincide with $\varphi^* \mathcal{E}$.

6.2 Grothendieck Cohomology

In the present section we will be concerned with $\mathcal{O}_X$-modules on rigid K-spaces X. As usual, the cohomology of such modules is defined via derived functors. The functors we want to consider are the *section functor*

$$\Gamma(X, \cdot)\colon \mathcal{F} \longmapsto \Gamma(X, \mathcal{F}) = \mathcal{F}(X),$$

which associates to an $\mathcal{O}_X$-module $\mathcal{F}$ the group of its global sections $\mathcal{F}(X)$ and, for a morphism of rigid K-spaces $\varphi\colon X \longrightarrow Y$, the *direct image functor*

$$\varphi_*\colon \mathcal{F} \longmapsto \varphi_* \mathcal{F},$$

which associates to an $\mathcal{O}_X$-module $\mathcal{F}$ its direct image $\varphi_* \mathcal{F}$. Both functors are left-exact. To define their right-derived functors we need injective resolutions. For shortness, let us write $\mathfrak{C}$ for the category of $\mathcal{O}_X$-modules.

Definition 1. *An object $\mathcal{F} \in \mathfrak{C}$ is called* injective *if the functor $\mathrm{Hom}(\cdot, \mathcal{F})$ is exact, i.e. if for each short exact sequence*

$$0 \longrightarrow \mathcal{E}' \longrightarrow \mathcal{E} \longrightarrow \mathcal{E}'' \longrightarrow 0$$

in $\mathfrak{C}$ also the sequence

$$0 \longrightarrow \mathrm{Hom}(\mathcal{E}'', \mathcal{F}) \longrightarrow \mathrm{Hom}(\mathcal{E}, \mathcal{F}) \longrightarrow \mathrm{Hom}(\mathcal{E}', \mathcal{F}) \longrightarrow 0$$

is exact.

As $\mathrm{Hom}(\cdot, \mathcal{F})$ is left-exact, the sequence

$$0 \longrightarrow \mathrm{Hom}(\mathcal{E}'', \mathcal{F}) \longrightarrow \mathrm{Hom}(\mathcal{E}, \mathcal{F}) \longrightarrow \mathrm{Hom}(\mathcal{E}', \mathcal{F})$$

will always be exact, and we see that $\mathcal{F}$ is injective if and only if for a given monomorphism $\mathcal{E}' \hookrightarrow \mathcal{E}$ any morphism $\mathcal{E}' \longrightarrow \mathcal{F}$ admits a (not necessarily unique) extension $\mathcal{E} \longrightarrow \mathcal{F}$. Without proof we will use:

Proposition 2. *The category $\mathfrak{C}$ of $\mathcal{O}_X$-modules on a rigid K-space X contains enough injectives, i.e. for each object $\mathcal{F} \in \mathfrak{C}$ there is a monomorphism $\mathcal{F} \hookrightarrow \mathcal{J}$ into an injective object $\mathcal{J} \in \mathfrak{C}$.*

The assertion of Proposition 2 is true for quite general categories $\mathfrak{C}$; cf. Grothendieck [Gr], Thm. 1.10.1.

Corollary 3. *Every object $\mathcal{F} \in \mathfrak{C}$ admits an injective resolution, i.e. there is an exact sequence*

$$0 \longrightarrow \mathcal{F} \longrightarrow \mathcal{J}^0 \longrightarrow \mathcal{J}^1 \longrightarrow \ \cdots$$

with injective objects $\mathcal{J}^i$, $i = 0, 1, \ldots$.

Recall that, more precisely, the above exact sequence has to be viewed as a quasi-isomorphism of complexes

$$
\begin{array}{ccccccccc}
0 & \longrightarrow & \mathcal{F} & \longrightarrow & 0 & & & & \\
 & & \downarrow & & & & & & \\
0 & \longrightarrow & \mathcal{J}^0 & \longrightarrow & \mathcal{J}^1 & \longrightarrow & \mathcal{J}^2 & \longrightarrow & \cdots
\end{array}
$$

where the lower row is referred to as an *injective resolution* of $\mathcal{F}$.

Proof of Corollary 3. We choose an embedding $\mathcal{F} \hookrightarrow \mathcal{I}^0$ of $\mathcal{F}$ into an injective object $\mathcal{I}^0$, an embedding $\mathcal{I}^0/\mathcal{F} \hookrightarrow \mathcal{I}^1$ into an injective object $\mathcal{I}^1$, then an embedding $\mathcal{I}^1/\operatorname{im}\mathcal{I}^0 \hookrightarrow \mathcal{I}^2$ into an injective object $\mathcal{I}^2$, and so on.   □

Now let us define right derived functors of the section functor $\Gamma = \Gamma(X, \cdot)$ and of the direct image functor φ_*, the latter for a morphism of rigid K-spaces $\varphi: X \longrightarrow Y$. To apply these functors to an $\mathcal{O}_X$-module $\mathcal{F}$, choose an injective resolution

$$0 \longrightarrow \mathcal{I}^0 \xrightarrow{\alpha^0} \mathcal{I}^1 \xrightarrow{\alpha^1} \mathcal{I}^2 \xrightarrow{\alpha^2} \cdots$$

of $\mathcal{F}$, apply the functor Γ to it, thereby getting a complex of abelian groups

$$0 \longrightarrow \Gamma(X, \mathcal{I}^0) \xrightarrow{\Gamma(\alpha^0)} \Gamma(X, \mathcal{I}^1) \xrightarrow{\Gamma(\alpha^1)} \Gamma(X, \mathcal{I}^2) \xrightarrow{\Gamma(\alpha^0)} \cdots,$$

and take the cohomology of this complex. Then

$$R^q \Gamma(X, \mathcal{F}) = H^q(X, \mathcal{F}) = \ker \Gamma(\alpha^q) / \operatorname{im} \Gamma(\alpha^{q-1})$$

is called the qth *cohomology group* of X with values in $\mathcal{F}$. Using the technique of homotopies, one can show that these cohomology groups are independent of the chosen injective resolution of $\mathcal{F}$, and that $R^q \Gamma(X, \cdot) = H^q(X, \cdot)$ is a functor on $\mathfrak{C}$; it is the so-called qth *right-derived* functor of the section functor $\Gamma(X, \cdot)$. Note that $R^0 \Gamma(X, \cdot) = \Gamma(X, \cdot)$, since the section functor is left-exact. For $\mathcal{F} = \mathcal{O}_X$, the cohomology groups $H^q(X, \mathcal{F})$ may be viewed as certain invariants of the rigid K-space X.

Similarly one proceeds with the direct image functor φ_*, which might be viewed as a relative version of the section functor. Applying φ_* to the above injective resolution of $\mathcal{F}$, we get the complex of $\mathcal{O}_Y$-modules

$$0 \longrightarrow \varphi_*\mathcal{I}^0 \xrightarrow{\varphi_*\alpha^0} \varphi_*\mathcal{I}^1 \xrightarrow{\varphi_*\alpha^1} \varphi_*\mathcal{I}^2 \xrightarrow{\varphi_*\alpha^2} \cdots$$

and

$$R^q\varphi_*(\mathcal{F}) = \ker \varphi_*\alpha^q / \operatorname{im} \varphi_*\alpha^{q-1}$$

is an $\mathcal{O}_Y$-module, which is called the qth *direct image* of $\mathcal{F}$. Clearly, $R^0\varphi_*(\mathcal{F})$ equals $\varphi_*(\mathcal{F})$, and one can show that $R^q\varphi_*(\mathcal{F})$ is the sheaf associated to the presheaf

$$Y \supset V \longmapsto H^q\big(\varphi^{-1}(V), \mathcal{F}|_{\varphi^{-1}(V)}\big).$$

Let us mention the existence of long exact cohomology sequences, writing Φ for a left-exact functor on $\mathfrak{C}$, such as the section functor or a direct image functor:

Theorem 4. *Let*

$$0 \longrightarrow \mathcal{F}' \stackrel{\alpha}{\longrightarrow} \mathcal{F} \stackrel{\beta}{\longrightarrow} \mathcal{F}'' \longrightarrow 0$$

be an exact sequence of objects in $\mathfrak{C}$. *Then there is an associated long exact sequence*:

$$
\begin{aligned}
0 \longrightarrow \quad \Phi(\mathcal{F}') & \stackrel{\Phi(\alpha)}{\longrightarrow} \quad \Phi(\mathcal{F}) \stackrel{\Phi(\beta)}{\longrightarrow} \quad \Phi(\mathcal{F}'') \\
\stackrel{\partial}{\longrightarrow} \quad R^1\Phi(\mathcal{F}') & \stackrel{R^1\Phi(\alpha)}{\longrightarrow} R^1\Phi(\mathcal{F}) \stackrel{R^1\Phi(\beta)}{\longrightarrow} R^1\Phi(\mathcal{F}'') \\
\stackrel{\partial}{\longrightarrow} \quad R^2\Phi(\mathcal{F}') & \stackrel{R^2\Phi(\alpha)}{\longrightarrow} R^2\Phi(\mathcal{F}) \stackrel{R^2\Phi(\beta)}{\longrightarrow} R^2\Phi(\mathcal{F}'') \\
\stackrel{\partial}{\longrightarrow} \quad & \cdots
\end{aligned}
$$

There is, of course, the problem of computing derived functors or cohomology groups. For example, for an injective object $\mathscr{I} \in \mathfrak{C}$ we have $R^0\Phi(\mathscr{I}) = \Phi(\mathscr{I})$ and $R^q\Phi(\mathscr{I}) = 0$ for $q > 0$ since we can use $0 \longrightarrow \mathscr{I} \longrightarrow 0$ as an injective resolution of $\mathscr{I}$. In general, one can try to compute cohomology groups via Čech cohomology. Below we give some details on this method, but for more information one may consult Artin [A], Grothendieck [Gr], or Godement [Go].

If $\mathcal{F}$ is an $\mathcal{O}_X$-module, we define the Čech cohomology groups $H^q(\mathfrak{U}, \mathcal{F})$ for any admissible covering $\mathfrak{U}$ of X as in Sect. 4.3. Then

$$\check{H}^q(X, \mathcal{F}) = \varinjlim_{\mathfrak{U}} H^q(\mathfrak{U}, \mathcal{F})$$

where the limit runs over all admissible coverings of X, is called the qth *Čech cohomology group* of X with values in $\mathcal{F}$. There is always a canonical morphism

$$\check{H}^q(X, \mathcal{F}) \longrightarrow H^q(X, \mathcal{F})$$

that it is bijective for $q = 0, 1$ and injective for $q = 2$. To compute higher cohomology groups via Čech cohomology, one needs special assumptions.

Theorem 5. *Let* $\mathfrak{U}$ *be an admissible covering of a rigid K-space* X *and let* $\mathcal{F}$ *be an* $\mathcal{O}_X$-*module. Assume* $H^q(U, \mathcal{F}) = 0$ *for* $q > 0$ *and* U *any finite intersection of sets in* $\mathfrak{U}$. *Then the canonical map*

$$H^q(\mathfrak{U}, \mathcal{F}) \longrightarrow H^q(X, \mathcal{F})$$

is bijective for all $q \geq 0$.

Theorem 6. *Let* X *be a rigid K-space,* $\mathcal{F}$ *an* $\mathcal{O}_X$-*module, and* $\mathfrak{S}$ *a system of admissible open subsets of* X *satisfying the following conditions:*

(i) *The intersection of two sets in $\mathfrak{S}$ is in $\mathfrak{S}$ again.*
(ii) *Each admissible covering of an admissible open subset of X admits an admissible refinement consisting of sets in $\mathfrak{S}$.*
(iii) *$\check{H}^q(U,\mathscr{F}) = 0$ for $q > 0$ and $U \in \mathfrak{S}$.*

Then the canonical homomorphism

$$\check{H}^q(X,\mathscr{F}) \longrightarrow H^q(X,\mathscr{F})$$

is bijective for $q \geq 0$.

For example, let us look at an affinoid K-space X and let $\mathfrak{S}$ be the system of all affinoid subdomains of X. Then the conditions of Theorem 6 are satisfied for the structure sheaf $\mathscr{F} = \mathscr{O}_X$ or for any $\mathscr{O}_X$-module associated to an $\mathscr{O}_X(X)$-module; for condition (iii), see Tate's Acyclicity Theorem 4.3/10 and Corollary 4.3/11. Thus, we can conclude:

Corollary 7. *Let X be an affinoid K-space. Then*

$$H^q(X,\mathscr{O}_X) = 0 \quad \text{for} \quad q > 0.$$

The same is true for any $\mathscr{O}_X$-module $\mathscr{F}$ in place of $\mathscr{O}_X$ that is associated to an $\mathscr{O}_X(X)$-module.

6.3 The Proper Mapping Theorem

We end the first part of these lectures by an advanced topic, Kiehl's Proper Mapping Theorem; its proof will follow in Sect. 6.4. The theorem requires the notions of properness and, in particular, of separatedness for morphisms of rigid spaces. In order to introduce the latter concept, we adapt the definition of closed immersions, as given in 6.1/1 for affinoid spaces, to the setting of global rigid spaces.

Definition 1. *A morphism of rigid K-spaces $\varphi: X \longrightarrow Y$ is called a* closed immersion *if there exists an admissible affinoid covering $(V_j)_{j \in J}$ of Y such that, for all $j \in J$, the induced morphism $\varphi_j: \varphi^{-1}(V_j) \longrightarrow V_j$ is a closed immersion of affinoid K-spaces in the sense of 4.2/1. The latter means that φ_j is a morphism of affinoid spaces, say $\varphi^{-1}(V_j) = \operatorname{Sp} A_j$ and $V_j = \operatorname{Sp} B_j$, and that the corresponding morphism of affinoid K-algebras $B_j \longrightarrow A_j$ is an epimorphism.*

If $\varphi: X \longrightarrow Y$ is a closed immersion in the sense of the definition, we can view $\varphi_* \mathscr{O}_X$ as a coherent $\mathscr{O}_Y$-module as characterized in 6.1/3. Using Kiehl's Theorem 6.1/4 in conjunction with 6.1/1, one can show that the condition in Definition 1 is independent of the chosen admissible affinoid covering $(V_j)_{j \in J}$.

In fact, a morphism of affinoid K-spaces $\mathrm{Sp}\, A \longrightarrow \mathrm{Sp}\, B$ is a closed immersion if and only if the corresponding morphism $B \longrightarrow A$ is an epimorphism. In particular, we thereby see that Definition 1 extends the notion of closed immersions for affinoid K-spaces, as given in 4.2/1.

Definition 2.

(i) *A rigid K-space X is called* quasi-compact *if it admits a finite admissible affinoid cover. A morphism of rigid K-spaces $\varphi\colon X \longrightarrow Y$ is called* quasi-compact *if for each quasi-compact open subspace $Y' \subset Y$ its inverse image $\varphi^{-1}(Y')$ is quasi-compact.*

(ii) *A morphism of rigid K-spaces $\varphi\colon X \longrightarrow Y$ is called* separated (*resp.* quasi-separated) *if the diagonal morphism $\Delta\colon X \longrightarrow X \times_Y X$ is a closed immersion (resp. a quasi-compact morphism).*

(iii) *A rigid K-space X is called* separated (*resp.* quasi-separated) *if the structural morphism $X \longrightarrow \mathrm{Sp}\, K$ is separated (resp. quasi-separated).*

Of course, every separated morphism of rigid K-spaces is quasi-separated since closed immersions are quasi-compact. As in algebraic geometry, one shows:

Proposition 3. *Every morphism of affinoid K-spaces $\varphi\colon \mathrm{Sp}\, A \longrightarrow \mathrm{Sp}\, B$ is separated.*

Proposition 4. *Let $\varphi\colon X \longrightarrow Y$ be a separated (resp. quasi-separated) morphism of rigid K-spaces and assume that Y is affinoid. Then, for any open affinoid subspaces $U, V \subset X$, the intersection $U \cap V$ is affinoid (resp. quasi-compact).*

In algebraic geometry, one knows for a morphism of schemes $\varphi\colon X \longrightarrow Y$ that the diagonal morphism $\Delta\colon X \longrightarrow X \times_Y X$ is always a locally closed immersion. Furthermore, Δ is a closed immersion and, hence, φ is separated, if and only if the image of Δ is closed in $X \times_Y X$. In rigid analytic geometry the diagonal morphism $\Delta\colon X \longrightarrow X \times_Y X$ is still a locally closed immersion, but the characterization of separated morphisms is a bit more complicated; see [BGR], 9.6.1/7 in conjunction with [BGR], 9.6.1/3:

Proposition 5. *A morphism of rigid K-spaces $\varphi\colon X \longrightarrow Y$ is separated if and only if the following hold:*

(i) *φ is quasi-separated.*

(ii) *The image of the diagonal morphism $\Delta\colon X \longrightarrow X \times_Y X$ is a closed analytic subset in $X \times_Y X$, i.e., locally on open affinoid parts $W \subset X \times_Y X$, it is a Zariski closed subset of W.*

Considering a rigid K-space Y as a base space, a morphism of rigid K-spaces $X \longrightarrow Y$ is quite often referred to as a rigid Y-space. We need to introduce a notion of relative compactness over such a base Y.

Definition 6. *Let X be a rigid Y-space where the base space Y is affinoid, and let $U \subset U' \subset X$ be open affinoid subspaces. We say that U is* relatively compact *in U' and write $U \Subset_Y U'$ if there exist affinoid generators $f_1, \ldots, f_r$ of $\mathcal{O}_X(U')$ over $\mathcal{O}_Y(Y)$ (in the sense that the structural morphism $\mathcal{O}_Y(Y) \longrightarrow \mathcal{O}_X(U')$ extends to an epimorphism $\mathcal{O}_Y(Y)\langle \zeta_1, \ldots, \zeta_r \rangle \longrightarrow \mathcal{O}_X(U')$ mapping ζ_i to f_i) such that*

$$U \subset \{ x \in U' \, ; \, | f_i(x) | < 1 \}$$

or, in equivalent terms, such that there is an $\varepsilon \in \sqrt{|K^*|}, 0 < \varepsilon < 1$, satisfying

$$U \subset U'(\varepsilon^{-1} f_1, \ldots, \varepsilon^{-1} f_r).$$

The notion of relative compactness behaves in a quite reasonable way:

Lemma 7. *Let X_1, X_2 be affinoid spaces over an affinoid K-space Y and consider affinoid subdomains $U_i \subset X_i, i = 1, 2$. Then:*

(i) *$U_1 \Subset_Y X_1 \implies U_1 \times_Y X_2 \Subset_{X_2} X_1 \times_Y X_2$.*
(ii) *$U_i \Subset_Y X_i, i = 1, 2, \implies U_1 \times_Y U_2 \Subset_Y X_1 \times_Y X_2$.*
(iii) *$U_i \Subset_Y X_i, i = 1, 2, \implies U_1 \cap U_2 \Subset_Y X_1 \cap X_2$ where, slightly different from the above, X_1, X_2 are open affinoid subspaces of an ambient rigid K-space X over Y and the morphism $X \longrightarrow Y$ is separated.*

Now we can introduce proper morphisms of rigid K-spaces. The definition is inspired from compact complex Riemann surfaces that are viewed as manifolds without boundary.

Definition 8. *A morphism of rigid K-spaces $\varphi \colon X \longrightarrow Y$, or X as a rigid Y-space, is called* proper *if the following hold:*

(i) *φ is separated.*
(ii) *There exist an admissible affinoid covering $(Y_i)_{i \in I}$ of Y and for each $i \in I$ two finite admissible affinoid coverings $(X_{ij})_{j=1 \ldots n_i}, (X'_{ij})_{j=1 \ldots n_i}$ of $\varphi^{-1}(Y_i)$ such that $X_{ij} \Subset_{Y_i} X'_{ij}$ for all i and j.*

It can easily be shown that properness, just like separateness, behaves well with respect to base change on Y and with respect to fiber products over Y; cf. Lemma 7. However, it is quite difficult to see that the composition of two proper morphisms is proper again. To deduce this result, one uses the characterization of properness in terms of properness on the level of formal models as we will study them in Sect. 8.4; for details see Lütkebohmert [L] if K carries a discrete valuation, as well as Temkin [Te] in the general case.

Of course, finite morphisms of rigid K-spaces are examples of proper morphisms. Furthermore the projective space $\mathbb{P}^n_K$, viewed as the rigid analytification of the corresponding K-scheme, is a prototype of a proper rigid K-space. More generally, if $\varphi: X \longrightarrow Y$ is a morphism of K-schemes of locally finite type, one can show that the corresponding rigid analytification $\varphi^{\text{rig}}: X^{\text{rig}} \longrightarrow Y^{\text{rig}}$ is proper if and only if φ is proper in the sense of algebraic geometry; see Köpf [Kö], Satz 2.16. On the other hand, an affinoid K-space will never be proper over K, unless it is finite over K, as can be read from Kiehl's theorem below.

We want to present now Kiehl's version of the Proper Mapping Theorem, see Kiehl [K2], as well as some of its applications. The proof of this theorem will be postponed until the next section.

Theorem 9 (Kiehl). *Let $\varphi: X \longrightarrow Y$ be a proper morphism of rigid K-spaces and $\mathcal{F}$ a coherent $\mathcal{O}_X$-module. Then the higher direct images $R^q \varphi_*(\mathcal{F})$, $q \geq 0$, are coherent $\mathcal{O}_Y$-modules.*

A basic lemma that has to be established on the way is the following one:

Lemma 10. *If, in the situation of Theorem 9, Y is affinoid, say $Y = \operatorname{Sp} B$, and if $Y' = \operatorname{Sp} B' \subset Y$ is an affinoid subdomain, then*

$$\Gamma\left(Y', R^q \varphi_*(\mathcal{F})\right) = H^q\left(\varphi^{-1}(Y'), \mathcal{F}\right) = H^q(X, \mathcal{F}) \otimes_B B', \qquad q \geq 0.$$

There are a lot of applications of the Proper Mapping Theorem, and before concluding this section, we want to discuss some of them. Let $\varphi: X \longrightarrow Y$ be a proper morphism of rigid K-spaces. Then, for any closed analytic subset $A \subset X$ (i.e., locally on open affinoid parts of X, one requires that A is Zariski closed in X), the image $\varphi(A)$ is a closed analytic subset of Y. Furthermore, there is the so-called *Stein Factorization* of φ: The coherent $\mathcal{O}_Y$-module $\varphi_*(\mathcal{O}_X)$ gives rise to a rigid K-space Y' that is finite over Y. Thus, φ splits into a proper morphism $X \longrightarrow Y'$ with connected fibers and a finite morphism $Y' \longrightarrow Y$.

Finally, we want to present the subsequent theorems applying to the GAGA-functor, dealt with in Sect. 5.4. Note that, for a K-scheme of locally finite type X, any $\mathcal{O}_X$-module $\mathcal{F}$ gives rise to an $\mathcal{O}_{X^{\text{rig}}}$-module $\mathcal{F}^{\text{rig}}$ on the rigid analytification X^{rig} of X, and one can show that $\mathcal{F}^{\text{rig}}$ is coherent if and only if the same is true for $\mathcal{F}$.

Theorem 11. *Let X be a proper K-scheme and $\mathcal{F}$ a coherent $\mathcal{O}_X$-module. Then the canonical maps*

$$H^q(X, \mathcal{F}) \longrightarrow H^q(X^{\text{rig}}, \mathcal{F}^{\text{rig}}), \qquad q \geq 0,$$

are isomorphisms.

Theorem 12. *Let X be a proper K-scheme and $\mathcal{F}, \mathcal{G}$ coherent $\mathcal{O}_X$-modules. Then the canonical map*

$$\mathrm{Hom}_{\mathcal{O}_X}(\mathcal{F}, \mathcal{G}) \longrightarrow \mathrm{Hom}_{\mathcal{O}_{X^{\mathrm{rig}}}}(\mathcal{F}^{\mathrm{rig}}, \mathcal{G}^{\mathrm{rig}})$$

is an isomorphism.

Theorem 13. *Let X be a proper K-scheme and $\mathcal{F}'$ a coherent $\mathcal{O}_{X^{\mathrm{rig}}}$-module. Then there is a coherent $\mathcal{O}_X$-module $\mathcal{F}$ satisfying $\mathcal{F}^{\mathrm{rig}} = \mathcal{F}'$; furthermore, $\mathcal{F}$ is unique up to canonical isomorphism.*

It should be mentioned that the last three theorems generalize to the relative GAGA-functor where one works over an affinoid K-algebra as base instead of K. For details, see Köpf [Kö].

One may apply Theorem 13 to the case where X equals the projective n-space $\mathbb{P}^n_K$ and where $\mathcal{F}'$ is a coherent ideal $\mathcal{J}' \subset \mathcal{O}_{X^{\mathrm{rig}}}$. As the zero sets of such coherent ideals are precisely the closed analytic subsets of X^{rig}, we obtain the analog of Chow's Theorem, namely that each analytic subset of $\mathbb{P}^{n,\mathrm{rig}}_K$ is algebraic.

6.4   Proof of the Proper Mapping Theorem

In this section we will prove Kiehl's Theorem 6.3/9, which states that all higher direct images of a coherent sheaf under a proper morphism are coherent again. To give a short preview on the method we will use, consider a proper morphism of rigid K-spaces $\varphi \colon X \longrightarrow Y$ where Y is affinoid, and assume that the following (slightly stronger) condition for φ is satisfied:

(†)   *There exist two finite admissible affinoid coverings $\mathfrak{U} = (U_i)_{i=1,\dots,s}$ as well as $\mathfrak{V} = (V_i)_{i=1,\dots,s}$ of X such that $V_i \Subset_Y U_i$ for all i.*

Note that a separated morphism φ is proper if and only if there is an admissible affinoid covering of Y such that condition (†) is satisfied for the inverse images of the members of this covering.

Now let $\mathcal{F}$ be a coherent $\mathcal{O}_X$-module. As a main step of proof, we will show that $H^q(X, \mathcal{F})$ is a finite module over $B = \mathcal{O}_Y(Y)$ for all $q \geq 0$. Applying 6.2/5 in conjunction with 6.2/7, we may look at Čech cohomology and use the fact that the canonical morphisms

$$H^q(\mathfrak{U}, \mathcal{F}) \xrightarrow{\ \mathrm{res}\ } H^q(\mathfrak{V}, \mathcal{F}) \longrightarrow H^q(X, \mathcal{F}), \qquad q \geq 0,$$

are isomorphisms. Thus, writing $Z^q(\mathfrak{V}, \mathcal{F})$ for the kernel of the coboundary map $d^q \colon C^q(\mathfrak{V}, \mathcal{F}) \longrightarrow C^{q+1}(\mathfrak{V}, \mathcal{F})$, it is enough to show that the maps

$$f^q : C^{q-1}(\mathfrak{V}, \mathcal{F}) \longrightarrow Z^q(\mathfrak{V}, \mathcal{F}), \qquad q \geq 0,$$

(with $C^{-1}(\mathfrak{V}, \mathcal{F}) = 0$) induced by the coboundary maps of the Čech complex $C^\bullet(\mathfrak{V}, \mathcal{F})$ have finite B-modules as cokernels. Let

$$r^q : Z^q(\mathfrak{U}, \mathcal{F}) \longrightarrow Z^q(\mathfrak{V}, \mathcal{F}), \qquad q \geq 0,$$

be the morphisms induced from the restriction map $C^\bullet(\mathfrak{U}, \mathcal{F}) \longrightarrow C^\bullet(\mathfrak{V}, \mathcal{F})$ on the kernels of coboundary maps. Then, relying on the fact that associated maps between cohomology groups are isomorphisms, as mentioned above, all maps

$$f^q + r^q : C^{q-1}(\mathfrak{V}, \mathcal{F}) \oplus Z^q(\mathfrak{U}, \mathcal{F}) \longrightarrow Z^q(\mathfrak{V}, \mathcal{F}), \qquad q \geq 0,$$

will be surjective. At this point a subtle approximation argument comes in. It says that the map r^q is suitably "nice" such that, when we disturb $f^q + r^q$ by subtracting r^q, the resulting map f^q, although not necessarily surjective any more, will still have finite cokernel. It is this approximation step that we will discuss first.

In order to make the notion of "nice" maps more explicit, we introduce some notation. As before, let B be an affinoid K-algebra that is equipped with a fixed residue norm $|\cdot|$. On B we will consider normed modules M that are complete. For any such M let

$$M^\circ = \{x \in M \; ; \; |x|_M \leq 1\},$$

and, for any B-linear continuous homomorphism $f : M \longrightarrow N$ between two such B-modules, set

$$|f| = \sup \left\{ \frac{|f(x)|_N}{|x|_M} \; ; \; x \in M - \{0\} \right\}.$$

Using Lemma 1 of Appendix B we see that $|f|$ is finite. In particular, we thereby get a complete B-module norm on the space of all B-linear homomorphisms from M to N. As usual, let R be the valuation ring of K.

Definition 1. *A continuous B-linear homomorphism $f : M \longrightarrow N$ is called completely continuous if it is the limit of a sequence $(f_i)_{i \in \mathbb{N}}$ of continuous B-linear homomorphisms such that $\mathrm{im}(f_i)$ is a finite B-module for all $i \in \mathbb{N}$. Furthermore, if there is an element $c \in R - \{0\}$ such that for all integers $i \in \mathbb{N}$ the B°-module $c f_i (M^\circ)$ is contained in a finite B°-submodule of N°, which may depend on i, then f is called* strictly completely continuous.

We want to give a basic example of a strictly completely continuous homomorphism.

Proposition 2. *Let* $f: B\langle\zeta\rangle \longrightarrow A$ *be a K-homomorphism where A and B are affinoid K-algebras and* $\zeta = (\zeta_1, \ldots, \zeta_n)$ *a system of variables. Consider on* $B\langle\zeta\rangle$ *the Gauß norm derived from a given residue norm on B and on A any residue norm* $|\cdot|$ *such that* $f|_B: B \longrightarrow A$ *is contractive. Then, if* $|f(\zeta_i)|_{\text{sup}} < 1$ *for all i, the map f is a strictly completely continuous homomorphism of complete normed B-modules.*

Proof. Since $|f(\zeta_i)|_{\text{sup}} < 1$, we see from 3.1/18 that $f(\zeta_i)$ is topologically nilpotent in A for all i, and it follows that $(f(\zeta^\nu))_{\nu \in \mathbb{N}^n}$ is a zero sequence in A.

For $i \in \mathbb{N}$ set $M_i = \bigoplus_{|\nu|=i} B\zeta^\nu$ so that $B\langle\zeta\rangle$ equals the complete direct sum $M = \hat{\bigoplus}_{i \in \mathbb{N}} M_i$. Furthermore, let $f_i: M \longrightarrow A$ be the B-module homomorphism that equals f on M_i and is trivial on the complement $\hat{\bigoplus}_{j \in \mathbb{N}, j \neq i} M_j$. Then, since $f(\zeta^\nu)$ is a zero sequence in A, we can conclude that $f = \sum_{i \in \mathbb{N}} f_i$ and, hence, since the M_i are finite B-modules, that f is completely continuous. In fact, choosing $c \in R - \{0\}$ such that $|f(\zeta^\nu)| \leq |c|^{-1}$ for all ν, we get $cf_i(M^\circ) \subset A^\circ$, and we see that f is strictly completely continuous, since $f_i(M^\circ) = f_i(M_i^\circ)$ and since each M_i° is a finite B°-module. $\quad\square$

We start the approximation process alluded to above by establishing a Theorem of L. Schwarz.

Theorem 3. *Let* $f, g: M \longrightarrow N$ *be continuous homomorphisms of complete normed B-modules where, as above, B is an affinoid K-algebra equipped with a certain residue norm. Assume that*

(i) *f is surjective, and*
(ii) *g is completely continuous.*

Then the image $\text{im}(f + g)$ *is closed in N, and the cokernel* $N/\text{im}(f + g)$ *is a finite B-module.*

Proof. We can view f as a continuous surjective linear map between K-Banach spaces. Thus, by Banach's Theorem, see [EVT], f is open and there exists a constant $t \in K^*$ such that $tN^\circ \subset f(M^\circ)$. In other words, replacing t by ct for some $c \in K^*$ with $|c| < 1$, we see for any $y \in N$ that there is some $x \in M$ satisfying

$$f(x) = y \qquad \text{and} \qquad |x| \leq |t|^{-1}|y|.$$

Now consider the special case where $|g| = \alpha|t|$ for some $\alpha < 1$. We claim that, under such an assumption, $f + g$ is still surjective. Indeed, given $y \in N - \{0\}$, we can pick $x \in M$ as before with $f(x) = y$, $|x| \leq |t|^{-1}|y|$, and write

$$(f + g)(x) = y + y'$$

where $y' = g(x) \in N$ satisfies $|y'| \leq \alpha |y|$. Then, proceeding with y' in the same way as we did with y, an iteration argument in combination with a limit process shows that, indeed, $f + g$ is surjective.

To deal with the general case, we use the fact that g is completely continuous and, hence, can be uniformly approximated by an infinite sum of continuous B-linear maps having module-finite image. By the just considered special case, we may assume that this sum is, in fact, finite and, hence, that g has module-finite image. Then $M/\ker g$ may be viewed as a finite B-module, and we can consider the commutative diagram

$$
\begin{array}{ccc}
M & \xrightarrow{\ f\ } & N \\
\downarrow & & \downarrow \\
M/\ker g & \xrightarrow{\ \overline{f}\ } & N/f(\ker g)
\end{array}
$$

where the lower row is induced from f and, hence, all arrows are epimorphisms. It follows that $N/f(\ker g)$ is a finite B-module and, since

$$f(\ker g) = (f + g)(\ker g) \subset (f + g)(M),$$

that the same is true for $N/(f + g)(M)$.

To show that $\mathrm{im}(f + g)$ is closed in M, observe that $\ker g$ is closed in M and that we can provide $M/\ker g$ with the canonical residue norm derived from the norm of M. Using the assertion of 2.3/10, one can show that any submodule of such a finite B-module is closed. In particular, $\ker \overline{f}$ is closed, and we can consider the residue norm via $\overline{f}$ on $N/f(\ker g)$. On the other hand, we can assume, due to Banach's Theorem (see above), that the norm of N coincides with the residue norm via f. Then it follows that the norm on $N/f(\ker g)$ coincides with the residue norm via $N \longrightarrow N/f(\ker g)$. In particular, the latter map is continuous. Since $(f + g)(M)$ can be interpreted as the inverse of a submodule of $N/f(\ker g)$, and since any such submodule is closed, as we have seen, it follows that $(f + g)(M)$ is closed in N.   □

Recalling the maps

$$f^q + r^q : C^{q-1}(\mathfrak{V}, \mathcal{F}) \oplus Z^q(\mathfrak{U}, \mathcal{F}) \longrightarrow Z^q(\mathfrak{V}, \mathcal{F}), \qquad q \geq 0,$$

as introduced in the beginning of the section, we would like to apply Theorem 3 to the maps $f = f^q + r^q$ and $g = -r^q$, for all q. Certainly, f is surjective then, but we do not know if g will be completely continuous. Basing our information about complete continuity upon the example given in Proposition 2, we need a slight generalization of the Theorem of Schwarz, as follows:

Theorem 4. *Let* $f, g: M \longrightarrow N$ *be continuous homomorphisms of complete normed B-modules where, as above, B is an affinoid K-algebra equipped with a certain residue norm. Assume that*

(i) *f is surjective, and*

(ii) *g is part of a sequence* $M^{\flat} \xrightarrow{\ p\ } M \xrightarrow{\ g\ } N \xrightarrow{\ j\ } N^{\sharp}$ *of continuous morphisms of complete normed B-modules where p is an epimorphism and j identifies N with a closed submodule of $N^{\sharp}$, and where the composed map $j \circ g \circ p$ is strictly completely continuous.*

Then the image $\mathrm{im}(f + g)$ *is closed in N, and the cokernel* $N/\mathrm{im}(f + g)$ *is a finite B-module.*

The proof of Theorem 4 requires some preparations.

Lemma 5. *Let E be a finite B°-module and $E' \subset E$ a B°-submodule. Then, for any constant $0 < \alpha < 1$, there is a finite B°-submodule $E'' \subset E'$ such that $a E' \subset E''$ for all $a \in R$ with $|a| \leq \alpha$.*

Proof. If $\pi: T_n = K\langle \zeta \rangle \longrightarrow B$ with a system of variables $\zeta = (\zeta_1, \ldots, \zeta_n)$ is an epimorphism defining the chosen residue norm on B, then the induced morphism $\pi^{\circ}: T_n{}^{\circ} = R\langle \zeta \rangle \longrightarrow B^{\circ}$ is surjective by 2.3/9. Thus, we may assume that B° coincides with the algebra $R\langle \zeta \rangle$ of all restricted power series in ζ having coefficients in R. If R is a discrete valuation ring, $R\langle \zeta \rangle$ is Noetherian by Grothendieck and Dieudonné [EGA I], Chap. 0, Prop. 7.5.2. Thus we are done in this case, since E is Noetherian then.

To deal with the general case, assume that the valuation on K is not discrete. There is a more or less obvious reduction step:

Let $0 \longrightarrow E_1 \longrightarrow E \xrightarrow{\ \psi\ } E_2 \longrightarrow 0$ *be an exact sequence of finite B°-modules. Then the assertion of Lemma 5 holds for E if and only if it holds for E_1 and E_2.*

In fact, the only-if part being trivial, assume that the assertion of the lemma holds for E_1 and E_2. Consider a submodule $E' \subset E$, and set $E_1' = E' \cap E_1$, as well as $E_2' = \psi(E')$. Then, given a constant $0 < \alpha < 1$, fix some constant γ satisfying $\sqrt{\alpha} < \gamma < 1$. There are finite submodules $E_1'' \subset E_1'$ and $E_2'' \subset E_2'$ such that $c E_1' \subset E_1''$ and $c E_2' \subset E_2''$ for all $c \in R$ with $|c| \leq \gamma$. Lifting E_2'' to a finite submodule $\widetilde{E}_2'' \subset E'$, we claim that the submodule $E'' = E_1'' + \widetilde{E}_2'' \subset E'$ satisfies the assertion of the Lemma. To justify this, pick some $a \in R$ with $|a| \leq \alpha$ and choose a constant $c \in R$ such that $\sqrt{|a|} \leq |c| \leq \gamma$. It follows $\psi(c E') = c E_2' \subset E_2''$ and, hence, $c E' \subset E_1' + \widetilde{E}_2''$. But then we have

$$a E' \subset c^2 E' \subset c(E_1' + \widetilde{E}_2'') \subset E_1'' + \widetilde{E}_2'' = E'',$$

as required.

Now, applying the above reduction step, we may assume that E is a finite free $R\langle\zeta\rangle$-module and, applying it again in a recursive way, that E coincides with $R\langle\zeta\rangle$ itself. Then E' is an ideal in $R\langle\zeta\rangle$. We will proceed by induction on n, the number of variables. The case $n = 0$ is trivial, and the same is true for $E' = 0$. Therefore assume $n > 0$ and $E' \neq 0$. Write $\beta = \sup\{|h| \,;\, h \in E'\}$ where $|\cdot|$ denotes the Gauß norm on $R\langle\zeta\rangle$, and consider some $g \in E'$ such that $|g| > \alpha\beta$, for a fixed constant $0 < \alpha < 1$. There is a constant $c \in R$ satisfying $|c| = |g|$, and we see that $f = c^{-1}g$ is a well-defined element of Gauß norm 1 in $R\langle\zeta\rangle$. Using 2.2/7, we may apply a change of variables to $K\langle\zeta\rangle$ and thereby can assume that f is ζ_n-distinguished of some order $s \geq 0$. Then, by Weierstraß Division 2.2/8, $R\langle\zeta\rangle/(f)$ is a finite $R\langle\zeta'\rangle$-module where $\zeta' = (\zeta_1, \ldots, \zeta_{n-1})$, and we can consider the exact sequence

$$0 \longrightarrow (f) \longrightarrow R\langle\zeta\rangle \longrightarrow R\langle\zeta\rangle/(f) \longrightarrow 0.$$

As a finite $R\langle\zeta'\rangle$-module, $R\langle\zeta\rangle/(f)$ satisfies the assertion of the lemma by the induction hypothesis. Thus, by the argument given in the above reduction step, it is enough to show that the assertion of the lemma holds for the submodule $E_1' = E' \cap (f) \subset (f)$. However, the latter is obvious from our construction. Indeed, consider the submodule $E_1'' = (g) \subset E_1'$. Any $h \in E_1'$ has Gauß norm $|h| \leq \beta$ and, hence, any $h \in aE_1'$ has Gauß norm $|h| \leq |a|\beta \leq \alpha\beta \leq |g|$ by the choice of $g \in E'$. But then, as we are working within the free monogenous $R\langle\zeta\rangle$-module $fR\langle\zeta\rangle \simeq R\langle\zeta\rangle$, we see that $aE_1' \subset (g) = E_1''$ as required. $\square$

Lemma 6. *Let* $M \xrightarrow{\;g\;} N \xrightarrow{\;j\;} N^\sharp$ *be a homomorphism of complete normed B-modules where j identifies N with a closed submodule of $N^\sharp$. Assume that M is topologically free in the sense that there exists a system $(e_\lambda)_{\lambda\in\Lambda}$ of elements in M such that every $x \in M$ can be written as a converging series $x = \sum_{\lambda\in\Lambda} b_\lambda e_\lambda$ with coefficients $b_\lambda \in B$ satisfying $\max_{\lambda\in\Lambda} |b_\lambda| = |x|$ (and, hence, where the coefficients b_λ are unique).*

Then, if $j \circ g$ is strictly completely continuous, the same is true for g.

Proof. We may assume that the norm of $N^\sharp$ restricts to the one of N. Furthermore, if $j \circ g$ is strictly completely continuous, it is, in particular continuous, and we may assume that $j \circ g$ and g are contractive. Then g and j restrict to morphisms of B°-modules

$$M^\circ \longrightarrow N^\circ \hookrightarrow N^{\sharp\circ}.$$

Since $j \circ g$ is strictly completely continuous, there exist continuous B-linear maps $h_i : M \longrightarrow N^\sharp$, $i \in \mathbb{N}$, satisfying $j \circ g = \lim_{i\in\mathbb{N}} h_i$, and there is a non-zero constant $c \in R$ such that $ch_i(M^\circ)$ is contained in a finite B°-submodule of $N^{\sharp\circ}$ for each i. Adjusting norms on N and $N^\sharp$ by the factor $|c|^{-1}$, we may assume $c = 1$ and, hence, that $h_i(M^\circ)$, for each i, is contained in a finite B°-submodule of $N^{\sharp\circ}$.

Now consider a constant α, $0 < \alpha < 1$, and assume that there is an element $a \in R$ satisfying $|a| = \alpha$. In order to show that g is strictly completely continuous, it is enough to construct for each ε, $0 < \varepsilon < 1$, a B-linear continuous map $g' \colon M \longrightarrow N$ such that

(i) $|g - g'| \leq \varepsilon \alpha^{-1}$, and
(ii) $ag'(M^\circ)$ is contained in a finite B°-submodule of N°.

To construct such an approximation g' of g, let $i \in \mathbb{N}$ be big enough such that $h = h_i$ satisfies $|j \circ g - h| \leq \varepsilon$. By our assumption, $h(M^\circ)$ is contained in a finite B°-submodule $E \subset N^{\sharp\circ}$. Thus, using Lemma 5, there is a finite B°-submodule $E'' \subset h(M^\circ)$ such that $ah(M^\circ) \subset E''$.

We will obtain the desired approximation $g' \colon M \longrightarrow N$ of g by modifying the approximation $h \colon M \longrightarrow N^\sharp$ in a suitable way. Fix generators $y_1, \ldots, y_r$ of E'', let $x_1, \ldots, x_r \in M^\circ$ be inverse images with respect to h such that $y_j = h(x_j)$, and set $z_j = g(x_j)$ for $j = 1, \ldots, r$. Then $z_j \in N^\circ$ and $|y_j - z_j| \leq \varepsilon$ for all j. Thus, we have approximated the elements $y_j \in N^{\sharp\circ}$ suitably well by certain elements $z_j \in N^\circ$. Now, using the fact that $ah(M^\circ) \subset E''$, there are elements $b_{j\lambda} \in B$, $j = 1, \ldots, r$, $\lambda \in \Lambda$, such that

$$h(e_\lambda) = \sum_{j=1}^{r} b_{j\lambda} y_j, \qquad |b_{j\lambda}| \leq \alpha^{-1},$$

and we can define a continuous B-linear map $g' \colon M \longrightarrow N$ by setting

$$g'(e_\lambda) = \sum_{j=1}^{r} b_{j\lambda} z_j.$$

Then, since $|y_j - z_j| \leq \varepsilon$ for all j, we have

$$\left| g(e_\lambda) - g'(e_\lambda) \right| \leq \max\left\{ \left| g(e_\lambda) - h(e_\lambda) \right|, \left| h(e_\lambda) - g'(e_\lambda) \right| \right\}$$

$$\leq \max\left\{ \varepsilon, \alpha^{-1}\varepsilon \right\} = \alpha^{-1}\varepsilon$$

for all λ and, hence, $|g - g'| \leq \alpha^{-1}\varepsilon$. Since $ag'(M^\circ) \subset \sum_{j=1}^{r} B^\circ z_j \subset N^\circ$, by the construction of g', we are done.   □

After these preparations, the *proof of Theorem* 4 is easy to achieve. First observe that the epimorphism $p \colon M^\flat \longrightarrow M$ is not really relevant, since composition with such a continuous (and, hence, by Banach's Theorem, open) B-linear map p does not change the image of $f + g$. Assume first that $M^\flat$ is topologically free, as needed in Lemma 6. Then, with the help of this lemma, the assertion follows from Theorem 3. If $M^\flat$ is not topologically free, we can compose our situation with a continuous B-linear epimorphism $M^{\flat\flat} \longrightarrow M^\flat$ and apply the reasoning used before. To obtain such an epimorphism, consider a bounded generating system

$(x_\lambda)_{\lambda \in \Lambda}$ for M^b as ordinary B-module, and let M^{bb} be the completion of $B^{(\Lambda)}$, the free B-module generated by Λ, with respect to the canonical maximum norm. Then M^{bb} is topologically free, and there is a canonical continuous epimorphism $M^{bb} \longrightarrow M^b$, as required. $\square$

Going back to Kiehl's Theorem 6.3/9, we consider a proper morphism of rigid K-spaces $\varphi \colon X \longrightarrow Y$ and a coherent $\mathcal{O}_X$-module $\mathcal{F}$. Then the higher direct image $R^q \varphi_*(\mathcal{F})$ is the sheaf associated to the presheaf

$$Y \supset Y' \longmapsto H^q\big(\varphi^{-1}(Y'), \mathcal{F}\big).$$

In order to show that $R^q \varphi_*(\mathcal{F})$ is a coherent $\mathcal{O}_Y$-module in the sense of 6.1/3, we may work locally on Y. In other words, we may assume that Y is affinoid and that, as in the beginning of the present section, the following condition is satisfied:

(†) *There exist two finite admissible affinoid coverings* $\mathfrak{U} = (U_i)_{i=1,\dots,s}$ *as well as* $\mathfrak{V} = (V_i)_{i=1,\dots,s}$ *of X such that* $V_i \Subset_Y U_i$ *for all i.*

As a first step we show:

Proposition 7. *Let $\varphi \colon X \longrightarrow Y$ be a proper morphism of rigid K-spaces where Y is affinoid, and where condition (†) is satisfied. Let $\mathcal{F}$ be a coherent $\mathcal{O}_X$-module. Then $H^q(X, \mathcal{F})$ is a finite module over $B = \mathcal{O}_Y(Y)$ for all $q \geq 0$.*

Proof. Looking at the maps $V_i \hookrightarrow U_i \overset{\varphi}{\longrightarrow} Y$, we can fix a residue norm on B, as well as residue norms on $\mathcal{O}_X(U_i)$ and $\mathcal{O}_X(V_i)$ for $i = 1, \dots, s$ in such a way that the canonical maps $B \longrightarrow \mathcal{O}_X(U_i) \longrightarrow \mathcal{O}_X(V_i)$ are contractive. As a result, we may view $\mathcal{O}_X(U_i) \longrightarrow \mathcal{O}_X(V_i)$ as a continuous homomorphism of complete normed B-modules. Furthermore, we can extend $B \longrightarrow \mathcal{O}_X(U_i)$ to an epimorphism $B\langle \zeta_1, \dots, \zeta_n \rangle \longrightarrow \mathcal{O}_X(U_i)$, for a number of variables $\zeta_1, \dots, \zeta_n$. Using the Gauß norm derived from the residue norm of B, we view $E_i = B\langle \zeta_1, \dots, \zeta_n \rangle$ as a topologically free complete normed B-module. Since we have $V_i \Subset_Y U_i$, we may even assume that the image of each variable ζ_j under the composition $E_i \longrightarrow \mathcal{O}_X(U_i) \longrightarrow \mathcal{O}_X(V_i)$ has supremum norm < 1. Then it follows from Proposition 2 that the latter composition is strictly completely continuous. From this we can conclude:

For each $q \in \mathbb{N}$, there exists a topologically free complete normed B-module E^q together with a continuous epimorphism $p \colon E^q \longrightarrow C^q(\mathfrak{U}, \mathcal{F})$ such that the composition

$$E^q \overset{p}{\longrightarrow} C^q(\mathfrak{U}, \mathcal{F}) \overset{\mathrm{res}}{\longrightarrow} C^q(\mathfrak{V}, \mathcal{F})$$

is completely continuous.

Indeed, to settle the case $q = 0$, we consider the cartesian product of the maps $E_i \longrightarrow \mathcal{O}_X(U_i) \longrightarrow \mathcal{O}_X(V_i)$ as introduced above. Since any intersection of type $V_{i_0} \cap \dots \cap V_{i_q}$ lies relatively compact in the intersection $U_{i_0} \cap \dots \cap U_{i_q}$

by 6.3/7 (iii), the same reasoning works for $q > 0$. Also note that the restriction of the above composition to the inverse image $p^{-1}(Z^q(\mathfrak{U}, \mathcal{F}))$ remains strictly completely continuous for trivial reasons.

In the beginning of the section, we have introduced the maps

$$f^q \colon C^{q-1}(\mathfrak{V}, \mathcal{F}) \longrightarrow Z^q(\mathfrak{V}, \mathcal{F}), \qquad q \geq 0,$$

(with $C^{-1}(\mathfrak{V}, \mathcal{F}) = 0$) given by coboundary maps, which are continuous. Also we have shown that the maps

$$f^q + r^q \colon C^{q-1}(\mathfrak{V}, \mathcal{F}) \oplus Z^q(\mathfrak{U}, \mathcal{F}) \longrightarrow Z^q(\mathfrak{V}, \mathcal{F}), \qquad q \geq 0,$$

are surjective where $r^q \colon Z^q(\mathfrak{U}, \mathcal{F}) \longrightarrow Z^q(\mathfrak{V}, \mathcal{F})$ is the canonical restriction map induced from the restriction map $\mathrm{res} \colon C^q(\mathfrak{U}, \mathcal{F}) \longrightarrow C^q(\mathfrak{V}, \mathcal{F})$, as considered above. We view r^q, in a more precise manner, as the map

$$r^q \colon C^{q-1}(\mathfrak{V}, \mathcal{F}) \oplus Z^q(\mathfrak{U}, \mathcal{F}) \longrightarrow Z^q(\mathfrak{V}, \mathcal{F})$$

that is zero on the first component and given by restriction on the second. Then we can conclude from the above statement that the composition

$$C^{q-1}(\mathfrak{V}, \mathcal{F}) \oplus p^{-1}\big(Z^q(\mathfrak{U}, \mathcal{F})\big) \xrightarrow{\ \mathrm{id} \times p\ } C^{q-1}(\mathfrak{V}, \mathcal{F}) \oplus Z^q(\mathfrak{U}, \mathcal{F})$$

$$\xrightarrow{\ r^q\ } Z^q(\mathfrak{V}, \mathcal{F}) \xhookrightarrow{\ j\ } C^q(\mathfrak{V}, \mathcal{F})$$

is strictly completely continuous, with $\mathrm{id} \times p$ a continuous epimorphism and j the canonical inclusion. But then, applying Theorem 4 to the epimorphism $f^q + r^q$ in place of f and to $-r^q$ in place of g, the cokernel of $f^q = (f^q + r^q) - r^q$, which coincides with $H^q(X, \mathcal{F})$, is a finite B-module. $\qquad\square$

As a next step, we want to show that in the situation of the above proposition the higher direct image sheaf $R^q \varphi_* \mathcal{F}$ is the sheaf associated to the finite B-module $H^q(X, \mathcal{F})$. The proof of this fact is based on a formal function type result. To explain it, choose an element $b \in B$ and consider the composition $\mathcal{F} \xrightarrow{\ [b^i]\ } \mathcal{F} \longrightarrow \mathcal{F}/b^i \mathcal{F}$ where $[b^i]$ for some exponent i is given by multiplication with $\varphi^\#(b^i)$. This composition is zero and so is the attached composition

$$H^q(X, \mathcal{F}) \xrightarrow{\ H^q([b^i])\ } H^q(X, \mathcal{F}) \longrightarrow H^q(X, \mathcal{F}/b^i \mathcal{F})$$

on the level of cohomology groups. Since $H^q([b^i])$ is just multiplication with b^i in the sense of B-modules, as is easily checked, we get a canonical map $\sigma_i \colon H^q(X, \mathcal{F})/b^i H^q(X, \mathcal{F}) \longrightarrow H^q(X, \mathcal{F}/b^i \mathcal{F})$ and then, varying i, a canonical map between associated inverse limits.

Proposition 8. *As in the situation of Proposition* 7, *let* $\varphi \colon X \longrightarrow Y$ *be a proper morphism of rigid K-spaces where Y is affinoid and where condition* (†) *is satisfied. Let $\mathcal{F}$ be a coherent $\mathcal{O}_X$-module and fix an element $b \in B = \mathcal{O}_Y(Y)$. Then the canonical morphism*

$$\sigma \colon \varprojlim_i H^q(X, \mathcal{F})/b^i H^q(X, \mathcal{F}) \longrightarrow \varprojlim_i H^q(X, \mathcal{F}/b^i \mathcal{F})$$

is an isomorphism for all $q \geq 0$.

For the proof of the proposition, we need to recall some notions applying to projective systems. Let $(M_i)_{i \in \mathbb{N}}$ be a projective system with connecting morphisms $f_{ij} \colon M_j \longrightarrow M_i$ for $i \leq j$. The system $(M_i)_{i \in \mathbb{N}}$ is said to satisfy the condition of *Mittag–Leffler* if for every $i \in \mathbb{N}$ there exists an index $j_0 \geq i$ in $\mathbb{N}$ such that $f_{ij}(M_j) = f_{ij_0}(M_{j_0})$ for all $j \geq j_0$. Furthermore, $(M_i)_{i \in \mathbb{N}}$ is called a *null system* if, more specifically, for every $i \in \mathbb{N}$ there is an index $j_0 \geq i$ in $\mathbb{N}$ such that $f_{ij}(M_j) = 0$ for all $j \geq j_0$. Note that any null system $(M_i)_{i \in \mathbb{N}}$ satisfies the condition of Mittag–Leffler and yields $\varprojlim_{i \in \mathbb{N}} M_i = 0$. Furthermore, an exact sequence of projective systems $0 \longrightarrow M_i' \longrightarrow M_i \longrightarrow M_i'' \longrightarrow 0$ induces an exact sequence of projective limits

$$0 \longrightarrow \varprojlim_{i \in \mathbb{N}} M_i' \longrightarrow \varprojlim_{i \in \mathbb{N}} M_i \longrightarrow \varprojlim_{i \in \mathbb{N}} M_i'' \longrightarrow 0,$$

provided the system $(M_i')_{i \in \mathbb{N}}$ satisfies the condition of Mittag–Leffler. In general, the functor $\varprojlim$ is only left exact.

In the situation of the proposition, the canonical morphisms

$$H^q(\mathcal{F}) \longrightarrow H^q(\mathcal{F}/b^i \mathcal{F}), \qquad H^q(\mathcal{F})/b^i H^q(\mathcal{F}) \longrightarrow H^q(\mathcal{F}/b^i \mathcal{F}),$$

(where we have suppressed the rigid space X, as it won't change for the moment) can be inserted into exact sequences

$$0 \longrightarrow D_i \longrightarrow H^q(\mathcal{F}) \longrightarrow H^q(\mathcal{F}/b^i \mathcal{F}) \longrightarrow E_i \longrightarrow 0,$$

$$0 \longrightarrow \overline{D}_i \longrightarrow H^q(\mathcal{F})/b^i H^q(\mathcal{F}) \longrightarrow H^q(\mathcal{F}/b^i \mathcal{F}) \longrightarrow E_i \longrightarrow 0 \quad (*)$$

by adding kernels and cokernels. Then the kernels D_i, resp. $\overline{D}_i$, form projective systems again, and the same is true for the cokernels E_i. Furthermore, we have $\overline{D}_i = D_i/b^i H^q(\mathcal{F})$. We claim that the assertion of Proposition 8 will be a consequence of the following facts:

Lemma 9. *In the above situation, the projective systems $(\overline{D}_i)_{i \in \mathbb{N}}$ and $(E_i)_{i \in \mathbb{N}}$ are null systems. Furthermore, that $(\overline{D}_i)_{i \in \mathbb{N}}$ is a null system follows from the fact that* *the filtration $H^q(\mathcal{F}) \supset D_0 \supset D_1 \supset \ldots$ becomes b-stable in the sense that there is an index $i_0 \in \mathbb{N}$ satisfying $D_i = b^{i-i_0} D_{i_0}$ for all $i \in \mathbb{N}, i \geq i_0$.*

Let us first show how to deduce the *proof of Proposition 8* from the lemma. Let H_i be the image of the canonical map

$$\sigma_i \colon H^q(\mathcal{F})/b^i H^q(\mathcal{F}) \longrightarrow H^q(\mathcal{F}/b^i \mathcal{F}).$$

Then we can split σ_i into the composition

$$\sigma_i \colon H^q(\mathcal{F})/b^i H^q(\mathcal{F}) \longrightarrow H_i \lhook\joinrel\longrightarrow H^q(\mathcal{F}/b^i \mathcal{F})$$

and deduce from the above exact sequence $(*)$ the short exact sequences

$$0 \longrightarrow \overline{D}_i \longrightarrow H^q(\mathcal{F})/b^i H^q(\mathcal{F}) \longrightarrow H_i \longrightarrow 0,$$
$$0 \longrightarrow H_i \longrightarrow H^q(\mathcal{F}/b^i \mathcal{F}) \longrightarrow E_i \longrightarrow 0.$$

Passing to inverse limits, the first of these remains exact, since $(\overline{D}_i)_{i \in \mathbb{N}}$, as a null system, satisfies the condition of Mittag–Leffler. The same is true for the second sequence, since all morphisms of the projective system $(H_i)_{i \in \mathbb{N}}$ are surjective so that, also in this case, the condition of Mittag–Leffler is satisfied. Since $(\overline{D}_i)_{i \in \mathbb{N}}$ and $(E_i)_{i \in \mathbb{N}}$ are null systems, we see that

$$\sigma \colon \varprojlim_i H^q(\mathcal{F})/b^i H^q(\mathcal{F}) \longrightarrow \varprojlim_i H_i \longrightarrow \varprojlim_i H^q(\mathcal{F}/b^i \mathcal{F})$$

is a composition of isomorphisms and, thus, an isomorphism, as claimed. $\qquad\square$

It remains to *prove Lemma 9*. First, assume that the filtration $D_0 \supset D_1 \supset \ldots$ is *b*-stable. Choosing $i_0 \in \mathbb{N}$ such that $D_i = b^{i-i_0} D_{i_0}$ for all $i \geq i_0$, we see that the image of $\overline{D}_i = D_i/b^i H^q(\mathcal{F})$ is trivial in $H^q(\mathcal{F})/b^{i-i_0} H^q(\mathcal{F})$ for all $i \geq i_0$ so that, indeed, $(\overline{D}_i)_{i \in \mathbb{N}}$ is a null system. Thus, it remains to show that the filtration of the D_i is *b*-stable and $(E_i)_{i \in \mathbb{N}}$ is a null system.

To do this, let $S = \bigoplus_{i \in \mathbb{N}} S_i = \bigoplus_{i \in \mathbb{N}} b^i B$ be the graded ring generated by the ideal $S_1 = bB \subset B$. The latter is Noetherian, since, as an algebra over the Noetherian ring B, it is generated by b, viewed as a homogeneous element of degree 1. Now consider the direct sum

$$M^q(\mathcal{F}) = \bigoplus_{i \in \mathbb{N}} H^q(b^i \mathcal{F})$$

as a graded S-module, where the multiplication by $b \in S_1 = bB$ is given by the maps $H^q(b^i \mathcal{F}) \longrightarrow H^q(b^{i+1} \mathcal{F})$ derived from the maps $b^i \mathcal{F} \longrightarrow b^{i+1} \mathcal{F}$ that, in turn, are given by multiplication with $b \in B$. We claim that:

$M^q(\mathcal{F})$ *is a finite S-module.*

If $\mathcal{F}$ does not admit b-torsion, the assertion is trivial, since then multiplication by $b \in B$ yields isomorphisms $b^i \mathcal{F} \xrightarrow{\sim} b^{i+1} \mathcal{F}$ and therefore isomorphisms $H^q(b^i \mathcal{F}) \xrightarrow{\sim} H^q(b^{i+1} \mathcal{F})$. Then $M^q(\mathcal{F})$, as an S-module, is generated by $H^q(\mathcal{F})$, the part of degree 0 in $M^q(\mathcal{F})$, where $H^q(\mathcal{F})$ is a finite B-module by Proposition 7. It follows that $M^q(\mathcal{F})$ is a finite S-module.

If there is non-trivial b-torsion in $\mathcal{F}$, the situation is slightly more complicated. In this case, consider the kernels of the morphisms $[b^i] : \mathcal{F} \longrightarrow \mathcal{F}$, which form an increasing sequence of submodules of the coherent $\mathcal{O}_X$-module $\mathcal{F}$. By a Noetherian and quasi-compactness argument, the sequence becomes stationary at a certain coherent submodule $\mathcal{T} \subset \mathcal{F}$. It follows that $\mathcal{T}$ is annihilated by a power of b, and that the quotient $\mathcal{F}/\mathcal{T}$ is without b-torsion. Now let $\mathcal{T}_i = \mathcal{T} \cap b^i \mathcal{F}$. Then, by the Lemma of Artin–Rees, see 7.1/4, the filtration $\mathcal{T}_0 \supset \mathcal{T}_1 \supset \ldots$ is b-stable. Thus, there is an index $i_0 \in \mathbb{N}$ such that $\mathcal{T}_i = 0$ for all $i \geq i_0$. Since $H^q(\mathcal{T}_i)$ is trivial for such i, it follows with the help of Proposition 7 that the graded S-module

$$N^q = \bigoplus_{i \in \mathbb{N}} H^q(\mathcal{T}_i)$$

is finitely generated over B and, hence, also over S.

Now observe that the short exact sequence

$$0 \longrightarrow \mathcal{T}_i \longrightarrow b^i \mathcal{F} \longrightarrow b^i(\mathcal{F}/\mathcal{T}) \longrightarrow 0$$

induces an exact sequence

$$N^q \longrightarrow M^q(\mathcal{F}) \longrightarrow M^q(\mathcal{F}/\mathcal{T}).$$

By construction, $\mathcal{F}/\mathcal{T}$ is without b-torsion. Therefore $M^q(\mathcal{F}/\mathcal{T})$ is a finite S-module, as we have seen above. Since also N^q is a finite S-module, it follows that $M^q(\mathcal{F})$ is a finite S-module, as claimed.

In order to justify the remaining assertions of the lemma, consider the exact sequence

$$0 \longrightarrow b^i \mathcal{F} \longrightarrow \mathcal{F} \longrightarrow \mathcal{F}/b^i \mathcal{F} \longrightarrow 0,$$

as well as the attached long cohomology sequence

$$\ldots \longrightarrow H^q(b^i \mathcal{F}) \longrightarrow H^q(\mathcal{F}) \longrightarrow H^q(\mathcal{F}/b^i \mathcal{F})$$
$$\longrightarrow H^{q+1}(b^i \mathcal{F}) \longrightarrow H^{q+1}(\mathcal{F}) \longrightarrow \ldots$$

Then we have

$$D_i = \ker\big(H^q(\mathcal{F}) \longrightarrow H^q(\mathcal{F}/b^i \mathcal{F})\big) = \mathrm{im}\big(H^q(b^i \mathcal{F}) \longrightarrow H^q(\mathcal{F})\big),$$
$$E_i = \mathrm{im}\big(H^q(\mathcal{F}/b^i \mathcal{F}) \longrightarrow H^{q+1}(b^i \mathcal{F})\big),$$

and it follows that $D = \bigoplus_{i \in \mathbb{N}} D_i$, as an image of the finite S-module $M^q(\mathcal{F})$, is a finite S-module itself. The latter means that the filtration $D_0 \supset D_1 \supset \ldots$ is b-stable.

Thus, it remains to show that $(E_i)_{i \in \mathbb{N}}$ is a null system. The characterization of E_i above says that the graded S-module $E = \bigoplus_{i \in \mathbb{N}} E_i$, as a submodule of the finite S-module $M^{q+1}(\mathcal{F})$, is finite itself. Furthermore, each E_i is annihilated by b^i, as it is an image of $H^q(\mathcal{F}/b^i\mathcal{F})$. Since E is a finite S-module, there is some $r \in \mathbb{N}$ such that $b^r E = 0$, viewing b^r as an element of $B = S_0$. On the other hand, using the fact that E is a finite S-module and writing b_1 instead of b for the corresponding element in $S_1 = bB$, one can find integers i_0 and $s \geq r$ with $b_1^s E_i = E_{i+s}$ for all $i \geq i_0$. Let $p_{i,s} \colon E_{i+s} \longrightarrow E_i$ be the map induced by the canonical map $H^q(\mathcal{F}/b^{i+s}\mathcal{F}) \longrightarrow H^q(\mathcal{F}/b^i\mathcal{F})$, i.e. the canonical map given by the projective system $(E_i)_{i \in \mathbb{N}}$. Then $p_{i,s}(b_1^s E_i) = b^s E_i$, as is easily checked, and we see that

$$p_{i,s}(E_{i+s}) = p_{i,s}(b_1^s E_i) = b^s E_i = 0$$

for $i \geq i_0$. Thus, $(E_i)_{i \in \mathbb{N}}$ is a null system. $\qquad\square$

Finally, using the characterization of coherent modules as given in 6.1/3, the assertion of Kiehl's Theorem 6.3/9 will be a consequence of the following result:

Theorem 10. *As in the situation of Proposition 7, let $\varphi \colon X \longrightarrow Y$ be a proper morphism of rigid K-spaces where Y is affinoid and where condition* (†) *is satisfied. Let $\mathcal{F}$ be a coherent $\mathcal{O}_X$-module. Then, for any $q \in \mathbb{N}$, the higher direct image $R^q\varphi_*(\mathcal{F})$ equals the $\mathcal{O}_Y$-module associated to the finite B-module $H^q(X,\mathcal{F})$, for $B = \mathcal{O}_Y(Y)$.*

Proof. We will proceed by induction on the Krull dimension d of B. The case $d = 0$ is trivial, since then Y is a finite disjoint union of rigid K-spaces supported at a single point each.

Therefore assume $d > 0$, and consider an affinoid subdomain $Y' = \operatorname{Sp} B'$ in $Y = \operatorname{Sp} B$. Let $X' = X \times_Y Y'$. We have to show that the canonical morphism

$$H^q(X, \mathcal{F}) \otimes_B B' \longrightarrow H^q(X', \mathcal{F})$$

is an isomorphism. In order to do this, it is enough to show that all localizations $H^q(X, \mathcal{F}) \otimes_B B'_{\mathfrak{m}'} \longrightarrow H^q(X', \mathcal{F}) \otimes_{B'} B'_{\mathfrak{m}'}$ at maximal ideals $\mathfrak{m}' \subset B'$ are isomorphisms or, since the $\mathfrak{m}'$-adic completion $\hat{B}'_{\mathfrak{m}'}$ of $B_{\mathfrak{m}'}$ is faithfully flat over $B'_{\mathfrak{m}'}$ (see [AC], Chap. III, § 3, no. 4, Thm. 3 and no. 5, Prop. 9), that all morphisms $H^q(X, \mathcal{F}) \otimes_B \hat{B}'_{\mathfrak{m}'} \longrightarrow H^q(X', \mathcal{F}) \otimes_{B'} \hat{B}'_{\mathfrak{m}'}$ are isomorphisms.

Now consider a maximal ideal $\mathfrak{m}' \subset B'$. Then it follows from 3.3/10, that there is a (unique) maximal ideal $\mathfrak{m} \subset B$ satisfying $\mathfrak{m}' = \mathfrak{m}B'$. Furthermore, by 2.2/11, there is a finite monomorphism $T_d \hookrightarrow B$, and we see that $\mathfrak{n} = \mathfrak{m} \cap T_d$ is a maximal ideal in T_d. Choosing a non-zero element $b \in \mathfrak{n} \subset T_d$, we conclude from 2.2/9 in conjunction with 2.2/7 that $B/(b^i)$ has Krull dimension $< d$ for all $i \in \mathbb{N}$. Thus, writing $Y_i = \operatorname{Sp} B/(b^i)$, we may apply the induction hypothesis to the morphisms

$$\varphi_i : X \times_Y Y_i \longrightarrow Y_i, \qquad i \in \mathbb{N},$$

and the induced coherent sheaves $\mathcal{F}/b^i \mathcal{F}$ on $X \times_Y Y_i$. Using identifications of type $H^q(X \times_Y Y_i, \mathcal{F}/b^i \mathcal{F}) = H^q(X, \mathcal{F}/b^i \mathcal{F})$, the canonical morphisms

$$H^q(X, \mathcal{F}/b^i \mathcal{F}) \otimes_B B' \longrightarrow H^q(X \times_Y Y', \mathcal{F}/b^i \mathcal{F})$$

are isomorphisms by induction hypothesis.

Next we recall the exact sequence $(*)$ from the proof of Proposition 8. Tensoring it with B' over B, we get the upper square of the following commutative diagram where we have written X' as an abbreviation for $X \times_Y Y'$:

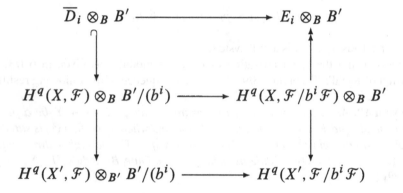

The lower vertical maps are induced by restriction from X to X', whereas the lower horizontal morphism is the equivalent of the middle morphism in $(*)$, with X replaced by X'. Taking inverse limits for $i \to \infty$, we get the commutative diagram

$$
\begin{array}{ccc}
H^q(X, \mathcal{F}) \otimes_B \widehat{B}' & \longrightarrow & \varprojlim_{i \in \mathbb{N}} \left[H^q(X, \mathcal{F}/b^i \mathcal{F}) \otimes_B B' \right] \\
\downarrow & & \downarrow \\
H^q(X', \mathcal{F}) \otimes_{B'} \widehat{B}' & \longrightarrow & \varprojlim_{i \in \mathbb{N}} H^q(X', \mathcal{F}/b^i \mathcal{F})
\end{array}
$$

where $\widehat{B}'$ is the b-adic completion of B'. Here we have used the fact that the b-adic completion of a finite B'-module M' is canonically isomorphic to $M' \otimes_B \widehat{B}'$; see [AC], Chap. III, § 3, no. 4, Thm. 3 (ii), or use the method of proof applied in 7.3/14. Now observe that the projective systems $(\overline{D}_i \otimes_B B')_{i \in \mathbb{N}}$ and $(E_i \otimes_B B')_{i \in \mathbb{N}}$ are null systems, since the same is true for $(\overline{D}_i)_{i \in \mathbb{N}}$ and $(E_i)_{i \in \mathbb{N}}$. Thus, the proof of Proposition 8 shows that the upper morphism is an isomorphism. Similarly, by the same proposition again, the lower morphism is an isomorphism. Since the right vertical map is an inverse limit of isomorphisms, it is an isomorphism, too. Therefore we can conclude that the left vertical map is an isomorphism. Thus, the canonical map

$$H^q(X, \mathcal{F}) \otimes_B B' \longrightarrow H^q(X', \mathcal{F})$$

yields an isomorphism when we tensor with $\widehat{B}'$ over B'. But then, since b belongs to the maximal ideal $\mathfrak{m}' = mB \subset B'$, the map from B' to the $\mathfrak{m}'$-adic completion $\widehat{B}'_{\mathfrak{m}'}$ of B' factors through the b-adic completion $\widehat{B}'$ of B', and we see that the above map gives rise to an isomorphism

$$H^q(X, \mathcal{F}) \otimes_B \widehat{B}'_{\mathfrak{m}'} \overset{\sim}{\longrightarrow} H^q(X', \mathcal{F}) \otimes_{B'} \widehat{B}'_{\mathfrak{m}'}.$$

Thus, we are done. $\qquad\qquad\qquad\qquad\qquad\qquad\qquad\qquad\qquad\qquad\qquad\quad$ $\square$

Part II
Formal Geometry

Chapter 7
Adic Rings and Their Associated Formal Schemes

7.1 Adic Rings

In classical rigid geometry, one works over a field K, carrying a non-Archimedean absolute value. The strategy of the formal approach to rigid geometry is to replace K by its valuation ring R. For example, one starts with R-algebras $R\langle \zeta_1, \ldots, \zeta_n \rangle$ of restricted power series having coefficients in R and considers quotients with respect to finitely generated ideals. This way one obtains R-algebras that may be viewed as R-models of affinoid K-algebras. In fact, taking the generic fiber of such an R-model, i.e. tensoring it with K over R, yields an affinoid K-algebra.

We want to look at rings R that are more general than just valuation rings as occurring above. Let us call a ring R together with a topology on it a *topological ring* if addition and multiplication on R yield continuous maps $R \times R \longrightarrow R$; of course, $R \times R$ is endowed with the product topology. There is a fundamental example. Let R be an arbitrary ring (commutative, and with identity) and $\mathfrak{a} \subset R$ an ideal. There is a unique topology on R making it a topological ring such that the ideals $\mathfrak{a}^n$, $n \in \mathbb{N}$, form a basis of neighborhoods of 0 in R. Just call a subset $U \subset R$ open if for each $x \in U$ there is an $n \in \mathbb{N}$ such that $x + \mathfrak{a}^n \subset U$. The resulting topology is called the $\mathfrak{a}$-*adic topology* on R. (In Grothendieck's terminology [EGA I], this is the $\mathfrak{a}$-*preadic* topology; the latter is called *adic* if it is separated and complete.) Note that all ideals $\mathfrak{a}^n$ are open and, being subgroups of R, also closed in R. A topological ring R is called an *adic ring* if its topology coincides with the $\mathfrak{a}$-adic one for some ideal $\mathfrak{a} \subset R$. Any such ideal $\mathfrak{a}$ is called an *ideal of definition*.

There are similar notions for modules. A module M over a topological ring R, together with a topology on M, is called a *topological R-module* if the addition map $M \times M \longrightarrow M$ and the multiplication map $R \times M \longrightarrow M$ are continuous. Furthermore, for any R-module M and an ideal $\mathfrak{a} \subset R$, we can define the $\mathfrak{a}$-*adic topology* on M: we endow R with its $\mathfrak{a}$-adic topology as described above and consider on M the unique topology making it a topological R-module, for which

S. Bosch, *Lectures on Formal and Rigid Geometry*, Lecture Notes in Mathematics 2105, DOI 10.1007/978-3-319-04417-0_7,
© Springer International Publishing Switzerland 2014

the submodules $\mathfrak{a}^n M$, $n \in \mathbb{N}$, form a basis of neighborhoods. Again, all these submodules are open and closed in M.

Remark 1. *Consider a ring R and an R-module M with $\mathfrak{a}$-adic topologies for some ideal $\mathfrak{a} \subset R$.*

(i) *R is separated (i.e. Hausdorff) if and only if $\bigcap_{n=0}^{\infty} \mathfrak{a}^n = 0$.*
(ii) *M is separated if and only if $\bigcap_{n=0}^{\infty} \mathfrak{a}^n M = 0$.*

Proof. We have $\bigcap_{n=0}^{\infty} \mathfrak{a}^n = 0$ if and only if, for each $x \in R - \{0\}$, there is an $n \in \mathbb{N}$ such that $x \notin \mathfrak{a}^n$. As $\mathfrak{a}^n$ is open and closed in R, assertion (i) follows by a translation argument; (ii) is derived in the same way. $\qquad\square$

For Noetherian rings, adic topologies have nice properties. Let us recall the basic facts from Commutative Algebra.

Theorem 2 (Krull's Intersection Theorem). *Let R be a Noetherian ring, $\mathfrak{a} \subset R$ an ideal, and M a finite R-module. Then:*

$$\bigcap_{n=0}^{\infty} \mathfrak{a}^n M = \{x \in M \; ; \; \text{there exists } r \in 1 + \mathfrak{a} \text{ with } rx = 0\}$$

Proof. Let $M' = \bigcap_{n=0}^{\infty} \mathfrak{a}^n M$, and let $x_1, \ldots, x_r \in M'$ be a generating system of M' as an R-module. By the Lemma of Artin–Rees below, there is some integer $n_0 \in \mathbb{N}$, such that

$$M' = \mathfrak{a}^n M \cap M' = \mathfrak{a}^{n-n_0}\left((\mathfrak{a}^{n_0} M) \cap M'\right) = \mathfrak{a}^{n-n_0} M'$$

for $n \geq n_0$. In particular, we have $M' = \mathfrak{a}M'$, and there are coefficients $a_{ij} \in \mathfrak{a}$ such that

$$x_i = \sum_{j=1}^{r} a_{ij} x_j, \qquad i = 1, \ldots, r.$$

Interpreting $\Lambda = (\delta_{ij} - a_{ij})_{ij}$ as a matrix in $R^{r \times r}$ and $x = (x_i)_i$ as a column vector in M', the above equations can be written in matrix form as $\Lambda \cdot x = 0$. Multiplying from the left with the adjoint matrix Λ^* of Λ yields

$$\det(\Lambda) \cdot x = \Lambda^* \cdot \Lambda \cdot x = 0$$

and, therefore, $\det(\Lambda) \cdot M' = 0$. By construction, we have $\det(\Lambda) \in 1 + \mathfrak{a}$ so that any element of M' is annihilated by an element in $1 + \mathfrak{a}$.

Conversely, assume that $u \in M$ is an element which is annihilated by some element of type $1 - a$ for $a \in \mathfrak{a}$. Then

$$u = au = a^2 u = \ldots \in \bigcap_{n=0}^{\infty} \mathfrak{a}^n M$$

and, hence, $u \in M'$. Thus, $M' = \bigcap_{n=0}^{\infty} \mathfrak{a}^n M$ is characterized as claimed. $\square$

Corollary 3. *Let R be a local Noetherian ring with maximal ideal $\mathfrak{m}$. Then R is $\mathfrak{m}$-adically separated. The same is true for any finitely generated R-module M.*

Lemma 4 (Artin–Rees). *Let R be a Noetherian ring, $\mathfrak{a} \subset R$ an ideal, M a finite R-module, and $M' \subset M$ an R-submodule. Then there is an integer $n_0 \in \mathbb{N}$ such that*

$$(\mathfrak{a}^n M) \cap M' = \mathfrak{a}^{n-n_0}\big((\mathfrak{a}^{n_0} M) \cap M'\big)$$

for all integers $n \geq n_0$.

Proof. Consider $R_* = \bigoplus_{n \in \mathbb{N}} \mathfrak{a}^n$ as a graded ring and $M_* = \bigoplus_{n \in \mathbb{N}} \mathfrak{a}^n M$ as a graded R_*-module. The ideal $\mathfrak{a} \subset R$ is finitely generated, since R is Noetherian, and any such generating system will generate R_* as an R-algebra, when viewed as a system of homogeneous elements of degree 1 in R_*. Thus, by Hilbert's Basis Theorem, R_* is Noetherian. Furthermore, any system of generators for M as an R-module, will generate M_* as an R_*-module. In particular, M_* is a finite R_*-module and, thus, Noetherian.

Now let $M'_n = \mathfrak{a}^n M \cap M'$ for $n \in \mathbb{N}$ and consider

$$\bigoplus_{n=0}^{m} M'_n \oplus \bigoplus_{n>m} \mathfrak{a}^{n-m} M'_m, \qquad m \in \mathbb{N},$$

as an ascending sequence of graded submodules of M_*. Since M_* is Noetherian, the sequence becomes stationary. Thus, there is an index $m = n_0 \in \mathbb{N}$ such that

$$M'_n = \mathfrak{a}^{n-n_0} M'_{n_0} \qquad \text{for all } n \geq n_0.$$

But then $(\mathfrak{a}^n M) \cap M' = \mathfrak{a}^{n-n_0}\big((\mathfrak{a}^{n_0} M) \cap M'\big)$ for $n \geq n_0$, as required. $\square$

Corollary 5. *In the situation of Lemma 4, the $\mathfrak{a}$-adic topology of M restricts to the $\mathfrak{a}$-adic topology of M'.*

Proof. We have

$$\mathfrak{a}^n M' \subset (\mathfrak{a}^n M) \cap M' \qquad \text{and} \qquad (\mathfrak{a}^{n+n_0} M) \cap M' \subset \mathfrak{a}^n M'$$

in the situation of Lemma 4. $\square$

Apart from Noetherian rings we will look at *valuation rings*. Recall that an integral domain R with field of fractions K is called a valuation ring if we have $x \in R$ or $x^{-1} \in R$ for every $x \in K$.

Remark 6. *Let R be a valuation ring.*

(i) *Every finitely generated ideal in R is principal.*
(ii) *For two ideals $\mathfrak{a},\mathfrak{b} \subset R$ we have $\mathfrak{a} \subset \mathfrak{b}$ or $\mathfrak{b} \subset \mathfrak{a}$.*

In particular, R is a local ring.

Proof. For two non-trivial elements $a,b \in R$ we have $ab^{-1} \in R$ or $a^{-1}b \in R$, i.e. b divides a or a divides b in R. This shows (i). To verify (ii) assume $\mathfrak{a} \not\subset \mathfrak{b}$ and $\mathfrak{b} \not\subset \mathfrak{a}$. Then there are elements $a \in \mathfrak{a} - \mathfrak{b}$ and $b \in \mathfrak{b} - \mathfrak{a}$. If a divides b, we have $b \in \mathfrak{a}$, and if b divides a, we must have $a \in \mathfrak{b}$. However, both is excluded, and we get a contradiction.   □

The length of a maximal chain of prime ideals in a valuation ring R is called the *height* of R. For example, starting with a non-Archimedean absolute value on a field K, the corresponding valuation ring $R = \{x \in K \,;\, |x| \leq 1\}$ is of height 1. However, there are valuation rings of higher, even infinite height. For any prime ideal $\mathfrak{p}$ of a valuation ring R, the localization $R_\mathfrak{p}$ is a valuation ring again. In fact, the map $\mathfrak{p} \longmapsto R_\mathfrak{p}$ defines a bijection between prime ideals of R and intermediate rings between R and its field of fractions K. Let us mention without proof that the concept of valuations and absolute values carries over to the field of fractions of valuation rings of arbitrary height. Then $\Gamma = K^*/R^*$, with its attached canonical ordering, serves as the value group of K^*, and the canonical maps $v \colon K \longrightarrow \Gamma \cup \{\infty\}$, resp. $|\cdot| \colon K \longrightarrow \Gamma \cup \{0\}$ are viewed as a valuation, resp. an absolute value on K. The valuation ring R is of height 1 if and only if Γ, together with its ordering, can be realized as a subgroup of the additive group $\mathbb{R}$, resp. the multiplicative group $\mathbb{R}_{>0}$. In precisely these cases, the valuation corresponds to a valuation or a non-Archimedean absolute value on K, as we have defined them in Sect. 2.1.

Any valuation ring R may be viewed as a topological ring by taking the system of its non-zero ideals as a basis of neighborhoods of 0. Then R is automatically *separated*, unless R is a field. We are only interested in valuation rings that are adic.

Remark 7. *Let R be a valuation ring and assume that R is not a field. Then the following are equivalent:*

(i) *R is adic with a finitely generated ideal of definition.*
(ii) *There exists a minimal non-trivial prime ideal $\mathfrak{p} \subset R$.*

If the conditions are satisfied, the topology of R coincides with the t-adic one for any non-zero element $t \in \mathfrak{p}$.

Proof. To begin, let us show that, for any non-unit $t \in R$, the ideal $\mathrm{rad}(t) \subset R$ is prime. To verify this, consider elements $a,b \in R$ satisfying $ab \in \mathrm{rad}(t)$, and look at the ideals $\mathrm{rad}(a)$ and $\mathrm{rad}(b)$. Using Remark 6 (ii), we may assume that $\mathrm{rad}(a) \subset \mathrm{rad}(b)$. Then b divides some power of a, and $ab \in \mathrm{rad}(t)$ implies $a \in \mathrm{rad}(t)$. Thus, $\mathrm{rad}(t)$ is prime.

Now assume condition (i). Due to Remark 6 (i), the topology of R coincides with the t-adic one for some non-zero element $t \in R$. As any non-zero ideal of R must contain a power of t, we see that any non-zero prime ideal in R will contain the ideal $\mathrm{rad}(t)$. The latter is prime by what we have shown and, thus, it is minimal among all non-zero prime ideals in R.

Conversely, assume (ii), i.e. that there is a minimal non-zero prime ideal $\mathfrak{p} \subset R$. Let $t \in \mathfrak{p}$ be a non-zero element and let $\mathfrak{a} \subset R$ be any non-zero ideal. We have to show that $\mathfrak{a}$ contains a power of t. To do this, we may assume that $\mathfrak{a}$ is principal, say $\mathfrak{a} = (a)$. Comparing $\mathrm{rad}(t)$ with $\mathrm{rad}(a)$, both ideals are prime. Thus, we must have $\mathrm{rad}(t) \subset \mathrm{rad}(a)$, and it follows that a power of t is contained in $(a) = \mathfrak{a}$. $\square$

Now let us turn to general adic rings again; let R be a such a ring with $\mathfrak{a} \subset R$ as ideal of definition. As the $\mathfrak{a}$-adic topology on R is invariant under translation, convergence in R can be defined in a natural way. We say that a sequence $c_\nu \in R$ *converges* to an element $c \in R$ if, for each $n \in \mathbb{N}$, there is an integer $\nu_0 \in \mathbb{N}$ such that $c_\nu - c \in \mathfrak{a}^n$ for all $\nu \geq \nu_0$. Similarly, c_ν is called a *Cauchy sequence* if, for each $n \in \mathbb{N}$, there is an integer $\nu_0 \in \mathbb{N}$ such that $c_\nu - c_{\nu'} \in \mathfrak{a}^n$ for all $\nu, \nu' \geq \nu_0$. As usual, R is called *complete* if every Cauchy sequence in R is convergent. A separated completion $\hat{R}$ of R can be constructed by dividing the ring of all Cauchy sequences in R by the ideal of all zero sequences.

For adic rings there is a nice description of completions, which we will explain. Consider the projective system

$$\cdots \longrightarrow R/\mathfrak{a}^n \longrightarrow \cdots \longrightarrow R/\mathfrak{a}^2 \longrightarrow R/\mathfrak{a}^1 \longrightarrow 0$$

where $\mathfrak{a}$ is an ideal of definition of R. Then its projective limit

$$\hat{R} = \varprojlim_n R/\mathfrak{a}^n$$

is seen to be the (separated) completion of R. The topology on this limit is the coarsest one such that all canonical projections $\pi_n \colon \hat{R} \longrightarrow R/\mathfrak{a}^n$ are continuous where $R/\mathfrak{a}^n$ carries the discrete topology (the one for which all subsets of $R/\mathfrak{a}^n$ are open). Thus, a subset of $\hat{R}$ is open if and only if it is a union of certain fibers of the π_n, with varying n, and it follows that a basis of neighborhoods of $0 \in \hat{R}$ is given by the ideals $\ker \pi_n \subset \hat{R}$. We claim that $\ker \pi_n$ equals the closure of $\mathfrak{a}^n$ in $\hat{R}$. In fact, $\ker \pi_n$ is closed in $\hat{R}$ by the definition of the topology on $\hat{R}$, and $\mathfrak{a}^n$ is dense in $\ker \pi_n$, as for any $f \in \ker \pi_n$ and any $m \in \mathbb{N}$ there is an element $f_m \in \mathfrak{a}^n$ such that $f - f_m \in \ker \pi_{m+n}$. Just choose $f_m \in R$ as a representative of the image $\pi_{m+n}(f) \in R/\mathfrak{a}^{m+n}$. If the ideal $\mathfrak{a} \subset R$ is finitely generated, say $\mathfrak{a} = (a_1, \ldots, a_r)$, it is easy to see that its closure in $\hat{R}$ equals $\mathfrak{a}\hat{R}$. First, $\mathfrak{a}$ is clearly dense in $\mathfrak{a}\hat{R}$, since $\mathfrak{a}\hat{R} \subset \ker \pi_1$ and $\mathfrak{a}$ is dense in $\ker \pi_1$. Furthermore, if $f = \sum_{i=1}^{\infty} f_i$ is an infinite sum with $f_i \in \mathfrak{a}^i$, then each f_i can be written as a combination $f_i = \sum_{j=1}^{r} f_{ij} a_i$

with coefficients $f_{ij} \in \mathfrak{a}^{i-1}$, which yields $f = \sum_{j=1}^{r}(\sum_{i=1}^{\infty} f_{ij})a_j$ and, hence, $f \in \mathfrak{a}\widehat{R}$. Thus, we have shown:

Remark 8. *If the ideal of definition $\mathfrak{a} \subset R$ is finitely generated, then $\mathfrak{a}\widehat{R}$ is the closure of $\mathfrak{a}$ in $\widehat{R}$ and it follows that $\widehat{R}$ is adic again with ideal of definition $\mathfrak{a}\widehat{R}$.*

On the other hand, if $\mathfrak{a}$ is not finitely generated, it can happen that $\widehat{R}$ fails to be $\mathfrak{a}\widehat{R}$-adically complete so that in this case the topology of $\widehat{R}$ will be different from the $\mathfrak{a}\widehat{R}$-adic one.

In the following we will always assume that R is complete and separated under its $\mathfrak{a}$-adic topology. In particular, the canonical homomorphism

$$R \longrightarrow \varprojlim_{n} R/\mathfrak{a}^n$$

is an isomorphism then. For $f \in R$ we set

$$R\langle f^{-1}\rangle = \varprojlim_{n}\big((R/\mathfrak{a}^n)[f^{-1}]\big)$$

and call it the *complete localization* of R by (the multiplicative system generated by) f. There is a canonical map $R \longrightarrow R\langle f^{-1}\rangle$, and the maps

$$R[f^{-1}] \longrightarrow (R/\mathfrak{a}^n)[f^{-1}]$$

give rise to a canonical map $R[f^{-1}] \longrightarrow R\langle f^{-1}\rangle$ showing that the image of f is invertible in $R\langle f^{-1}\rangle$.

Remark 9. *The canonical homomorphism $R[f^{-1}] \longrightarrow R\langle f^{-1}\rangle$ exhibits $R\langle f^{-1}\rangle$ as the adic completion of $R[f^{-1}]$ with respect to the ideal $\mathfrak{a}R[f^{-1}]$ generated by $\mathfrak{a}$ in $R[f^{-1}]$. If $\mathfrak{a}$ is finitely generated, the topology on $R\langle f^{-1}\rangle$ coincides with the $\mathfrak{a}R\langle f^{-1}\rangle$-adic one.*

Proof. Tensoring the exact sequence

$$0 \longrightarrow \mathfrak{a}^n \longrightarrow R \longrightarrow R/\mathfrak{a}^n \longrightarrow 0$$

with $R[f^{-1}]$, which is flat over R, yields the exact sequence

$$0 \longrightarrow \mathfrak{a}^n R[f^{-1}] \longrightarrow R[f^{-1}] \longrightarrow (R/\mathfrak{a}^n)[f^{-1}] \longrightarrow 0$$

and, hence, an isomorphism

$$R[f^{-1}]/(\mathfrak{a}^n) \overset{\sim}{\longrightarrow} (R/\mathfrak{a}^n)[f^{-1}].$$

Thus $R\langle f^{-1}\rangle = \varprojlim R[f^{-1}]/(\mathfrak{a}^n)$ is the $\mathfrak{a}R[f^{-1}]$-adic completion of $R[f^{-1}]$. As we have explained in Remark 8, the topology on the latter is the $\mathfrak{a}R\langle f^{-1}\rangle$-adic one if $\mathfrak{a}$ is finitely generated. $\square$

To give a more explicit description of $R\langle f^{-1}\rangle$, we consider the R-algebra $R\langle\zeta\rangle$ of restricted power series with coefficients in R and with a variable ζ, i.e. of all power series $\sum_{\nu=0}^{\infty} c_\nu \zeta^\nu$ satisfying $\lim c_\nu = 0$, a condition that is meaningful, as we have explained. Note that $R\langle\zeta\rangle$ is complete and separated under the $(\mathfrak{a})$-adic topology and that, in fact, $R\langle\zeta\rangle = \varprojlim_n R/\mathfrak{a}^n[\zeta]$. Thus, there is a canonical continuous homomorphism $R\langle\zeta\rangle \longrightarrow R\langle f^{-1}\rangle$ mapping ζ to f^{-1}.

Remark 10. *The canonical homomorphism* $R\langle\zeta\rangle \longrightarrow R\langle f^{-1}\rangle$ *induces an isomorphism*

$$R\langle\zeta\rangle/(1 - f\zeta) \overset{\sim}{\longrightarrow} R\langle f^{-1}\rangle.$$

Proof. To abbreviate, let us write $R_n = R/\mathfrak{a}^n$ for $n \in \mathbb{N}$. Then consider the projective system of exact sequences:

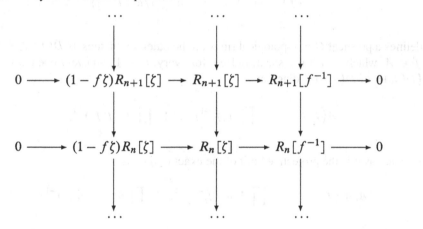

As $\varprojlim$ is left exact, it gives rise to a left exact sequence

$$0 \longrightarrow \varprojlim (1 - f\zeta)R_n[\zeta] \longrightarrow \varprojlim R_n[\zeta] \longrightarrow \varprojlim R_n[f^{-1}] \longrightarrow 0, \quad (*)$$

which is, in fact, exact, since all maps

$$(1 - f\zeta)R_{n+1}[\zeta] \longrightarrow (1 - f\zeta)R_n[\zeta]$$

are surjective so that the system on the left-hand side in $(*)$ satisfies the condition of Mittag–Leffler. Thus, as $1 - f\zeta$ is not a zero divisor in $R_n[\zeta]$, we get an exact sequence

$$0 \longrightarrow (1 - f\zeta)R\langle\zeta\rangle \longrightarrow R\langle\zeta\rangle \longrightarrow R\langle f^{-1}\rangle \longrightarrow 0$$

as claimed. $\qquad\qquad\qquad\qquad\qquad\qquad\qquad\qquad\qquad\qquad\qquad\qquad\qquad\square$

7.2 Formal Schemes

Formal schemes are locally topologically ringed spaces where all occurring rings
have to be viewed as objects of the category of topological rings. Just as ordinary
schemes, they are built from local affine parts. To define such affine formal schemes,
consider an adic ring A; from now on, we will always assume that adic rings are
complete and *separated*. Let $\mathfrak{a}$ be an ideal of definition of A. We denote by $\mathrm{Spf}\, A$
the set of all open prime ideals $\mathfrak{p} \subset A$. As a prime ideal in A is open if and only
if it contains some power of $\mathfrak{a}$ and, hence, $\mathfrak{a}$ itself, we see that $\mathrm{Spf}\, A$ is canonically
identified with the closed subset $\mathrm{Spec}\, A/\mathfrak{a} \subset \mathrm{Spec}\, A$, for any ideal of definition $\mathfrak{a}$
of A.

This way the Zariski topology on $\mathrm{Spec}\, A$ induces a topology on $\mathrm{Spf}\, A$. As usual,
let $D(f)$ for $f \in A$ be the open subset in $\mathrm{Spf}\, A$ where f does not vanish. Then

$$D(f) \longmapsto A\langle f^{-1}\rangle = \varprojlim_n \left(A/\mathfrak{a}^n[f^{-1}]\right)$$

defines a presheaf $\mathcal{O}$ of topological rings on the category of subsets $D(f) \subset \mathrm{Spf}\, A$,
$f \in A$, which in fact is a sheaf. Indeed, for every $f \in A$ and every open covering
$\left(D(f_i)\right)_i$ of $D(f)$, the diagram

$$A\langle f^{-1}\rangle \longrightarrow \prod_i A\langle f_i^{-1}\rangle \rightrightarrows \prod_{i,j} A\langle (f_i f_j)^{-1}\rangle$$

is exact, as it is the projective limit of the exact diagrams

$$A/\mathfrak{a}^n[f^{-1}] \longrightarrow \prod_i A/\mathfrak{a}^n[f_i^{-1}] \rightrightarrows \prod_{i,j} A/\mathfrak{a}^n[(f_i f_j)^{-1}],$$

and as $\varprojlim$ is left exact. By the usual procedure, the sheaf $\mathcal{O}$ can be extended to the
category of all Zariski open subsets of $\mathrm{Spf}\, A$ and we will use the notation $\mathcal{O}$ for
it again. In fact, if $U \subset \mathrm{Spf}\, A$ is Zariski open and $U = \bigcup_{i \in J} D(f_i)$ is an open
covering by basic open subsets $D(f_i) \subset \mathrm{Spf}\, A$, $f_i \in A$, then the exact diagram

$$\mathcal{O}(U) \longrightarrow \prod_i A\langle f_i^{-1}\rangle \rightrightarrows \prod_{i,j} A\langle (f_i f_j)^{-1}\rangle$$

is obtained by taking the projective limit of the exact diagrams

$$\mathcal{O}_{\mathrm{Spec}\, A/\mathfrak{a}^n}(U) \longrightarrow \prod_i A/\mathfrak{a}^n[f_i^{-1}] \rightrightarrows \prod_{i,j} A/\mathfrak{a}^n[(f_i f_j)^{-1}].$$

Thus, it makes sense to write $\mathcal{O} = \varprojlim \mathcal{O}_{\mathrm{Spec}\, A/\mathfrak{a}^n}$, i.e. $\mathcal{O}$ is the projective limit of the
sheaves $\mathcal{O}_{\mathrm{Spec}\, A/\mathfrak{a}^n}$.

However, let us point out, although this is only of minor importance, that the interpretation of the sheaf $\mathcal{O}$ on Spf A as the projective limit of the sheaves $\mathcal{O}_{\mathrm{Spec}\, A/\mathfrak{a}^n}$ has actually to be carried out in a more specific setting. To get a projective limit topology on $\mathcal{O}(U)$ for $U \subset$ Spf A open, which is in accordance with the definition of a sheaf with values in the category of topological rings ([EGA I], Chap. 0, 3.3.1), especially if U is not quasi-compact, we have to view the sheaves $\mathcal{O}_{\mathrm{Spec}\, A/\mathfrak{a}^n}$ as sheaves of pseudo-discrete topological rings ([EGA I], Chap. 0, 3.9.1).

Remark 1. *If in the above situation a point $x \in$ Spf A corresponds to the open prime ideal $\mathfrak{j}_x \subset A$, then the stalk $\mathcal{O}_x = \varinjlim_{x \in D(f)} A\langle f^{-1} \rangle$ is a local ring with a maximal ideal $\mathfrak{m}_x$ containing $\mathfrak{j}_x \mathcal{O}_x$. Furthermore, $\mathfrak{m}_x = \mathfrak{j}_x \mathcal{O}_x$ if $\mathfrak{a}$ is finitely generated.*

Proof. For each $f \in A - \mathfrak{j}_x$, there are canonical exact sequences

$$0 \longrightarrow \mathfrak{j}_x A/\mathfrak{a}^n [f^{-1}] \longrightarrow A/\mathfrak{a}^n [f^{-1}] \longrightarrow A/\mathfrak{j}_x [f^{-1}] \longrightarrow 0$$

where $n \geq 1$. Taking projective limits over n and using the fact that the projective system on the left-hand side is surjective and, hence, satisfies the condition of Mittag–Leffler, we get an exact sequence

$$0 \longrightarrow \mathfrak{j}_x \langle f^{-1} \rangle \longrightarrow A\langle f^{-1} \rangle \longrightarrow A/\mathfrak{j}_x [f^{-1}] \longrightarrow 0$$

where we have used the abbreviation $\mathfrak{j}_x \langle f^{-1} \rangle = \varprojlim \mathfrak{j}_x A/\mathfrak{a}^n [f^{-1}]$ for the completion of $\mathfrak{j}_x A[f^{-1}]$ with respect to the topology induced from the $(\mathfrak{a})$-adic topology on $A[f^{-1}]$. Then, taking the direct limit over all $f \in A - \mathfrak{j}_x$ and writing $\mathfrak{m}_x = \varinjlim \mathfrak{j}_x \langle f^{-1} \rangle$, we get an exact sequence

$$0 \longrightarrow \mathfrak{m}_x \longrightarrow \mathcal{O}_x \longrightarrow Q(A/\mathfrak{j}_x) \longrightarrow 0$$

showing that $\mathfrak{m}_x$ is a maximal ideal in $\mathcal{O}_x$ containing $\mathfrak{j}_x \mathcal{O}_x$.

To see that $\mathfrak{m}_x$ is the only maximal ideal in $\mathcal{O}_x$, we show that $\mathcal{O}_x - \mathfrak{m}_x$ consists of units. To do this, fix an element $g_x \in \mathcal{O}_x - \mathfrak{m}_x$, say represented by an element $g \in A\langle f^{-1} \rangle$ for some $f \in A$ satisfying $f \notin \mathfrak{j}_x$. Then g cannot belong to $\mathfrak{j}_x \langle f^{-1} \rangle$ and, hence, using $\mathfrak{a} \subset \mathfrak{j}_x$, its residue class $\overline{g} \in A/\mathfrak{a}[f^{-1}]$ cannot belong to $\mathfrak{j}_x A/\mathfrak{a}[f^{-1}]$. Multiplying g by a suitable power of f, we can even assume that $\overline{g}$ belongs to $A/\mathfrak{a}$ and, thus, admits a representative $g' \in A$ where $g' \notin \mathfrak{j}_x$. Then $fg' \in A - \mathfrak{j}_x$, and we claim that the image of g is invertible in $A\langle (fg')^{-1} \rangle$, which implies that it is invertible in $\mathcal{O}_x$ as well. To see this, consider the equation $g = g'(1 - d)$ in $A\langle (fg')^{-1} \rangle$ with $d = 1 - g'^{-1}g$ where we have written g, g' again for the corresponding images in $A\langle (fg')^{-1} \rangle$. Thus, in order to show that g is invertible in $A\langle (fg')^{-1} \rangle$, we need to know that $1 - d$ is invertible. However, using the geometric series, the latter is clear since d^n is a zero sequence in $A\langle (fg')^{-1} \rangle$, due to the fact that the image of d is trivial in $A/\mathfrak{a}[(fg')^{-1}]$ and, hence, the image of d^n is trivial in $A/\mathfrak{a}^n [(fg')^{-1}]$ for all $n \in \mathbb{N}$.

Finally, if $\mathfrak{a}$ is finitely generated, we can conclude $j_x\langle f^{-1}\rangle = j_x A\langle f^{-1}\rangle$ from 7.1/8 and, hence, that $\mathfrak{m}_x = j_x \mathcal{O}_x$.   □

Definition 2. *For an adic ring A with ideal of definition $\mathfrak{a} \subset A$, set $X = \operatorname{Spf} A$ and let $\mathcal{O}_X$ be the sheaf of topological rings we have constructed above. Then the locally ringed space $(X, \mathcal{O}_X)$ (where "ringed" has to be understood in the sense of topological rings) is called the* affine formal scheme *of A. It is denoted by $\operatorname{Spf} A$ again.*

There is a slight problem with this definition. If we consider an affine formal scheme $X = \operatorname{Spf} A$ and a basic open subset $U = D(f) \subset \operatorname{Spf} A$ for some $f \in A$, we would like to interpret $(U, \mathcal{O}_X|_U)$ as the affine formal scheme $\operatorname{Spf} A\langle f^{-1}\rangle$, although we do not know in general if $A\langle f^{-1}\rangle$, which is defined as the $\mathfrak{a}$-adic completion of $A[f^{-1}]$, is an adic ring again. Due to 7.1/9, no problems arise, when $\mathfrak{a}$ is finitely generated, since then the topology of $A\langle f^{-1}\rangle$ coincides with the $\mathfrak{a}$-adic one.

When such a finiteness condition is to be avoided, affine formal schemes $\operatorname{Spf} A$ should be constructed for slightly more general topological rings than just adic ones. One needs that A is an admissible ring in the sense of Grothendieck. This means that:

(i) A is linearly topologized, i.e. there is a basis of neighborhoods $(I_\lambda)_{\lambda \in \Lambda}$ of 0 consisting of ideals in A; such ideals are automatically open.

(ii) A has an ideal of definition, i.e. there is an *open* ideal $\mathfrak{a} \subset A$ such that $\mathfrak{a}^n$ tends to zero in the sense that, for each neighborhood $U \subset A$ of 0, there is an $n \in \mathbb{N}$ satisfying $\mathfrak{a}^n \subset U$. (This does not necessarily imply that $\mathfrak{a}^n$ is open for $n > 1$.)

(iii) A is separated and complete.

If A is an admissible ring with a family of ideals $(I_\lambda)_{\lambda \in \Lambda}$ forming a basis of neighborhoods of 0, then the canonical map $A \overset{\sim}{\longrightarrow} \varprojlim_\lambda A/I_\lambda$ is a topological isomorphism. Admissible rings can be dealt with in essentially the same way as adic ones, just replacing the system of powers $(\mathfrak{a}^n)_{n \in \mathbb{N}}$ for an ideal of definition $\mathfrak{a} \subset A$ by the system $(I_\lambda)_{\lambda \in \Lambda}$. However, for our purposes, it will be enough to restrict to complete and separated adic rings, as later we will always suppose that there is an ideal of definition that is *finitely generated*.

When working with affine formal schemes, morphisms are, of course, meant in the sense of morphisms of locally topologically ringed spaces. So all inherent ring homomorphisms are supposed to be continuous. Just as in the scheme case or in the case of affinoid K-spaces, one shows that morphisms of locally topologically ringed spaces $\operatorname{Spf} A \longrightarrow \operatorname{Spf} B$ correspond bijectively to continuous homomorphisms $B \longrightarrow A$.

Definition 3. *A* formal scheme *is a locally topologically ringed space $(X, \mathcal{O}_X)$ such that each point $x \in X$ admits an open neighborhood U where $(U, \mathcal{O}_X|_U)$ is isomorphic to an affine formal scheme $\operatorname{Spf} A$, as constructed above.*

As usual, global formal schemes can be constructed by gluing local ones. In particular, fiber products can be constructed by gluing local affine ones. Similarly as for schemes or rigid K-spaces, the fiber product of two affine formal schemes Spf A and Spf B over a third one Spf R is given by Spf($A \hat{\otimes}_R B$) where

$$A \hat{\otimes}_R B = \varprojlim A/\mathfrak{a}^n \otimes_R B/\mathfrak{b}^n$$

with ideals of definition $\mathfrak{a}$ of A and $\mathfrak{b}$ of B is the *complete tensor product* of A and B over R. The latter is the $(\mathfrak{a}, \mathfrak{b})$-adic completion of the ordinary tensor product $A \otimes_R B$. If $\mathfrak{a}$ and $\mathfrak{b}$ are finitely generated, we see from 7.1/8 that $A \hat{\otimes}_R B$ is an adic ring again with ideal of definition generated by the image of $\mathfrak{a} \otimes_R B + A \otimes_R \mathfrak{b}$.

We end this section by a fundamental example of a formal scheme, the so-called *formal completion of a scheme X along a closed subscheme $Y \subset X$.*

Example 4. Let X be a scheme and $Y \subset X$ a closed subscheme, defined by a quasi-coherent ideal $\mathcal{J} \subset \mathcal{O}_X$. Then consider the sheaf $\mathcal{O}_Y$ obtained by restricting the projective limit $\varprojlim_n \mathcal{O}_X/\mathcal{J}^n$ to Y. It follows that $(Y, \mathcal{O}_Y)$ is a locally topologically ringed space, the desired *formal completion of X along Y*. Locally, the construction looks as follows: Let $X = \operatorname{Spec} A$ and assume that $\mathcal{J}$ is associated to the ideal $\mathfrak{a} \subset A$. Then

$$(Y, \mathcal{O}_Y) = \operatorname{Spf}\left(\varprojlim_n A/\mathfrak{a}^n\right) = \operatorname{Spf} \hat{A}$$

where $\hat{A}$ is the $\mathfrak{a}$-adic completion of A.

For example, assume $A = R[\zeta]$ where ζ is a system of n variables, R a complete valuation ring of height 1, and where $\mathfrak{a} = (t)$ for some non-unit $t \in R - \{0\}$. So X coincides with the affine n-space $\mathbb{A}_R^n$ and Y (pointwise) with its special fiber $\mathbb{A}_k^n$ where k is the residue field of R. The formal completion of X along Y then yields the formal affine n-space Spf $R\langle\zeta\rangle$. The latter admits the affinoid unit ball $\mathbb{B}_K^n = \operatorname{Sp} K\langle\zeta\rangle = \operatorname{Sp}(R\langle\zeta\rangle \otimes_R K)$ for $K = Q(R)$ as "generic fiber", as we will explain later in Sect. 7.4, and there is a canonical open immersion $\mathbb{B}_K^n \hookrightarrow \mathbb{A}_K^{n, \mathrm{rig}}$ into the rigid analytification of $\mathbb{A}_K^n$.

A canonical open immersion of this type exists on a more general scale. Let X be an R-scheme of locally finite type that is flat over R, and denote by $\hat{X}$ its formal completion along the special fiber. Then $\hat{X}$ is an admissible formal R-scheme using the terminology of 7.3/3 and 7.4/1, and its generic fiber $\hat{X}_{\mathrm{rig}}$ in the sense of Sect. 7.4 admits canonically an open immersion $\hat{X}_{\mathrm{rig}} \hookrightarrow (X_K)^{\mathrm{rig}}$ into the rigid analytification via the GAGA-functor of the generic fiber $X_K = X \otimes_R K$ of X. As we have seen above, this immersion is not necessarily an isomorphism. But in case X is proper over R, one can show $\hat{X}_{\mathrm{rig}} = (X_K)^{\mathrm{rig}}$ relying on the valuative criterion of properness.

7.3 Algebras of Topologically Finite Type

Let R be a (complete and separated) adic ring with a finitely generated ideal of definition $I \subset R$. We will assume that R does not have I-torsion, i.e. that the ideal

$$(I\text{-torsion})_R = \{r \in R \,;\, I^n r = 0 \text{ for some } n \in \mathbb{N}\}$$

is trivial, a condition that, apparently, is independent of the choice of I. Choosing generators $g_1, \dots, g_r$ of I, we see that R does not have I-torsion if and only if the canonical map

$$R \longrightarrow \prod_{i=1}^{r} R[g_i^{-1}]$$

is injective. We will admit only the following two classes of rings:

(V) R is an adic valuation ring with a finitely generated ideal of definition (which automatically is principal by 7.1/6).

(N) R is a Noetherian adic ring with an ideal of definition I such that R does not have I-torsion.

These classes of adic rings R have been chosen bearing in mind that topological R-algebras with certain finiteness conditions, for example as we will set them up in Definition 3, should be accessible in a satisfactory way. Of course, the Noetherian hypothesis in class (N) is quite convenient and useful, especially since there are interesting objects such as Raynaud's universal Tate curve that live over a non-local base of this type; see Sect. 9.2. On the other hand, even if the Noetherian hypothesis is not present, it turns out that adic valuation rings of class (V) can still be handled reasonably well. Indeed, this class allows the extension of several important results on R-algebras that otherwise are only valid in the Noetherian situation. A good example for this is Gabber's flatness result 8.2/2. Also note that class (V) includes all classical valuation rings that are obtained from a field with a complete non-Archimedean absolute value, especially in the non-discrete case where the Noetherian hypothesis is not available.

In the following, let R be of type (V) or (N). As usual, we define the R-algebra $R\langle \zeta_1, \dots, \zeta_n \rangle$ of restricted power series in the variables $\zeta_1, \dots, \zeta_n$ as the subalgebra of the R-algebra $R[\![\zeta_1, \dots, \zeta_n]\!]$ of formal power series, consisting of all series $\sum_{\nu \in \mathbb{N}^n} c_\nu \zeta^\nu$ with coefficients $c_\nu \in R$ constituting a zero sequence in R. Of course, $R\langle \zeta_1, \dots, \zeta_n \rangle$ equals the I-adic completion of the ring of polynomials $R[\zeta_1, \dots, \zeta_n]$.

Remark 1. $R\langle \zeta_1, \dots, \zeta_n \rangle$ *is Noetherian if R is of class* (N).

Proof. If R is Noetherian, the polynomial ring $(R/I)[\zeta_1, \dots, \zeta_n]$ is Noetherian and the assertion follows from [AC], Chap. III, § 2, no. 11, Cor. 2 of Prop. 14. $\square$

Remark 2. $R\langle \zeta_1, \ldots, \zeta_n \rangle$ *is flat over* R.

Proof. A module M over a ring R is flat if and only if, for each finitely generated ideal $\mathfrak{a} \subset R$, the canonical map $\mathfrak{a} \otimes_R M \longrightarrow M$ is injective. If R is an integral domain and if every finitely generated ideal in R is principal, the latter condition is equivalent to the fact that M does not admit R-torsion. Thus, if R is of class (V), we see from 7.1/6 that $R\langle \zeta_1, \ldots, \zeta_n \rangle$ is flat over R.

On the other hand, if R is of class (N), the map $R \longrightarrow R[\zeta_1, \ldots, \zeta_n]$ is flat being module-free. Furthermore, the map from $R[\zeta_1, \ldots, \zeta_n]$ into its I-adic completion is flat by Bourbaki [AC], Chap. III, § 5, no. 4, Cor. of Prop. 3. $\square$

Having defined restricted power series with coefficients in R, let us introduce now the analogs of affinoid algebras.

Definition 3. *A topological R-algebra A is called*

(i) *of* topologically finite type *if it is isomorphic to an R-algebra of type* $R\langle \zeta_1, \ldots, \zeta_n \rangle / \mathfrak{a}$ *that is endowed with the I-adic topology and where $\mathfrak{a}$ is an ideal in* $R\langle \zeta_1, \ldots, \zeta_n \rangle$,

(ii) *of* topologically finite presentation *if, in addition to* (i), $\mathfrak{a}$ *is finitely generated,*

(iii) admissible *if, in addition to* (i) *and* (ii), *A does not have I-torsion.*

It is a fundamental fact, which will be used extensively in the sequel, that an R-algebra of topologically finite type that is flat over R, is automatically of topologically finite presentation. Properties of this type are proved using the flattening techniques of Raynaud and Gruson; see [RG], Part I, 3.4.6.

Theorem 4 (Raynaud–Gruson). *Let A be an R-algebra of topologically finite type and M a finite A-module that is flat over R. Then M is an A-module of finite presentation, i.e. M is isomorphic to the cokernel of some A-linear map* $A^r \longrightarrow A^s$.

Proof. As an R-algebra of topologically finite type, A is a quotient of some algebra of restricted power series $R\langle \zeta_1, \ldots, \zeta_n \rangle$. Viewing M as a module over such a power series ring, we may assume $A = R\langle \zeta_1, \ldots, \zeta_n \rangle$. In the Noetherian case (N), nothing has to be shown, since A is Noetherian then. If R is an adic valuation ring of type (V), we can choose an element t generating an ideal of definition of R. Then A/tA is an $R/(t)$-algebra of finite presentation, and M/tM is a finite A/tA-module that is flat over $R/(t)$. Furthermore, it follows from the above cited result of Raynaud and Gruson that M/tM is an A/tA-module of finite presentation. Now consider a short exact sequence of A-modules

$$0 \longrightarrow N \longrightarrow A^s \longrightarrow M \longrightarrow 0.$$

Since M is flat over R, the sequence remains exact when tensoring it with $R/(t)$ over R. Since M/tM is an A/tA-module of finite presentation, N/tN is a finite A/tA-module. But then, viewing N as a submodule of A^s for $A = R\langle \zeta_1, \ldots, \zeta_n \rangle$, a standard approximation argument in terms of the t-adic topology on A^s shows that N is a finite A-module and, hence, that M is an A-module of finite presentation.

In the most interesting case where R is an adic valuation ring of height 1, the Theorem is accessible by more elementary methods. First, one reduces to the case where $A = R\langle \zeta \rangle$, for a finite system of variables $\zeta = (\zeta_1, \ldots, \zeta_n)$, as indicated above. Then, as before, consider a short exact sequence

$$0 \longrightarrow N \longrightarrow (R\langle \zeta \rangle)^s \longrightarrow M \longrightarrow 0$$

of $R\langle \zeta \rangle$-modules. Since M is flat over R, there is no R-torsion in M and, consequently, looking at the inclusion map $(R\langle \zeta \rangle)^s \hookrightarrow (R\langle \zeta \rangle)^s \otimes_R K = T_n^s$ where K is the field of fractions of R, we get

$$\left(N \otimes_R K \right) \cap \left(R\langle \zeta \rangle \right)^s = N.$$

Applying 2.3/10 to the T_n-module $N \otimes_R K$, we see that N is a finite $R\langle \zeta \rangle$-module and, hence, that M is an $R\langle \zeta \rangle$-module of finite presentation. $\square$

Corollary 5. *Let A be an R-algebra of topologically finite type. If A has no I-torsion, A is of topologically finite presentation.*

Proof. The assertion is trivial in the Noetherian case. So assume that R is of class (V). Interpreting A as a residue algebra $R\langle \zeta \rangle/\mathfrak{a}$ with a system of variables ζ, we can view A as an $R\langle \zeta \rangle$-module via the canonical projection $R\langle \zeta \rangle \longrightarrow A$. If A has no I-torsion, it is flat over R and, thus, by Theorem 4, a finitely presented $R\langle \zeta \rangle$-module. But then $\mathfrak{a}$ must be finitely generated so that A is of topologically finite presentation. $\square$

Recall that, similarly as in 6.1/2, a module M over a ring A is called *coherent* if M is finitely generated and if every finite submodule of M is of finite presentation. A itself is called a *coherent ring* if it is coherent as a module over itself, i.e. if each finitely generated ideal $\mathfrak{a} \subset A$ is of finite presentation. One can show that all members of a short exact sequence of A-modules

$$0 \longrightarrow M' \longrightarrow M \longrightarrow M'' \longrightarrow 0$$

are coherent as soon as two of them are; see for example [Bo], 1.5/15.

Corollary 6. *Let A be an R-algebra of topologically finite presentation. Then A is a coherent ring. In particular, any A-module of finite presentation is coherent.*

Proof. We may assume that R is of class (V). Let us first consider the case where A does not have I-torsion and, hence, is flat over R. Then any finitely generated ideal in A is flat over R and, hence, of finite presentation by Theorem 4. Thus, A is coherent in this case.

In the general case, we can write A as a quotient $R\langle \zeta \rangle / \mathfrak{a}$ with a system of variables ζ and a finitely generated ideal $\mathfrak{a}$. The algebra $R\langle \zeta \rangle$ is coherent, as we have seen. Thus, $\mathfrak{a}$ is coherent, too, and it follows that $A = R\langle \zeta \rangle / \mathfrak{a}$ is coherent. $\square$

We want to draw some further conclusions from Theorem 4.

Lemma 7. *Let A be an R-algebra of topologically finite type, M a finite A-module, and $N \subset M$ a submodule. Then:*

(i) *If N is saturated in M in the sense that*

$$N_{\mathrm{sat}} = \{ x \in M \; ; \; \text{there is an } n \in \mathbb{N} \text{ such that } I^n x \subset N \}$$

coincides with N, then N is finitely generated.

(ii) *The I-adic topology of M restricts to the I-adic topology on N.*

Proof. If R is of class (N), assertion (i) is trivial, and assertion (ii) follows from the lemma of Artin–Rees; cf. 7.1/5. So assume that R is of class (V). If N is saturated in M, the quotient M/N does not admit I-torsion and, hence, is flat over R, since R is a valuation ring. Thus M/N, as a finite A-module that is flat over R, is of finite presentation by Theorem 4 and there is an exact sequence of A-modules

$$0 \longrightarrow K \longrightarrow F \longrightarrow M/N \longrightarrow 0$$

where F is finite free and K is finite. As M is finitely generated, we may assume that $F \longrightarrow M/N$ factors through M via an epimorphism $F \longrightarrow M$. But then this map restricts to an epimorphism $K \longrightarrow N$ and we see that N is finitely generated. This verifies (i).

To verify assertion (ii), we can consider the saturation $N_{\mathrm{sat}} \subset M$ of N; it is finitely generated by (i). Thus, there is an integer $n \in \mathbb{N}$ such that $I^n N_{\mathrm{sat}} \subset N$, and we have

$$I^{m+n} M \cap N \subset I^m N \subset I^m M \cap N$$

for all $m \in \mathbb{N}$. So we are done. $\square$

Proposition 8. *Let A be an R-algebra of topologically finite type and M a finite A-module. Then M is I-adically complete and separated.*

Proof. We may replace A by an algebra of restricted power series $R\langle \zeta \rangle$ and thereby assume that A is I-adically complete and separated. Then, viewing M as a quotient of a finite cartesian product of A and using Lemma 7 (ii), we see that M is I-adically complete for trivial reasons. To show that it is also I-adically separated, consider an element $x \in \bigcap_{n=0}^{\infty} I^n M$, and look at the submodule $N = Ax \subset M$. Using Lemma 7 (ii), there is an integer $n \in \mathbb{N}$ such that $N = I^n M \cap N \subset IN$. Hence there is an equation $(1 - c)x = 0$ for some $c \in I$. However, using the geometric series, we see that $1 - c$ is a unit in R and, hence, that x must be zero. □

Corollary 9. *Any R-algebra of topologically finite type is I-adically complete and separated.*

In particular, if A is an R-algebra of topologically finite type, we can identify A with the projective limit $\varprojlim_n A/I^n A$. To abbreviate, we will write $R_n = R/I^{n+1}$ and $A_n = A/I^{n+1} = A \otimes_R R_n$ for $n \in \mathbb{N}$. Similar notions will be used for R-modules.

Proposition 10. *Let A be an R-algebra, which is I-adically complete and separated. Then:*

(i) *A is of topologically finite type if and only if A_0 is of finite type over R_0.*
(ii) *A is of topologically finite presentation if and only if A_n is of finite presentation over R_n for all $n \in \mathbb{N}$.*

Proof. We need only to verify the if-parts. So assume that A_0 is of finite type over R_0. Then there is an epimorphism $\varphi_0: R_0[\zeta] \longrightarrow A_0$ for a finite system of variables $\zeta = (\zeta_1, \ldots, \zeta_m)$. Let $a_i \in A$ be a representative of $\varphi_0(\zeta_i)$ and define a continuous R-algebra homomorphism $\varphi: R\langle\zeta\rangle \longrightarrow A$ by mapping ζ_i onto a_i; the latter is possible, as A is I-adically complete and separated. Then $A = \operatorname{im} \varphi + IA$, and a limit argument shows that φ is surjective.

Now, setting $\mathfrak{a} = \ker \varphi$, consider the exact sequence

$$0 \longrightarrow \mathfrak{a} \longrightarrow R\langle\zeta\rangle \overset{\varphi}{\longrightarrow} A \longrightarrow 0$$

and assume that all algebras A_n are of finite presentation over R_n. Then, due to Lemma 7 (ii) there is an integer $n \in \mathbb{N}$ satisfying $\mathfrak{a} \cap I^{n+1} \subset I\mathfrak{a}$, and we get the exact sequence

$$0 \longrightarrow \mathfrak{a}/\mathfrak{a} \cap I^{n+1} R\langle\zeta\rangle \longrightarrow R_n[\zeta] \longrightarrow A_n \longrightarrow 0.$$

By our assumption $\mathfrak{a}/\mathfrak{a} \cap I^{n+1} R\langle\zeta\rangle$ and, hence, also $\mathfrak{a}/I\mathfrak{a}$ are finitely generated. Thus there is a finitely generated ideal $\mathfrak{a}' \subset \mathfrak{a}$ such that $\mathfrak{a} = \mathfrak{a}' + I\mathfrak{a}$. Again a limit argument yields $\mathfrak{a} = \mathfrak{a}'$ and, hence, that $\mathfrak{a}$ is finitely generated. □

Proposition 11. *Let $\varphi: A \longrightarrow B$ be a morphism of R-algebras of topologically finite type, and M a finite B-module. Then M is a flat (resp. faithfully flat) A-module if and only if M_n is a flat (resp. faithfully flat) A_n-module for all $n \in \mathbb{N}$.*

Proof. The only-if part is trivial, since flatness is preserved under base change. To verify the if part, we have to show that the canonical map $\mathfrak{a} \otimes_A M \longrightarrow M$ is injective for each finitely generated ideal $\mathfrak{a} \subset A$. This can be done similarly as in the proof of the Bourbaki criterion on flatness; see [AC], Chap. III, § 5, no. 2. Given an arbitrary $m \in \mathbb{N}$, there is an integer $n \in \mathbb{N}$ such that $I^{n+1} A \cap \mathfrak{a} \subset I^m \mathfrak{a}$; see Lemma 7 (ii). Setting $N = \mathfrak{a}/(I^{n+1} A \cap \mathfrak{a})$, we get a commutative diagram, whose upper row is exact:

$$
\begin{array}{ccccccc}
(I^{n+1} A \cap \mathfrak{a}) \otimes_A M & \longrightarrow & \mathfrak{a} \otimes_A M & \longrightarrow & N \otimes_A M & \longrightarrow & 0 \\
& & \Big\downarrow{\scriptstyle g} & & \Big\downarrow{\scriptstyle h} & & \\
& & M = A \otimes_A M & \longrightarrow & A_n \otimes_A M & &
\end{array}
$$

We may interpret h as the map obtained from $N \hookrightarrow A_n$ by tensoring with M_n over A_n. Therefore, due to our flatness assumption, h is injective, and this implies $\ker g \subset I^m(\mathfrak{a} \otimes_A M)$. Now, as a finitely generated B-module, $\mathfrak{a} \otimes_A M$ is I-adically separated by Proposition 8. Thus, varying m, we get

$$
\ker g \subset \bigcap_{m=0}^{\infty} I^m(\mathfrak{a} \otimes_A M) = 0,
$$

which shows that M is a flat A-module.

Next, assume for all $n \in \mathbb{N}$ that M_n is a faithfully flat A_n-module, and let N be a finitely generated A-module such that $M \otimes_A N = 0$. Then $M_n \otimes_{A_n} N_n = 0$ for all $n \in \mathbb{N}$ and, consequently, $N_n = 0$. In particular, we get $N = IN$ and, hence $N = 0$, as N is I-adically separated; see Proposition 8. $\square$

Corollary 12. *Let A be an R-algebra that is topologically of finite type, and let $f_1, \ldots, f_r \in A$ be elements generating the unit ideal. Then all canonical maps $A \longrightarrow A\langle f_i^{-1} \rangle$ are flat, and $A \longrightarrow \prod_{i=1}^{r} A\langle f_i^{-1} \rangle$ is faithfully flat.*

Proof. Use the corresponding facts for ordinary localizations in conjunction with Proposition 11. $\square$

Corollary 13. *Let A be an R-algebra that is I-adically complete and separated, and let $f_1, \ldots, f_r \in A$ be elements generating the unit ideal. Then the following are equivalent:*

(i) *A is of topologically finite type (resp. finite presentation, resp. admissible).*
(ii) *$A\langle f_i^{-1} \rangle$ is of topologically finite type (resp. finite presentation, resp. admissible) for each i.*

Proof. The assertion on "finite type" and "finite presentation" follows from the corresponding fact on ordinary localizations in conjunction with Proposition 10.

To extend the equivalence between (i) and (ii) to the condition "admissible", let $I = (g_1, \ldots, g_r)$. If A is admissible, the map $A \longrightarrow \prod_{j=1}^r A[g_j^{-1}]$ is injective. Tensoring it with $A\langle f_i^{-1} \rangle$, which is flat over A by Corollary 12, we see that $A\langle f_i^{-1} \rangle \longrightarrow \prod_{j=1}^r A\langle f_i^{-1} \rangle [g_j^{-1}]$ is injective and, hence, that $A\langle f_i^{-1} \rangle$ is admissible.

Conversely, assume that all $A\langle f_i^{-1} \rangle$ are admissible. Then consider the commutative diagram:

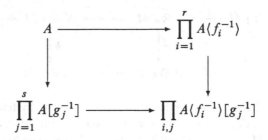

By assumption the right vertical map is injective. As the upper horizontal map is injective anyway due to the faithful flatness, see Corollary 12, the left vertical map must be injective as well.   □

We end by a lemma that will be useful later.

Lemma 14. *Let A be an R-algebra of topologically finite type, B an A-algebra of finite type, and M a finite B-module. Then, if $\widehat{B}$ and $\widehat{M}$ are the I-adic completions of B and M, the canonical map*

$$M \otimes_B \widehat{B} \longrightarrow \widehat{M}$$

is an isomorphism.

Proof. Choose an exact sequence of B-modules

$$0 \longrightarrow N \longrightarrow B^n \xrightarrow{p} M \longrightarrow 0$$

and consider the commutative diagram

having exact rows where the first row is obtained via tensoring with $\hat{B}$ over B and the second one via completion; $\overline{N}$ is the closure of N in $\hat{B}^n$. Since M is a finite B-module, a standard approximation argument shows that $\hat{p}$ and h are surjective. Furthermore, $\overline{N} \subset \ker \hat{p}$ holds by continuity. That, indeed, $\overline{N}$ equals the kernel of $\hat{p}$ is seen as follows. Let $(b_\nu)_{\nu \in \mathbb{N}}$ be a sequence in B^n converging I-adically towards an element $b \in \ker \hat{p} \subset \hat{B}^n$. Then $(p(b_\nu))_{\nu \in \mathbb{N}}$ is a zero sequence in M. Since p is surjective and, hence, satisfies $p(I^\nu B^n) = I^\nu M$, we can find a zero sequence $(b'_\nu)_{\nu \in \mathbb{N}}$ in B^n such that $p(b'_\nu) = p(b_\nu)$ for all ν. But then we have $b_\nu - b'_\nu \in \ker p = N$ for all ν and, hence,

$$b = \lim_{n \to \infty} b_\nu = \lim_{n \to \infty} (b_\nu - b'_\nu) \in \overline{N}.$$

Now use the fact that B is an A-algebra of finite type and A an R-algebra of topologically finite type. From this we may conclude using Proposition 10 that $\hat{B}$ is an R-algebra of topologically finite type. Then we see from Proposition 8 that any submodule $L \subset \hat{B}^n$ is closed, since $\hat{B}^n/L$ is I-adically separated. In particular, the image of $N \otimes_B \hat{B}$ in $\hat{B}^n$ is closed and therefore equals $\overline{N}$, since it must contain the image of N. Thus, $f \colon N \otimes_B \hat{B} \longrightarrow \overline{N}$ is surjective, and it follows by diagram chase that h is injective. Hence, being surjective as well, h is bijective.   □

7.4 Admissible Formal Schemes

Let A be an R-algebra that is I-adically complete and separated. We have seen in 7.3/13 that the condition of A being of topologically finite type, of topologically finite presentation, or admissible, can be tested locally on localizations of type $A\langle f^{-1} \rangle$. This enables us to extend these notions to formal R-schemes.

Definition 1. *Let X be a formal R-scheme. X is called locally of topologically finite type (resp. locally of topologically finite presentation, resp. admissible) if there is an open affine covering $(U_i)_{i \in J}$ of X with $U_i = \operatorname{Spf} A_i$ where A_i is an R-algebra of topologically finite type (resp. of topologically finite presentation, resp. an admissible R-algebra).*

As an immediate consequence we get from 7.3/13:

Remark 2. *Let A be an R-algebra that is I-adically complete and separated, and let $X = \operatorname{Spf} A$ be the associated formal R-scheme. Then the following are equivalent:*

(i) *X is locally of topologically finite type (resp. locally of topologically finite presentation, resp. admissible).*

(ii) *A is of topologically finite type (resp. of topologically finite presentation, resp. admissible) as R-algebra.*

Similarly as in the scheme case, a formal R-scheme X is called of *topologically finite type* if it is locally of topologically finite type and quasi-compact. It is called of *topologically finite presentation* if it is locally of topologically finite presentation, quasi-compact, and quasi-separated. Recall that X is called *quasi-separated* if the diagonal embedding $X \longrightarrow X \times_R X$ is quasi-compact. If X is locally of topologically finite type, the quasi-separateness of X is automatic if R is Noetherian, since X, as a topological space, is locally Noetherian then. The same is true for R a complete valuation ring of height 1. Indeed, if A is an R-algebra of topologically finite type and $\mathfrak{m}$ is the maximal ideal of R, then, as a topological space, Spf A coincides with $\mathrm{Spec}(A \otimes_R R/\mathfrak{m})$. Since $A \otimes_R R/\mathfrak{m}$ is of finite type over the field $R/\mathfrak{m}$, its spectrum $\mathrm{Spec}(A \otimes_R R/\mathfrak{m})$ is a Noetherian space.

Let X be a formal R-scheme that is locally of topologically finite type, and let $\mathcal{O}_X$ be its structure sheaf. Then we can look at the ideal $\mathcal{J} \subset \mathcal{O}_X$ representing the I-torsion of $\mathcal{O}_X$ where $\mathcal{J}(U)$, for any open subset $U \subset X$, consists of all sections $f \in \mathcal{O}_X(U)$ such that there is an open affine covering $(U_\lambda)_{\lambda \in \Lambda}$ of U with the property that each restriction $f|_{U_\lambda}$ is killed by some power I^n of the ideal of definition $I \subset R$. It is clear from the definition that $\mathcal{J}$ really is an ideal sheaf in $\mathcal{O}_X$. Furthermore, if $U \subset X$ is an affine open formal subscheme, say $U = \mathrm{Spf}\, A$, then one gets

$$\mathcal{J}(U) = (I\text{-torsion})_A = \{ f \in A \, ; \, I^n f = 0 \text{ for some } n \in \mathbb{N} \}.$$

Indeed, we clearly have $(I\text{-torsion})_A \subset \mathcal{J}(U)$, and the quotient $A/(I\text{-torsion})_A$ does not have I-torsion locally on Spf A, due to 7.3/13. In particular, we can replace the structure sheaf $\mathcal{O}_X$ by the quotient $\mathcal{O}_X/\mathcal{J}$ and restrict X to the support X_{ad} of $\mathcal{O}_X/\mathcal{J}$. Thereby we get a formal R-scheme X_{ad} that is still locally of topologically finite type and whose structure sheaf does not have I-torsion. Then X_{ad} is locally of topologically finite presentation by 7.3/5 and, thus, admissible. We call X_{ad} the *admissible formal R-scheme induced from X*.

For a moment, let us look at the classical rigid case where R consists of a complete valuation ring of height 1 with field of fractions K. To simplify our terminology, let us assume in the following that all formal R-schemes are at least *locally of topologically finite type*, unless stated otherwise. We want to define a functor "rig" from the category of formal R-schemes to the category of rigid K-spaces, which will be interpreted as associating to a formal R-scheme X its generic fiber X_{rig}. On affine formal R-schemes Spf A this functor is defined by

$$\mathrm{rig} \colon X = \mathrm{Spf}\, A \longmapsto X_{\mathrm{rig}} = \mathrm{Sp}(A \otimes_R K)$$

where we claim that $A \otimes_R K$ is an affinoid K-algebra. To justify this claim, we set $S = R - \{0\}$ and interpret $A \otimes_R K$ as the localization $S^{-1}A$. By our assumption, A is of topologically finite type and, thus, isomorphic to a quotient $R\langle \zeta \rangle / \mathfrak{a}$ where $R\langle \zeta \rangle$ is an algebra of restricted power series in finitely many variables $\zeta = (\zeta_1, \ldots, \zeta_n)$ and where $\mathfrak{a}$ is an ideal in $R\langle \zeta \rangle$. Since $A \otimes_R K = S^{-1}(R\langle \zeta \rangle)/(\mathfrak{a})$, it is enough to show $S^{-1}(R\langle \zeta \rangle) = K\langle \zeta \rangle$. However, the latter is clear by looking at the canonical inclusions

$$R\langle\zeta\rangle \subset S^{-1}\big(R\langle\zeta\rangle\big) \subset K\langle\zeta\rangle$$

and by observing that, for any series $f = \sum_{\nu\in\mathbb{N}^n} c_\nu \zeta^\nu \subset K\langle\zeta\rangle$ with coefficients $c_\nu \in K$, there is a constant $s \in S$ such that $s^{-1} f$ has coefficients in R, due to the fact that $\lim c_\nu = 0$. Thus $A \otimes_R K$ really is an affinoid K-algebra and, for any affine formal R-scheme $X = \mathrm{Spf}\, A$, the corresponding rigid K-space $X_{\mathrm{rig}} = \mathrm{Sp}\, A \otimes_R K$ is well-defined.

Next, if $\varphi\colon \mathrm{Spf}\, A \longrightarrow \mathrm{Spf}\, B$ is a morphism of affine formal R-schemes, we know from Sect. 7.2, as explained just before 7.2/3, that it is induced from a unique R-homomorphism $\varphi^*\colon B \longrightarrow A$. Then, by 5.3/2, the corresponding generic fiber

$$\varphi^*_{\mathrm{rig}}\colon B \otimes_R K \longrightarrow A \otimes_R K$$

determines a well-defined morphism of affinoid K-spaces

$$\varphi_{\mathrm{rig}}\colon \mathrm{Sp}(A \otimes_R K) \longrightarrow \mathrm{Sp}(B \otimes_R K),$$

which we define as the image of φ under the functor rig. Furthermore, let us observe that this functor commutes with complete localization. Indeed, for any R-algebra of topologically finite type A and any $f \in A$ we get

$$A\langle f^{-1}\rangle \otimes_R K = \big[A\langle\zeta\rangle/(1 - f\zeta)\big] \otimes_R K$$
$$= (A \otimes_R K)\langle\zeta\rangle/(1 - f\zeta) = (A \otimes_R K)\langle f^{-1}\rangle$$

where we have used 7.1/10 in conjunction with the fact that $A\langle\zeta\rangle \otimes_R K$ coincides with $(A \otimes_R K)\langle\zeta\rangle$; the latter is justified, similarly as above, by interpreting $[A\langle\zeta\rangle/(1 - f\zeta)] \otimes_R K$ as a localization of $A\langle\zeta\rangle/(1 - f\zeta)$ and by representing A as a quotient of an R-algebra of restricted power series by some ideal. Then we get a canonical commutative diagram

$$
\begin{array}{ccc}
A & \longrightarrow & A \otimes_R K \\
\downarrow & & \downarrow \\
A\langle f^{-1}\rangle & \longrightarrow & (A \otimes_R K)\langle f^{-1}\rangle
\end{array}
$$

showing that the functor rig produces from a basic open subspace of type

$$X(f^{-1}) = \mathrm{Spf}\, A\langle f^{-1}\rangle \subset X = \mathrm{Spf}\, A$$

the Laurent domain

$$X_{\mathrm{rig}}(f^{-1}) = \mathrm{Sp}(A \otimes_R K)\langle f^{-1}\rangle \subset X_{\mathrm{rig}} = \mathrm{Sp}(A \otimes_R K)$$

of the generic fiber associated to X. More generally, it follows that rig maps
any open immersion of affine formal R-schemes Spf A' $\longrightarrow$ Spf A to an open
immersion of affinoid K-spaces $\mathrm{Sp}(A' \otimes_R K)$ $\longrightarrow$ $\mathrm{Sp}(A \otimes_R K)$.

Now, to extend the functor rig to global formal R-schemes, let us look at such
a scheme X and assume first that X is separated and, hence, that the intersection
of two open affine formal subschemes of X is affine again. Fixing an open affine
covering $(U_i)_{i \in J}$ of X, all intersections $U_i \cap U_j$ are affine again. Hence, we
can glue the generic fibers $U_{i,\mathrm{rig}}$ via the "intersections" $(U_i \cap U_j)_{\mathrm{rig}}$ to produce a
global rigid K-space X_{rig}. It is easily checked that the latter is independent (up
to canonical isomorphism) of the chosen affine open covering $(U_i)_{i \in J}$ and that
any morphism of separated formal R-schemes X $\longrightarrow$ Y leads to a canonical
morphism X_{rig} $\longrightarrow$ Y_{rig} so that we really get a functor. In particular, as affine
formal R-schemes are separated, the functor rig is defined on all open formal
subschemes U of an affine formal R-scheme X. Furthermore, since such a U
is necessarily quasi-compact, the generic fiber U_{rig} is admissible open and, thus,
an open subspace of X_{rig}. Therefore, to extend the functor rig to the category of
all formal R-schemes, we can repeat the above construction, now interpreting an
arbitrary global formal R-scheme X by gluing open affine parts U_i via arbitrary
open subspaces of these. Hence, we have shown:

Proposition 3. *Let R be a complete valuation ring of height 1 with field of fractions
K. Then the functor $A \longmapsto A \otimes_R K$ on R-algebras A of topologically finite type
gives rise to a functor $X \longmapsto X_{\mathrm{rig}}$ from the category of formal R-schemes that are
locally of topologically finite type, to the category of rigid K-spaces.*

As indicated above, X_{rig} is called the *generic fiber* of the formal R-scheme X. In
an affine situation, say $X = \mathrm{Spf}\, A$, it coincides pointwise with the set of all closed
points of $\mathrm{Spec}(A \otimes_R K)$, the latter being the generic fiber of the ordinary scheme
$\mathrm{Spec}\, A$. This way the *generic fiber* of the formal scheme $\mathrm{Spf}\, A$ can be exhibited,
although, on the level of points, it is not visible in $\mathrm{Spf}\, A$.

In view of Proposition 3, one would like to describe all formal R-schemes X
whose generic fiber X_{rig} coincides with a given rigid K-space X_K. To answer this
question, observe first that the functor $X \longmapsto X_{\mathrm{rig}}$ factors through the category
of admissible formal R-schemes, since the tensor product with K over R kills any
R-torsion. In particular, the generic fiber of a formal R-scheme X coincides with
the one of its induced admissible formal R-scheme X_{ad}. Thus, we are reduced to the
problem of describing all *admissible* formal R-schemes X admitting a given rigid
K-space X_K as generic fiber. Such formal schemes will be referred to as formal
R-models:

Definition 4. *Given a rigid K-space X_K, any admissible formal R-scheme X
satisfying $X_{\mathrm{rig}} \simeq X_K$ is called a formal R-model of X_K.*

Thus, our problem consists in determining all formal R-models of a given rigid K-space X_K. To solve it, the notion of *admissible formal blowing-up*, which will be introduced in the Sect. 8.2, plays a central role.

Chapter 8
Raynaud's View on Rigid Spaces

8.1 Coherent Modules

Now, let us return to the general situation where R is an adic ring of type (V) or (N), with a finitely generated ideal of definition I. So R is a Noetherian adic ring or an adic valuation ring with a finitely generated ideal of definition.

Let A be an R-algebra of topologically finite type and $X = \operatorname{Spf} A$ the associated formal R-scheme. There is a functor $M \longmapsto M^{\Delta}$ that associates to any A-module M an $\mathcal{O}_X$-module M^{Δ} as follows: for a basic open subset $D_f = D(f) \subset X$, given by some $f \in A$, set

$$M^{\Delta}(D_f) = \varprojlim_{n \in \mathbb{N}} M \otimes_A A_n[f^{-1}]$$

where, as usual, $A_n = A/I^{n+1}A$. As $\varprojlim$ is left-exact, we get a sheaf which can be extended to all open subsets of X by the usual procedure. In fact we may say that M^{Δ} is the inverse limit of the modules $\widetilde{M}_n$ where the latter are the modules induced on $X_n = \operatorname{Spec} A_n$ from the A_n-modules $M_n = M \otimes_A A_n$. If M is a finite A-module, the sheaf M^{Δ} can be described in more convenient terms:

Proposition 1. *Let $X = \operatorname{Spf} A$ be a formal R-scheme of topologically finite type. Then, for any finite A-module M, the sheaf M^{Δ} coincides on basic open subsets $D_f \subset X$, $f \in A$, with the functor*

$$D_f \longmapsto M \otimes_A A\langle f^{-1}\rangle.$$

Proof. Since $A\langle f^{-1}\rangle$ is an R-algebra of topologically finite type, see 7.1/10 or 7.3/13, we know from 7.3/8 that $M \otimes_A A\langle f^{-1}\rangle$, which is a finite $A\langle f^{-1}\rangle$-module, is I-adically complete and separated. By the definition of $M^{\Delta}(D_f)$, we may view it as the I-adic completion of $M \otimes_A A[f^{-1}]$. However, since the latter is dense in $M \otimes_A A\langle f^{-1}\rangle$, we are done. $\square$

S. Bosch, *Lectures on Formal and Rigid Geometry*, Lecture Notes
in Mathematics 2105, DOI 10.1007/978-3-319-04417-0_8,
© Springer International Publishing Switzerland 2014

Corollary 2. *Let* $X = \mathrm{Spf}\, A$ *be a formal R-scheme of topologically finite type.*

(i) *The functor* $M \longmapsto M^{\Delta}$ *from the category of finite A-modules to the category of* $\mathcal{O}_X$*-modules is fully faithful and exact.*

(ii) *Assume that* X *is of topologically finite presentation and, hence by 7.3/6, that* A *is coherent. Then the functor* $M \longmapsto M^{\Delta}$ *commutes on the category of coherent A-modules with the formation of kernels, images, cokernels, and tensor products. Furthermore, a sequence of coherent A-modules*

$$0 \longrightarrow M' \longrightarrow M \longrightarrow M'' \longrightarrow 0$$

is exact if and only if the associated sequence of $\mathcal{O}_X$*-modules*

$$0 \longrightarrow M'^{\Delta} \longrightarrow M^{\Delta} \longrightarrow M''^{\Delta} \longrightarrow 0$$

is exact.

Proof. We use the same argument as the one given in 6.1/1. First, it is clear that the canonical map

$$\mathrm{Hom}_A(M, M') \longrightarrow \mathrm{Hom}_{\mathcal{O}_X}(M^{\Delta}, M'^{\Delta})$$

is bijective, since an $\mathcal{O}_X$-morphism $M^{\Delta} \longrightarrow M'^{\Delta}$ is uniquely determined by its inherent A-morphism between $M = M^{\Delta}(X)$ and $M' = M'^{\Delta}$. Next, if

$$0 \longrightarrow M' \longrightarrow M \longrightarrow M'' \longrightarrow 0$$

is an exact sequence of finite A-modules, then, for all $f \in A$, the associated sequence of $A\langle f^{-1}\rangle$-modules

$$0 \longrightarrow M' \otimes_A A\langle f^{-1}\rangle \longrightarrow M \otimes_A A\langle f^{-1}\rangle \longrightarrow M'' \otimes_A A\langle f^{-1}\rangle \longrightarrow 0$$

is exact, since $A\langle f^{-1}\rangle$ is flat over A by 7.3/12. Thus, the sequence

$$0 \longrightarrow M'^{\Delta} \longrightarrow M^{\Delta} \longrightarrow M''^{\Delta} \longrightarrow 0$$

is exact, showing that the functor $M \longmapsto M^{\Delta}$ is exact.

Now, let us consider the situation of (ii) and assume that X is of topologically finite presentation. Then A is coherent by 7.3/6, and the same is true for any finite A-module. If $M \longrightarrow N$ is a morphism of coherent A-modules, we know that its kernel, image, and cokernel are coherent again. Thus, we see from assertion (i) that the functor $M \longmapsto M^{\Delta}$ commutes with the formation of these modules. Furthermore, one can conclude from Proposition 1 that it commutes with tensor products.

Finally, look at a sequence of coherent A-modules $M' \xrightarrow{\varphi} M \xrightarrow{\psi} M''$ and assume that the corresponding sequence $M'^{\Delta} \xrightarrow{\varphi^{\Delta}} M^{\Delta} \xrightarrow{\psi^{\Delta}} M''^{\Delta}$ is exact. Then, using the just mentioned compatibility of the functor $M \longmapsto M^{\Delta}$, we get

$$(\ker \psi / \operatorname{im} \varphi)^{\Delta} = (\ker \psi)^{\Delta} / (\operatorname{im} \varphi)^{\Delta} = \ker(\psi^{\Delta}) / \operatorname{im}(\varphi^{\Delta}) = 0$$

and, hence, that $\ker \psi / \operatorname{im} \varphi$ is trivial. □

Next, we want to apply Corollary 2 in order to deal with coherent modules on formal R-schemes. The definition of such modules follows the general concept of coherent sheaves.

Definition 3. *Let X be a formal R-scheme and $\mathcal{F}$ an $\mathcal{O}_X$-module.*

(i) *$\mathcal{F}$ is called of* finite type, *if there exists an open covering $(X_i)_{i \in J}$ of X together with exact sequences of type*

$$\mathcal{O}_X^{s_i}|_{X_i} \longrightarrow \mathcal{F}|_{X_i} \longrightarrow 0, \quad i \in J.$$

(ii) *$\mathcal{F}$ is called of* finite presentation, *if there exists an open covering $(X_i)_{i \in J}$ of X together with exact sequences of type*

$$\mathcal{O}_X^{r_i}|_{X_i} \longrightarrow \mathcal{O}_X^{s_i}|_{X_i} \longrightarrow \mathcal{F}|_{X_i} \longrightarrow 0, \quad i \in J.$$

(iii) *$\mathcal{F}$ is called* coherent, *if $\mathcal{F}$ is of finite type and if for every open subscheme $U \subset X$ the kernel of any morphism $\mathcal{O}_X^s|_U \longrightarrow \mathcal{F}|_U$ is of finite type.*

For an affine formal R-scheme $X = \operatorname{Spf} A$, any power $\mathcal{O}_X^r$ may be viewed as the $\mathcal{O}_X$-module $(A^r)^{\Delta}$ associated to the A-module A^r. Furthermore, if A is of topologically finite presentation, A is coherent by 7.3/6, and we can conclude from Corollary 2 that kernels and cokernels of morphisms of type $\mathcal{O}_X^r \longrightarrow \mathcal{O}_X^s$ are associated to finite A-modules.

Remark 4. *Let X be a formal R-scheme that is locally of topologically finite presentation, and let $\mathcal{F}$ be an $\mathcal{O}_X$-module. Then the following are equivalent:*

(i) *$\mathcal{F}$ is coherent.*
(ii) *$\mathcal{F}$ is of finite presentation.*
(iii) *There is an open affine covering $(X_i)_{i \in J}$ of X such that $\mathcal{F}|_{X_i}$ is associated to a finite $\mathcal{O}_{X_i}(X_i)$-module for all $i \in J$.*

Proof. That (i) implies (ii) is immediately clear from the definitions. Next, assume that $\mathcal{F}$ is of finite presentation as in (ii). Then, in order to derive (iii), it is only necessary to consider the case where X is affine, say $X = \operatorname{Spf} A$ with an R-algebra A of topologically finite presentation. In addition, we may assume that there is an exact sequence

$$(A^r)^\Delta \longrightarrow (A^s)^\Delta \longrightarrow \mathcal{F} \longrightarrow 0.$$

Then it follows from Corollary 2 that the morphism $(A^r)^\Delta \longrightarrow (A^s)^\Delta$ corresponds to an A-linear map $A^r \longrightarrow A^s$ and that $\mathcal{F}$ is associated to its cokernel. The latter is a finite A-module so that (ii) implies (iii).

Finally, let $\mathcal{F}$ satisfy condition (iii). To show that $\mathcal{F}$ is coherent, we may assume, similarly as before, that X is affine, say $X = \operatorname{Spf} A$ with A of topologically finite presentation, and that $\mathcal{F}$ is associated to a finite A-module M. Let U be an open subscheme of X and $\varphi: \mathcal{O}_X^s|_U \longrightarrow \mathcal{F}|_U$ a morphism of $\mathcal{O}_X$-modules. To show that $\ker \varphi$ is of finite type, we may assume $U = X$. Then φ is associated to an A-linear map $A^s \longrightarrow M$. Since A is coherent by 7.3/6, the kernel of this map is, in particular, of finite type, and the same is true for its associated $\mathcal{O}_X$-module. As the latter coincides with $\ker \varphi$, we are done. $\square$

Just as in the scheme case or in the case of rigid K-spaces, one may ask if coherent modules on affine formal R-schemes $X = \operatorname{Spf} A$ are associated to coherent A-modules.

Proposition 5. *Let $X = \operatorname{Spf} A$ be an affine formal R-scheme of topologically finite presentation and let $\mathcal{F}$ be a coherent $\mathcal{O}_X$-module. Then $\mathcal{F}$ is associated to a coherent A-module M.*

Proof. There is a covering of X by basic open affine subschemes $U_i = \operatorname{Spf} A_i$, with i varying in a finite index set J, such that $\mathcal{F}|_{U_i}$ is associated to a coherent A_i-module M_i. Set $U_{ij} = U_i \cap U_j$ and let $U_{ij} = \operatorname{Spf} A_{ij}$. Then $\mathcal{F}|_{U_{ij}}$ is associated to the coherent A_{ij}-module $M_{ij} = M_i \otimes_{A_i} A_{ij} = M_j \otimes_{A_j} A_{ij}$.

Now observe that $\mathcal{F}$ induces for each $n \in \mathbb{N}$ a coherent module $\mathcal{F}_n$ on the scheme $X_n = \operatorname{Spec} A_n$ where, as usual, $A_n = A/I^{n+1}A$. This module sheaf is constructed by gluing the $\mathcal{O}_{U_{i,n}}$-modules that are associated to the coherent $A_{i,n}$-modules $M_{i,n} = M_i/I^{n+1}M_i$, $i \in J$. Then we can use the fact that $\mathcal{F}_n$ is associated to a coherent A_n-module M_n, thereby getting exact diagrams of type

$$M_n \longrightarrow \prod_{i \in J} M_{i,n} \rightrightarrows \prod_{ij \in J} M_{ij,n}, \qquad n \in \mathbb{N}.$$

Since $\mathcal{F}_n$ is derived from $\mathcal{F}_{n+1}$ via base change with X_{n+1} over X_n, we see that $M_n = M_{n+1} \otimes_{A_{n+1}} A_n$. Taking projective limits, the above diagrams give rise to an exact diagram

$$M \longrightarrow \prod_{i \in J} M_i \rightrightarrows \prod_{ij \in J} M_{ij}$$

where $M = \varprojlim_{n \in \mathbb{N}} M_n$. Let $K^{(n)} \subset M$ for $n \in \mathbb{N}$ be the kernel of the projection $M \longrightarrow M_{n-1}$, setting $M_{-1} = 0$. Then $I^n M \subset K^{(n)}$, and we claim that, in fact,

$K^{(n)} = I^n M$ and that $M = K^{(0)}$ is a finite A-module. Granting these facts, the topology on M, as a projective limit of the M_n, must coincide with the I-adic one, and it follows that the canonical map

$$M \otimes_A A_i = \varprojlim_{n \in \mathbb{N}} (M_n \otimes_{A_n} A_{i,n}) \longrightarrow \varprojlim_{n \in \mathbb{N}} M_{i,n} = M_i$$

is an isomorphism for all $i \in J$. But then $\mathcal{F}$ is associated to the finite and, hence, coherent A-module M.

To justify the above claim, observe that M_0 is a finite A_0-module. Since the projective system $(M_n)_{n \in \mathbb{N}}$ is surjective, there exist finitely many elements $x_1, \ldots, x_r$ in M with the property that their images generate M_0 as an A_0-module. Set $M' = \sum_{\rho=1}^{r} A x_i$ and let M'_n be the image of M' in M_n. Then

$$M_n = M'_n + I M_n, \quad n \in \mathbb{N},$$

and, hence, by finite induction, $M_n = M'_n$ for all n. From this we deduce that

$$K^{(n)} = I^n M' + K^{(n+1)}, \quad n \in \mathbb{N}. \tag{$*$}$$

Indeed, viewing $M \longrightarrow M_{n-1}$ as the composition of the projection $M \longrightarrow M_n$ and the canonical map $M_n \longrightarrow M_{n-1}$, the first map has kernel $K^{(n+1)}$, whereas the second one has kernel $I^n M_n = I^n M'_n$. Thus, the composition has a kernel $K^{(n)}$ equal to $I^n M' + K^{(n+1)}$ as stated. Now fix n and a finite set of generators $y_1, \ldots, y_s$ of $I^n M'$ as A-module. Then applying the equations $(*)$ inductively, we can write any element $z \in K^{(n)}$ as a limit of linear combinations of type $\sum_{\sigma=1}^{s} a_{\sigma v} y_\sigma$, $v \in \mathbb{N}$, with coefficients $a_{\sigma v} \in A$ where the sequences $(a_{\sigma v})_{v \in \mathbb{N}}$ have I-adic limits $a_\sigma \in A$. Since the I-adic topology on M is finer than the projective limit topology, we must have $z = \sum_{\sigma=1}^{s} a_\sigma y_\sigma$ and, thus, $K^{(n)} \subset I^n M'$ for all $n \in \mathbb{N}$. As the opposite inclusion holds anyway, the latter implies $K^{(n)} = I^n M'$. In particular, we see for $n = 0$ that M coincides with M' and therefore is finitely generated. Hence, we get $K^{(n)} = I^n M$, and it follows that the topology of M coincides with the I-adic one. $\qquad \square$

8.2 Admissible Formal Blowing-Up

In the following we will discuss the technique of admissible formal blowing-up on formal R-schemes X, as sort of a completed scheme theoretic blowing-up on the affine open parts of X. In order to control torsion submodules under such a completion process, for example I-torsion submodules, we need an auxiliary flatness result due to Gabber, which we will prove below in Lemma 2. It extends certain results on the flatness of adic completions, as contained in [AC], Chap. III, Sect. 5, no. 4, to the non-Noetherian situations we have to work with. Gabber's

Lemma will be used in the proof of Proposition 7 and is essential for showing that the formal blowing-up of an admissible formal R-scheme yields an admissible formal R-scheme again.

Lemma 1. *Let M be a module over some ring A, and let $\pi \in A$ be an element that is not a zero-divisor in A. Then the following are equivalent:*

(i) *M is flat over A.*
(ii) *The torsion*

$$(\pi\text{-torsion})_M = \{x \in M \; ; \pi^n x = 0 \text{ for some } n \in \mathbb{N}\}$$

of π in M is trivial, $M/\pi M$ is flat over $A/\pi A$, and $M \otimes_A A[\pi^{-1}]$ is flat over $A[\pi^{-1}]$.

Proof. Assume first that M is flat over A. Then the multiplication by π is injective on M, since it is injective on A. Furthermore, the flatness assertions in (ii) for $M/\pi M$ and $M \otimes_A A[\pi^{-1}]$ follow by base change.

Conversely, assume condition (ii). Proceeding step by step, we will show that the Tor modules $\mathrm{Tor}_q^A(M, N)$ are trivial for $q > 0$ and all A-modules N.

(a) Let $N = A/\pi A$. Then the short exact sequence

$$0 \longrightarrow A \overset{\pi}{\longrightarrow} A \longrightarrow N \longrightarrow 0$$

yields a free resolution of N. Tensoring it with M, we obtain the sequence

$$0 \longrightarrow M \overset{\pi}{\longrightarrow} M \longrightarrow M \otimes_A N \longrightarrow 0,$$

which is exact since $(\pi\text{-torsion})_M$ is supposed to be trivial. However, this implies $\mathrm{Tor}_1^A(M, N) = (\pi\text{-torsion})_M = 0$ and, hence, that $\mathrm{Tor}_q^A(M, N) = 0$ for $q > 0$.

(b) Next, assume $\pi N = 0$ and choose a projective resolution P_* of M. Since $\mathrm{Tor}_q^A(M, A/\pi A) = 0$ for $q > 0$ by step (a), the sequence

$$P_* \otimes_A A/\pi A \longrightarrow M/\pi M \longrightarrow 0$$

is seen to be exact. Thus, $P_* \otimes_A A/\pi A$ is a projective resolution of $M/\pi M$. Since $M/\pi M$ is flat over $A/\pi A$ by assumption, we have $\mathrm{Tor}_q^{A/\pi A}(M/\pi M, N) = 0$ for $q > 0$. Hence, the sequence

$$P_* \otimes_A A/\pi A \otimes_{A/\pi A} N \longrightarrow M/\pi M \otimes_{A/\pi A} N \longrightarrow 0$$

is exact. As the latter coincides with the sequence

$$P_* \otimes_A N \longrightarrow M \otimes_A N \longrightarrow 0,$$

it follows that $\operatorname{Tor}_q^A(M, N) = 0$ for $q > 0$.

(c) Assume that $\pi^n N = 0$ for some $n > 1$. We consider the long Tor sequence associated to the short exact sequence

$$0 \longrightarrow \pi N \longrightarrow N \longrightarrow N/\pi N \longrightarrow 0.$$

Since πN and $N/\pi N$ are killed by π^{n-1}, an inductive argument in conjunction with step (b) shows $\operatorname{Tor}_q^A(M, N) = 0$ for $q > 0$.

(d) Assume that $(\pi\text{-torsion})_N = N$, i.e. that each element of N is killed by a power of π. For $n \in \mathbb{N}$, let $N_n = \{x \in N ; \pi^n x = 0\}$. Then $N = \varinjlim_n N_n$. Since the formation of Tor is compatible with direct limits, we can conclude from step (c) that $\operatorname{Tor}_1^A(M, N) = 0$ for $q > 0$.

(e) Assume that $(\pi\text{-torsion})_N = 0$, i.e. that N does not admit π-torsion. Then consider the long Tor sequence associated to the short exact sequence

$$0 \longrightarrow N \longrightarrow N \otimes_A A[\pi^{-1}] \longrightarrow T \longrightarrow 0$$

where T is a π-torsion module, i.e. $(\pi\text{-torsion})_T = T$. Since

$$\operatorname{Tor}_q^A(M, N \otimes_A A[\pi^{-1}]) = \operatorname{Tor}^{A[\pi^{-1}]}(M \otimes_A A[\pi^{-1}], N \otimes_A A[\pi^{-1}]) = 0$$

for $q > 0$ by our assumption, we see from (d) that $\operatorname{Tor}_1^A(M, N) = 0$ for $q > 0$.

Finally, that condition (ii) of the lemma implies the flatness of M over A follows from steps (d) and (e) if we consider the long Tor sequence associated to the short exact sequence

$$0 \longrightarrow (\pi\text{-torsion})_N \longrightarrow N \longrightarrow N/(\pi\text{-torsion})_N \longrightarrow 0.$$

$\square$

Lemma 2 (Gabber). *As in Sect. 7.3, let R be an adic ring of type* (V) *or* (N). *Furthermore, let A be an R-algebra of topologically finite type and C an A-algebra of finite type. Then the I-adic completion $\widehat{C}$ of C is flat over C.*

Proof. If R is of type (N), then A and, hence, C are Noetherian, and the assertion of the lemma is well-known; see [AC], Chap. III, Sect. 5, no. 4, Cor. of Prop. 3. Therefore, we can assume that R is an adic valuation ring of type (V). Let $I = (\pi)$ be an ideal of definition of R.

We start with the special case where $C = A[\zeta]$, with a finite system of variables $\zeta = (\zeta_1, \dots, \zeta_r)$. Then the π-adic completion $\widehat{C}$ of C equals the algebra $A\langle\zeta\rangle$ of restricted power series in ζ with coefficients in A. Furthermore, let us assume that $\widehat{C}$ does not admit π-torsion. Then, by Lemma 1, we have only to show that

$\hat{C} \otimes_R R[\pi^{-1}]$ is flat over $C \otimes_R R[\pi^{-1}]$. Observing that $K = R[\pi^{-1}]$ is a field, we may interpret K as the field of fractions of the valuation ring obtained by localizing R at its minimal non-zero prime ideal $\mathrm{rad}(I) = \mathrm{rad}(\pi)$. Thus, K is the field of fractions of a valuation ring of height 1. Consequently, we may view $A_K = A \otimes_R K$ as an affinoid K-algebra, and we can use the identifications

$$C \otimes_R K = A_K[\zeta], \qquad \hat{C} \otimes_R K = A_K\langle\zeta\rangle,$$

where $A_K\langle\zeta\rangle$ is the K-algebra of strictly convergent power series over A_K in the sense of classical rigid geometry. In order to show that $A_K\langle\zeta\rangle$ is flat over $A_K[\zeta]$, it is enough to show that, for any maximal ideal $\mathfrak{m} \subset A_K\langle\zeta\rangle$ and its restriction $\mathfrak{n}$ to $A_K[\zeta]$, the canonical morphism $A_K[\zeta]_\mathfrak{n} \longrightarrow A_K\langle\zeta\rangle_\mathfrak{m}$ is flat. Since we are dealing with Noetherian rings, see 3.1/3, we may apply the above mentioned result of [AC] and thereby are reduced to showing that the preceding map induces an isomorphism between the $\mathfrak{n}$-adic completion of $A_K[\zeta]_\mathfrak{n}$ and the $\mathfrak{m}$-adic completion of $A_K\langle\zeta\rangle_\mathfrak{m}$. Thus, it is enough to show that the inclusion of $A_K[\zeta]$ into $A_K\langle\zeta\rangle$ induces isomorphisms $A_K[\zeta]/\mathfrak{n}^n \overset{\sim}{\longrightarrow} A_K\langle\zeta\rangle/\mathfrak{m}^n$ for $n \in \mathbb{N}$.

To do this, we proceed similarly as in the proof of 3.3/10. For any maximal ideal $\mathfrak{m} \subset A_K\langle\zeta\rangle$, we know from 3.1/4 that the quotient $A_K\langle\zeta\rangle/\mathfrak{m}$ is of finite vector space dimension over K. Then, being a subspace of $A_K\langle\zeta\rangle/\mathfrak{m}$, the same is true for $A_K[\zeta]/\mathfrak{n}$ and it follows that the latter is a field. Therefore $\mathfrak{n}$ is a maximal ideal in $A_K[\zeta]$. The same argument shows that $\mathfrak{n} \cap A_K$ is a maximal ideal in A_K. From this we can conclude that $\dim_K A_K[\zeta]/\mathfrak{n}^n < \infty$ for all n. Indeed, the restriction $\mathfrak{n}^n \cap A_K$ has radical $\mathfrak{n} \cap A_K$ in A_K, and the latter implies $\dim_K A_K/(\mathfrak{n}^n \cap A_K) < \infty$, again by 3.1/4. Hence, $A_K[\zeta]/\mathfrak{n}^n$ is a K-algebra of finite type, which is local, and it follows from Noether normalization, that $\dim_K A_K[\zeta]/\mathfrak{n}^n < \infty$.

Now look at the following commutative diagram

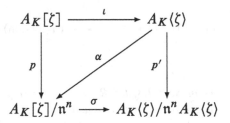

where the square consists of canonical maps and where the map α still has to be explained. Fixing a residue norm on the affinoid K-algebra A_K, we consider on $A_K[\zeta]$ and $A_K\langle\zeta\rangle$ the associated Gauß norms, as well as on the quotients $A_K[\zeta]/\mathfrak{n}^n$ and $A_K\langle\zeta\rangle/\mathfrak{n}^n A_K\langle\zeta\rangle$ the corresponding residue norms. Then all maps of the square are continuous, and $A_K[\zeta]/\mathfrak{n}^n$ is complete, since it is of finite vector space dimension over K and K is complete; use Theorem 1 of Appendix A. Thus, we can extend the projection $p : A_K[\zeta] \longrightarrow A_K[\zeta]/\mathfrak{n}^n$ to a continuous homomorphism $\alpha: A_K\langle\zeta\rangle \longrightarrow A_K[\zeta]/\mathfrak{n}^n$ such that the upper triangle of the diagram is commutative. By a density argument, the lower triangle will be commutative as well. But then the surjectivity of p' implies the surjectivity of σ. Furthermore, since $\ker \alpha$ must contain the ideal generated by $\mathfrak{n}^n$, it follows that σ is injective and, hence, bijective. In particular, we see that $A_K \langle \zeta \rangle / \mathfrak{n} A_K \langle \zeta \rangle$ is a field, since the same is true for $A_K [\zeta] / \mathfrak{n}$, and we get $\mathfrak{m} = \mathfrak{n} A_K \langle \zeta \rangle$, as well as $\mathfrak{m}^n = \mathfrak{n}^n A_K \langle \zeta \rangle$ for all $n \in \mathbb{N}$. Thus, $\sigma \colon A_K [\zeta] / \mathfrak{n}^n \overset{\sim}{\longrightarrow} A_K \langle \zeta \rangle / \mathfrak{m}^n$ is an isomorphism as claimed, settling the assertion of Gabber's Lemma in the special case where $C = A[\zeta]$ for some R-algebra of topologically finite type A that does not admit π-torsion.

In the general case we choose an epimorphism $A' \longrightarrow A$ where A' is an R-algebra of topologically finite type without π-torsion. For example, A' could be an R-algebra of restricted power series over R. The epimorphism can be extended to an epimorphism of type $\gamma \colon A'[\zeta] \longrightarrow C$, since C is of finite type over A. Then, by the above special case, $A'\langle \zeta \rangle$ is flat over $A'[\zeta]$, and we see by base change that $A'\langle \zeta \rangle \otimes_{A'[\zeta]} C$ is flat over C. It remains to exhibit the tensor product as the π-adic completion of C. To do this, let $\mathfrak{a} = \ker \gamma$ so that $C = A'[\zeta]/\mathfrak{a}$ and, hence, $A'\langle \zeta \rangle \otimes_{A'[\zeta]} C = A'\langle \zeta \rangle / \mathfrak{a} A'\langle \zeta \rangle$. Now look at the canonical map $\varphi \colon A'[\zeta]/\mathfrak{a} \longrightarrow A'\langle \zeta \rangle / \mathfrak{a} A'\langle \zeta \rangle$. Tensoring it with $R/(\pi^n)$ over R yields an isomorphism $\varphi \otimes_R R/(\pi^n)$, for any n. Since $A'\langle \zeta \rangle / \mathfrak{a} A'\langle \zeta \rangle$ is an R-algebra of topologically finite type, it is π-adically complete and separated by 7.3/8. It follows that $A'\langle \zeta \rangle / \mathfrak{a} A'\langle \zeta \rangle$ is the π-adic completion of $C = A'[\zeta]/\mathfrak{a}$ and we are done. $\square$

The notion of coherent modules applies, in particular, to ideals in the structure sheaf $\mathcal{O}_X$ of a formal R-scheme X. Such an ideal $\mathcal{A} \subset \mathcal{O}_X$ is called *open*, if locally on X, it contains powers of type $I^n \mathcal{O}_X$. In the following we will always assume that X is a formal R-scheme of *locally of topologically finite presentation* since then, by 8.1/5, a coherent open ideal $\mathcal{A} \subset \mathcal{O}_X$ is associated on any affine open part $\operatorname{Spf} A \subset X$ to a coherent open ideal $\mathfrak{a} \subset A$.

Definition 3. *Let X be a formal R-scheme that is locally of topologically finite presentation and let $\mathcal{A} \subset \mathcal{O}_X$ be a coherent open ideal. Then the formal R-scheme*

$$X_{\mathcal{A}} = \varinjlim_{n \in \mathbb{N}} \operatorname{Proj}\Big(\bigoplus_{d=0}^{\infty} \mathcal{A}^d \otimes_{\mathcal{O}_X} \big(\mathcal{O}_X / I^n \mathcal{O}_X \big) \Big)$$

together with the canonical projection $X_{\mathcal{A}} \longrightarrow X$ is called the formal blowing-up *of $\mathcal{A}$ on X. Any such blowing-up is referred to as an* admissible formal blowing-up *of X.*

To explain the construction of $X_{\mathcal{A}}$ in more detail, let $|X|$ be the topological space underlying the formal scheme X. Then $\mathcal{O}_X / I^n \mathcal{O}_X$ is a sheaf of rings on $|X|$ and the pair $(|X|, \mathcal{O}_X / I^n \mathcal{O}_X)$ may be viewed as an ordinary *scheme* over R or R/I^n. The latter is locally of finite presentation since X is supposed to be locally of topologically finite presentation; see 7.3/10. All schemes $X_n = (|X|, \mathcal{O}_X / I^{n+1} \mathcal{O}_X)$ for $n \in \mathbb{N}$ live on the same topological space $|X|$ and we will write $X = \varinjlim_{n \in \mathbb{N}} X_n$, which means that X consists of the topological space $|X|$ with the inverse limit $\mathcal{O}_X = \lim_{\substack{\longleftarrow \\ n \in \mathbb{N}}} \mathcal{O}_X / I^n \mathcal{O}_X$ as structure sheaf on it.

Next observe that the direct sum

$$\bigoplus_{d=0}^{\infty} \mathcal{A}^d \otimes_{\mathcal{O}_X} \left(\mathcal{O}_X / I^{n+1} \mathcal{O}_X \right)$$

is a quasi-coherent sheaf of graded $\mathcal{O}_{X_n}$-algebras on X_n and, hence, that

$$X_{\mathcal{A},n} = \mathrm{Proj}\left(\bigoplus_{d=0}^{\infty} \mathcal{A}^d \otimes_{\mathcal{O}_X} \left(\mathcal{O}_X / I^{n+1} \mathcal{O}_X \right) \right)$$

is a well-defined scheme over X_n. Since the tensor product commutes with localization and, in particular, homogeneous localization, we obtain

$$X_{\mathcal{A},n} = X_{\mathcal{A},n+1} \times_{X_{n+1}} X_n$$

for $n \in \mathbb{N}$. Thus, all $X_{\mathcal{A},n}$ live on the same topological space, say on $|X_{\mathcal{A}}|$, and the equation

$$X_{\mathcal{A}} = \lim_{\substack{\longrightarrow \\ n \in \mathbb{N}}} X_{\mathcal{A},n}$$

in Definition 3 expresses the fact that $X_{\mathcal{A}}$ consists of the topological space $|X_{\mathcal{A}}|$ with $\mathcal{O}_{X_{\mathcal{A}}} = \lim_{\substack{\longleftarrow \\ n \in \mathbb{N}}} \mathcal{O}_{X_{\mathcal{A},n}}$ as structure sheaf on it. That $\mathcal{O}_{X_{\mathcal{A}}}$ really is a sheaf follows along the lines of Sect. 7.2 from the fact that $\lim_{\substack{\longleftarrow}}$ is left exact. Also note that the structural morphisms $X_{\mathcal{A},n} \longrightarrow X_n$ give rise to a canonical morphism of formal R-schemes $X_{\mathcal{A}} \longrightarrow X$. As a caveat, let us point out that the components $\mathcal{A}^d \otimes_{\mathcal{O}_X} \left(\mathcal{O}_X / I^{n+1} \mathcal{O}_X \right)$ for $d \in \mathbb{N}$ cannot generally be viewed as powers of an ideal in $\mathcal{O}_X / I^{n+1} \mathcal{O}_X$. This is a clear hint for the fact that $X_{\mathcal{A},n}$ is not to be interpreted as a scheme theoretic blowing-up on X_n.

If X is affine, say $X = \mathrm{Spf}\, A$, an ideal $\mathcal{A} \subset \mathcal{O}_X$ is coherent open if and only if it is associated to a coherent open ideal $\mathfrak{a} \subset A$, see 8.1/5, where coherent may be replaced by finitely generated as A is a coherent ring by 7.3/6. Furthermore, if $\mathcal{A}$ is associated to the ideal $\mathfrak{a} \subset A$, the definition of $X_{\mathcal{A}}$ amounts to

$$X_{\mathcal{A}} = \lim_{\substack{\longrightarrow \\ n \in \mathbb{N}}} \mathrm{Proj}\left(\bigoplus_{d=0}^{\infty} \mathfrak{a}^d \otimes_R \left(R / I^n \right) \right).$$

It is easily deduced from this fact that admissible formal blowing-ups of coherent open ideals on formal R-schemes of locally topologically finite presentation yield formal R-schemes that are locally of topologically finite *type*; for example, this will be a consequence of Proposition 6 below. However, we are not able to show that admissible formal blowing-up maintains the property of a formal R-scheme to be locally of topologically finite *presentation*. The latter property will only come in via 7.3/5, when we blow up admissible formal R-schemes and show that the blowing-up does not admit I-torsion; see Proposition 7 and Corollary 8 below.

We want to establish some basic properties of admissible formal blowing-up. Let us call a morphism of formal R-schemes of topologically finite type $\varphi\colon X' \longrightarrow X$ *flat* if for every affine open part $U \subset X$ and every affine open part $U' \subset X'$ where $\varphi(U') \subset U$, the inherent morphism of R-algebras $\mathcal{O}_X(U) \longrightarrow \mathcal{O}_{X'}(U')$ is flat. It is easily checked using 7.3/11 that for φ to be flat it is enough to find affine open coverings $(U_i)_{i \in I}$ of X and $(U_i')_{i \in I}$ of X' such that $\varphi(U_i') \subset U_i$ and the attached morphisms of R-algebras $\mathcal{O}_X(U_i) \longrightarrow \mathcal{O}_{X'}(U_i')$ are flat for all $i \in I$. Also it is possible to characterize the flatness of a morphism φ in the usual way via the flatness of the local maps between stalks of structure sheaves.

Proposition 4. *Admissible formal blowing-up commutes with flat base change.*

Proof. It is enough to consider a situation where X is affine, say $X = \mathrm{Spf}\, A$, and where $\mathcal{A}$ is associated to a finitely generated open ideal $\mathfrak{a} \subset A$. Then

$$X_{\mathcal{A}} = \varinjlim_{n \in \mathbb{N}} \mathrm{Proj}\Big(\bigoplus_{d=0}^{\infty} \mathfrak{a}^d \otimes_R \left(R/I^n\right)\Big).$$

Now consider a base change morphism $\varphi\colon X' \longrightarrow X$ where we may assume X' to be affine, too, say $X' = \mathrm{Spf}\, A'$ with an R-algebra A' of topologically finite presentation. Then

$$X_{\mathcal{A}} \times_X X' = \varinjlim_{n \in \mathbb{N}} \mathrm{Proj}\Big(\bigoplus_{d=0}^{\infty} \mathfrak{a}^d \otimes_A A' \otimes_R \left(R/I^n\right)\Big).$$

If A' is flat over A, the canonical map $\mathfrak{a}^d \otimes_A A' \longrightarrow \mathfrak{a}^d A'$ is an isomorphism and, hence,

$$X_{\mathcal{A}} \times_X X' = \varinjlim_{n \in \mathbb{N}} \mathrm{Proj}\Big(\bigoplus_{d=0}^{\infty} (\mathfrak{a}A')^d \otimes_R \left(R/I^n\right)\Big)$$

equals the admissible blowing-up of the coherent open ideal $\mathcal{A}\mathcal{O}_{X'} \subset \mathcal{O}_{X'}$ on X'. Note that the same argument works if A' is replaced by a complete adic ring R' of type (V) or (N) over R such that IR' is an ideal of definition of R'. $\qquad\square$

In particular, it follows that the notion of admissible formal blowing-up is local on the base (although this can just as well be deduced directly from Definition 3, without the intervention of Proposition 4):

Corollary 5. *Let X be a formal scheme that is locally of topologically finite presentation, and let $\mathcal{A} \subset \mathcal{O}_X$ be a coherent open ideal. Then, for any open formal subscheme $U \subset X$, the restriction $X_{\mathcal{A}} \times_X U$ of the formal blowing-up $X_{\mathcal{A}}$ of $\mathcal{A}$ on X to U coincides with the formal blowing-up of the coherent open ideal $\mathcal{A}|_U \subset \mathcal{O}_U$ on U.*

Next, we want to relate admissible formal blowing-up to scheme theoretic blowing-up.

Proposition 6. *Let $X = \operatorname{Spf} A$ be an affine formal R-scheme of topologically finite presentation. Furthermore, let $\mathcal{A} = \mathfrak{a}^{\Delta}$ be a coherent open ideal in $\mathcal{O}_X$ that is associated to a coherent open ideal $\mathfrak{a} \subset A$. Then the formal blowing-up $X_{\mathcal{A}}$ equals the I-adic completion of the scheme theoretic blowing-up $(\operatorname{Spec} A)_{\mathfrak{a}}$ of $\mathfrak{a}$ on $\operatorname{Spec} A$. In other words, it equals the formal completion of $(\operatorname{Spec} A)_{\mathfrak{a}}$ along its subscheme defined by the ideal $IA \subset A$.*

Proof. The scheme theoretic blowing-up of $\mathfrak{a}$ on the affine scheme $\operatorname{Spec} A$ is given by

$$P = \operatorname{Proj}\left(\bigoplus_{d=0}^{\infty} \mathfrak{a}^d\right).$$

Since tensoring with R/I^n over R for $n \in \mathbb{N}$ is compatible with localization and, in particular, homogeneous localization of $\bigoplus_{d=0}^{\infty} \mathfrak{a}^d$, the I-adic completion of P is

$$\widehat{P} = \varinjlim_{n \in \mathbb{N}}(P \otimes_R R/I^n) = \varinjlim_{n \in \mathbb{N}} \operatorname{Proj}\left(\bigoplus_{d=0}^{\infty} \mathfrak{a}^d \otimes_R R/I^n\right)$$

and, thus, coincides with the formal blowing-up of $\mathcal{A}$ on X. $\square$

Relying on this result, we can describe admissible formal blowing-ups in quite precise terms, at least when X is admissible.

Proposition 7. *Let $X = \operatorname{Spf} A$ be an admissible formal R-scheme that is affine, and let $\mathcal{A} = \mathfrak{a}^{\Delta}$ be a coherent open ideal in $\mathcal{O}_X$ associated to a coherent open ideal $\mathfrak{a} = (f_0, \ldots, f_r) \subset A$. Then the following assertions hold for the formal blowing-up $X_{\mathcal{A}}$ of $\mathcal{A}$ on X:*

(i) *The ideal $\mathcal{A}\mathcal{O}_{X_{\mathcal{A}}} \subset \mathcal{O}_{X_{\mathcal{A}}}$ is invertible, i.e., in terms of $\mathcal{O}_{X_{\mathcal{A}}}$-modules, it is locally isomorphic to $\mathcal{O}_{X_{\mathcal{A}}}$.*

(ii) *Let U_i be the locus in $X_{\mathcal{A}}$ where $\mathcal{A}\mathcal{O}_{X_{\mathcal{A}}}$ is generated by f_i, $i = 0, \ldots, r$. Then the U_i define an open affine covering of $X_{\mathcal{A}}$.*

(iii) *Write*

$$C_i = A\left\langle \frac{f_j}{f_i} ; j \neq i \right\rangle = A\langle \zeta_j ; j \neq i \rangle \big/ (f_i \zeta_j - f_j ; j \neq i).$$

Then the I-torsion of C_i coincides with its f_i-torsion, and $U_i = \operatorname{Spf} A_i$ holds for $A_i = C_i / (I\text{-torsion})_{C_i}$.

Proof. Viewing $S = \bigoplus_{d=0}^{\infty} \mathfrak{a}^d$ as a graded ring, the scheme theoretic blowing-up of $\mathfrak{a}$ on $\widetilde{X} = \operatorname{Spec} A$ is given by

$$\widetilde{X}' = \operatorname{Proj} S = \operatorname{Proj} \bigoplus_{d=0}^{\infty} \mathfrak{a}^d.$$

The latter admits the canonical open covering $\widetilde{X}' = \bigcup_{i=0}^{r} D_+(f_i)$ with $D_+(f_i)$ the open set of all homogeneous prime ideals in S where f_i, viewed as a homogeneous element of degree 1 in $\mathfrak{a}^1 \subset S$, does not vanish. One knows that $D_+(f_i)$ is equipped with the structure of an affine open subscheme of $\operatorname{Proj} S$, namely $D_+(f_i) = \operatorname{Spec} S_{(f_i)}$ where $S_{(f_i)}$ is the homogeneous localization of S by f_i, i.e. the degree 0 part of the ordinary localization S_{f_i} of S by f_i.

The ideal $\mathfrak{a} \subset A$ induces an invertible ideal $\mathfrak{a}\mathcal{O}_{\widetilde{X}'}$ on $\widetilde{X}' = \operatorname{Proj} S$, since for any i, the ideal $\mathfrak{a} S_{(f_i)} \subset S_{(f_i)}$ is generated by f_i and the latter is not a zero divisor in $S_{(f_i)}$. Furthermore, from the construction of $\operatorname{Proj} S$ one knows that $D_+(f_i)$ coincides precisely with the locus in $\widetilde{X}'$ where the ideal $\mathfrak{a}\mathcal{O}_{\widetilde{X}'}$ is generated by f_i.

Now observe that the formal blowing-up $X_{\mathcal{A}}$ of $\mathcal{A}$ on X is covered by the I-adic completions $\operatorname{Spf} \widehat{S}_{(f_i)}$ of the affine schemes $D_+(f_i) = \operatorname{Spec} S_{(f_i)}$. Since $\widehat{S}_{(f_i)}$ is flat over $S_{(f_i)}$ by the Lemma of Gabber (Lemma 2), the ideal $\mathfrak{a}\widehat{S}_{(f_i)} \subset \widehat{S}_{(f_i)}$ is invertible. Thus, $\mathcal{A}\mathcal{O}_{X_{\mathcal{A}}}$ is an invertible ideal on $X_{\mathcal{A}}$, which settles assertion (i). Furthermore, (ii) follows from the fact that, in terms of sets, U_i is the restriction of $D_+(f_i)$ to $X_{\mathcal{A}}$. In fact, $U_i = \operatorname{Spf} \widehat{S}_{(f_i)}$. Thus, it remains to verify assertion (iii) for $A_i = \widehat{S}_{(f_i)}$.

To do this, we give a more specific description of $S_{(f_i)}$. Choose variables $\zeta_0, \ldots, \zeta_r$ and, for each i, look at the canonical epimorphism

$$A[\zeta_j ; j \neq i] \longrightarrow S_{(f_i)} \subset S_{f_i}, \qquad \zeta_j \longmapsto \frac{f_j}{f_i}.$$

The latter factors through the quotient

$$\widetilde{C}_i = A\left[\frac{f_j}{f_i} ; j \neq i \right] = A[\zeta_j ; j \neq i] \big/ (f_i \zeta_j - f_j ; j \neq i),$$

and it is easily seen that it induces an isomorphism

$$\widetilde{C}_i / (f_i\text{-torsion}) \overset{\sim}{\longrightarrow} S_{(f_i)},$$

since $S_{(f_i)}$, due to its nature as a localization by f_i, does not admit f_i-torsion.

Being an open ideal, $\mathfrak{a}$ contains a power of I. Thus, since $\mathfrak{a}\tilde{C}_i$ is generated by f_i, we must have

$$(f_i\text{-torsion})_{\tilde{C}_i} \subset (I\text{-torsion})_{\tilde{C}_i}.$$

Since X is admissible, A and, hence, the graded ring $S = \bigoplus_{d=0}^{\infty} \mathfrak{a}^d$, as well as its homogeneous localizations $S_{(f_i)}$ do not have I-torsion. Therefore the preceding inclusion must be an equality:

$$(f_i\text{-torsion})_{\tilde{C}_i} = (I\text{-torsion})_{\tilde{C}_i}$$

Now let us pass to the I-adic completion C_i of $\tilde{C}_i$. Applying 7.3/14, we see that $C_i = \tilde{C}_i \otimes_{A[\zeta_j ; j \neq i]} A\langle \zeta_j ; j \neq i\rangle$ and, hence, that

$$C_i = A\left\langle \frac{f_j}{f_i} ; j \neq i \right\rangle = A\langle \zeta_j ; j \neq i\rangle / (f_i\zeta_j - f_j ; j \neq i).$$

By the Lemma of Gabber (Lemma 2), the I-adic completion C_i of $\tilde{C}_i$ is flat over $\tilde{C}_i$. This implies that

$$(I\text{-torsion})_{C_i} = (I\text{-torsion})_{\tilde{C}_i} \otimes_{\tilde{C}_i} C_i$$

and, likewise,

$$(f_i\text{-torsion})_{C_i} = (f_i\text{-torsion})_{\tilde{C}_i} \otimes_{\tilde{C}_i} C_i,$$

so that both torsions coincide. But then, again by 7.3/14,

$$A_i = \hat{S}_{(f_i)} = A\left\langle \frac{f_j}{f_i} ; j \neq i \right\rangle / (I\text{-torsion}),$$

and $U_i = \mathrm{Spf}\, A_i$ is as claimed. $\square$

In particular, we see:

Corollary 8. *Let X be an admissible formal R-scheme and $\mathcal{A} \subset \mathcal{O}_X$ a coherent open ideal. Then the formal blowing-up $X_{\mathcal{A}}$ of $\mathcal{A}$ on X does not admit I-torsion and, thus, by 7.3/5, is an admissible formal R-scheme again.*

Next, let us show that admissible formal blowing-up is characterized by a certain universal property.

Proposition 9. *For an admissible formal R-scheme X and a coherent open ideal $\mathcal{A} \subset \mathcal{O}_X$ the formal blowing-up $X_{\mathcal{A}} \longrightarrow X$ satisfies the following universal property:*

Any morphism of formal R-schemes $\varphi: Y \longrightarrow X$ such that $\mathcal{A}\mathcal{O}_Y$ is an invertible ideal in $\mathcal{O}_Y$ factorizes uniquely through $X_{\mathcal{A}}$.

Proof. We may assume that X is affine, say $X = \operatorname{Spf} A$, and that $\mathcal{A}$ is associated to a finitely generated ideal $\mathfrak{a} = (f_0, \ldots, f_r) \subset A$. Then consider a morphism of formal schemes $\varphi: Y \longrightarrow X$ such that the ideal $\mathcal{A}\mathcal{O}_Y \subset \mathcal{O}_Y$ is invertible. We may assume that Y is affine, say $Y = \operatorname{Spf} B$, and that the ideal $\mathcal{A}\mathcal{O}_Y$ is generated by f_i, for some i. Then $\mathcal{A}\mathcal{O}_Y$ is associated to the ideal $f_i B = \mathfrak{a}B \subset B$.

Let $\varphi^*: A \longrightarrow B$ be the morphism of R-algebras, given by the morphism $\varphi: Y \longrightarrow X$. Since by our assumption, the ideal $\mathfrak{a}B$ is invertible, the fractions $f_j f_i^{-1}$ are well-defined in B. Therefore, using the terminology of the proof of Proposition 7, there is a unique homomorphism

$$A_i = A\left\langle \frac{f_j}{f_i} ; j \neq i \right\rangle / (f_i\text{-torsion}) \longrightarrow B$$

that extends $\varphi^*: A \longrightarrow B$ and maps the fractions $f_j f_i^{-1} \in A_i$ to the corresponding fractions in B. The attached morphism $Y \longrightarrow X_{\mathcal{A}}$ settles the existence part of the assertion.

To justify the uniqueness part, it is enough to show that, in the above considered special situation, any factorization $Y \longrightarrow X_{\mathcal{A}}$ of the morphism $\varphi: Y \longrightarrow X$ maps Y into $U_i = \operatorname{Spf} A_i$. However, this is easily checked, since U_i coincides with the locus in X where the ideal $\mathcal{A}\mathcal{O}_{X_{\mathcal{A}}} \subset \mathcal{O}_{X_{\mathcal{A}}}$ is generated by f_i. $\qquad\square$

We need to work out some basic properties of admissible formal blowing-up. Let us start with a simple observation.

Remark 10. *Let X be an admissible formal R-scheme and let $\mathcal{A}, \mathcal{B} \subset \mathcal{O}_X$ be coherent open ideals on X. Let $X_{\mathcal{A}}$ be the formal blowing-up of $\mathcal{A}$ on X, and set $\mathcal{B}' = \mathcal{B}\mathcal{O}_{X_{\mathcal{A}}}$. Then the composition*

$$(X_{\mathcal{A}})_{\mathcal{B}'} \longrightarrow X_{\mathcal{A}} \longrightarrow X$$

of the formal blowing-up of $\mathcal{B}'$ on $X_{\mathcal{A}}$ with the formal blowing-up of $\mathcal{A}$ on X is canonically isomorphic to the formal blowing-up of the ideal $\mathcal{A}\mathcal{B}$ on X.

Proof. The assertion is a direct consequence of the universal property of admissible blowing-up in Proposition 9, once we know that the ideal generated by $\mathcal{A}$ on $(X_{\mathcal{A}})_{\mathcal{B}'}$ is invertible. However, the latter follows from the construction of blowing-up. Consider an R-algebra A of topologically finite presentation and a coherent open ideal $\mathfrak{a} \subset A$. Then, if for some $g \in A$ the g-torsion of A is trivial, the same is true for all localizations of the graded ring $S = \bigoplus_{d \in \mathbb{N}} \mathfrak{a}^d$ and there is no g-torsion on the scheme theoretic blowing-up $\operatorname{Proj} S$ of $\mathfrak{a}$ on $\operatorname{Spec} A$. Using Gabber's Lemma (Lemma 2), the same holds for the formal blowing-up of $\mathfrak{a}$ on $\operatorname{Spf} A$. $\qquad\square$

Proposition 11. *Let X be an admissible formal R-scheme that is quasi-compact and quasi-separated, and consider two admissible formal blowing-ups $\varphi\colon X' \longrightarrow X$ and $\varphi'\colon X'' \longrightarrow X'$. Then the composition $\varphi \circ \varphi'\colon X'' \longrightarrow X$ is an admissible formal blowing-up again.*

Proof. Let $\mathcal{A} \subset \mathcal{O}_X$ and $\mathcal{A}' \subset \mathcal{O}_{X'}$ be coherent open ideals of the structure sheaves of X and X' such that $\varphi\colon X' \longrightarrow X$ is the formal blowing-up of $\mathcal{A}$ on X and $\varphi'\colon X'' \longrightarrow X'$ is the formal blowing-up of $\mathcal{A}'$ on X'. We start with the special case where X is affine, say $X = \operatorname{Spf} A$. Then $\mathcal{A}$ is associated to a coherent open ideal $\mathfrak{a} \subset A$ by 8.1/5. Setting $\tilde{X} = \operatorname{Spec} A$, let $\tilde{\varphi}\colon \tilde{X}' \longrightarrow \tilde{X}$ be the scheme theoretic blowing-up of the ideal $\mathfrak{a}$ on $\tilde{X}$. By Proposition 6, the formal blowing-up X' equals the I-adic completion of $\tilde{X}'$. More specifically, we choose a system of generators f_i of $\mathfrak{a}$, $i = 0, \ldots, r$, and consider for each i the affine open subscheme $\operatorname{Spec} \tilde{A}'_i \subset \tilde{X}'$ where f_i generates the invertible ideal $\mathfrak{a}\mathcal{O}_{\tilde{X}'}$. Then the schemes $\operatorname{Spec} \tilde{A}'_i$ cover $\tilde{X}'$ and the associated affine formal schemes $\operatorname{Spf} A'_i$ where A'_i is the I-adic completion of $\tilde{A}_i$, form an open covering of X'. For each i, the canonical map $\tilde{A}_i \longrightarrow A_i$ induces isomorphisms $\tilde{A}_i/(I^\ell) \xrightarrow{\sim} A_i/(I^\ell)$, $\ell \in \mathbb{N}$. Since the coherent ideal $\mathcal{A}' \subset \mathcal{O}_X$ is open and, thus, contains some power of I, we see that there exists canonically a coherent open ideal $\tilde{\mathcal{A}}' \subset \mathcal{O}_{\tilde{X}'}$ satisfying $\mathcal{A}' = \tilde{\mathcal{A}}'\mathcal{O}_{X'}$.

Now, writing $\tilde{\varphi}'\colon \tilde{X}'' \longrightarrow \tilde{X}'$ for the scheme theoretic blowing-up of $\tilde{\mathcal{A}}'$ on $\tilde{X}'$, it is enough to show that the composition $\tilde{\varphi} \circ \tilde{\varphi}'\colon \tilde{X}'' \longrightarrow \tilde{X}$ is the scheme theoretic blowing-up of a coherent open ideal $\tilde{\mathcal{A}}'' \subset \mathcal{O}_{\tilde{X}}$ on $\tilde{X}$, as then $\varphi \circ \varphi'\colon X'' \longrightarrow X$ will be the formal blowing-up of the ideal $\mathcal{A}'' = \tilde{\mathcal{A}}''\mathcal{O}_X$ on X. To exhibit such an ideal, note that $\mathcal{L} = \mathfrak{a}\mathcal{O}_{\tilde{X}'}$ is an ample invertible sheaf on $\tilde{X}'$. Thus, by Grothendieck and Dieudonné [EGA II], 4.6.8, or see [Bo], 9.4/14, there is an integer $n_0 \in \mathbb{N}$ such that, for all $n \geq n_0$, the sheaf $\tilde{\mathcal{A}}' \otimes \mathcal{L}^n$, which we may view as an ideal in $\mathcal{O}_{\tilde{X}'}$, is generated by its global sections.

We conclude from the universal property of blowing-up (or by direct computation) that the morphism $\tilde{\varphi}\colon \tilde{X}' \longrightarrow \tilde{X}$ is an isomorphism over the complement of the closed subscheme in $\tilde{X}$ defined by the sections $f_0, \ldots, f_r$. In particular, the canonical maps between localizations $A_{f_i} \longrightarrow \tilde{A}_{i,f_i}$ are isomorphisms. As a result, any given section in $\tilde{A}_i$ is induced from a section in A, provided we multiply it by a suitable power of f_i. Thus, we can take n_0, as introduced above, big enough such that, for all $n \geq n_0$, there is a (finite) set of global generators of $\tilde{\mathcal{A}}' \otimes \mathcal{L}^n$ that are induced from sections in A. Thus, if we choose some $n \geq n_0$ and define an $\mathcal{O}_{\tilde{X}}$-ideal $\tilde{\mathcal{A}}''$ via the canonical exact sequence

$$0 \longrightarrow \tilde{\mathcal{A}}'' \longrightarrow \mathcal{O}_{\tilde{X}} \longrightarrow \tilde{\varphi}_*\mathcal{O}_{\tilde{X}'}/\tilde{\varphi}_*(\tilde{\mathcal{A}}' \otimes \mathcal{L}^n),$$

we get an open ideal on $\tilde{X}$ that generates $\tilde{\mathcal{A}}' \otimes \mathcal{L}^n$ on $\tilde{X}'$. Due to its definition, $\tilde{\mathcal{A}}''$ is a quasi-coherent ideal on $\tilde{X}$ and therefore is associated to an open ideal $\mathfrak{a}'' \subset A$. Since $\tilde{\mathcal{A}}' \otimes \mathcal{L}^n$ is generated by finitely many global sections on $\tilde{X}$, there is a finitely generated open ideal $\mathfrak{a}''' \subset \mathfrak{a}''$ satisfying $\mathfrak{a}'''\mathcal{O}_{\tilde{X}'} = \tilde{\mathcal{A}}' \otimes \mathcal{L}^n$. Then $\mathfrak{a}'''$ induces a coherent open ideal on $\tilde{X}$, and it follows that the composition $\tilde{X}'' \longrightarrow \tilde{X}' \longrightarrow \tilde{X}$ is the scheme theoretic blowing-up of $\mathfrak{a}\mathfrak{a}'''$ on $\tilde{X}$. Likewise, $X'' \longrightarrow X' \longrightarrow X$ will be the formal blowing-up of $\mathfrak{a}\mathfrak{a}'''$ on X; see Remark 10 and its proof.

Next, in order to approach the general case, we want to show that the construction of the quasi-coherent ideal $\tilde{\mathcal{A}}'' \subset \mathcal{O}_{\tilde{X}}$ above is compatible with affine flat base change on X. Thus, let $Y \longrightarrow X$ be a morphism of affine formal R-schemes of topologically finite presentation where $X = \mathrm{Spf}\, A$ and $Y = \mathrm{Spf}\, B$. Then we know from Proposition 4 that

$$\psi: Y' = X' \times_X Y \longrightarrow Y$$

is the formal blowing-up of the ideal $\mathcal{B} = \mathcal{A}\mathcal{O}_Y$ on Y and, likewise,

$$\psi': Y'' = X'' \times_X Y = X'' \times_{X'} Y' \longrightarrow Y'$$

is the formal blowing-up of $\mathcal{B}' = \mathcal{A}'\mathcal{O}_{Y'}$ on Y'. Writing $\tilde{B}_i = \tilde{A}_i \otimes_A B$ and $B_i = A_i \,\hat{\otimes}_A\, B$, the commutative diagrams

$$
\begin{array}{ccc}
\tilde{A}_i/(I^\ell) & \xrightarrow{\sim} & A_i/(I^\ell) \\
\downarrow & & \downarrow \\
\tilde{B}_i/(I^\ell) & \xrightarrow{\sim} & B_i/(I^\ell)
\end{array}
$$

for $\ell \in \mathbb{N}$ show that $\tilde{\mathcal{B}}' = \tilde{\mathcal{A}}'\mathcal{O}_{\tilde{Y}'} \subset \mathcal{O}_{\tilde{Y}'}$ is the canonical ideal on $\tilde{Y}'$ generating the ideal $\mathcal{B}' \subset \mathcal{O}_{Y'}$. Now consider the ideal $\tilde{\mathcal{B}}'' \subset \mathcal{O}_{\tilde{Y}}$ given by the exact sequence

$$0 \longrightarrow \tilde{\mathcal{B}}'' \longrightarrow \mathcal{O}_{\tilde{Y}} \longrightarrow \tilde{\psi}_*\mathcal{O}_{\tilde{Y}'}/\tilde{\psi}_*(\tilde{\mathcal{B}}' \otimes \mathcal{L}^n|_{\tilde{Y}'})$$

where $\tilde{\psi}: \tilde{Y}' \longrightarrow \tilde{Y} = \mathrm{Spec}\, B$ is the scheme theoretic blowing-up of the coherent ideal $\mathfrak{b} \subset B$ corresponding to the coherent sheaf of ideals $\mathcal{B} = \mathcal{A}\mathcal{O}_Y \subset \mathcal{O}_Y$. Since for any quasi-coherent $\mathcal{O}_{\tilde{X}'}$-module $\mathcal{F}'$ on $\tilde{X}'$ and its pull-back $\mathcal{F}' \otimes_{\mathcal{O}_{\tilde{X}'}} \mathcal{O}_{\tilde{Y}'}$ on $\tilde{Y}'$, there is a canonical isomorphism

$$\psi_*(\mathcal{F}' \otimes_{\mathcal{O}_{\tilde{X}'}} \mathcal{O}_{\tilde{Y}'}) \xrightarrow{\sim} \varphi_*(\mathcal{F}') \otimes_{\mathcal{O}_{\tilde{X}}} \mathcal{O}_{\tilde{Y}}$$

by Grothendieck and Dieudonné [EGA III], 1.4.15, the flatness of B over A implies $\tilde{\mathcal{B}}'' = \tilde{\mathcal{A}}'' \otimes_{\mathcal{O}_{\tilde{X}}} \mathcal{O}_{\tilde{Y}}$ and, hence, that the construction of $\tilde{\mathcal{A}}''$, respectively $\tilde{\mathcal{B}}''$, is compatible with flat base change on X.

Finally, to conclude the proof of the proposition for an arbitrary admissible formal R-scheme X that is quasi-compact, we can consider a covering of X by finitely many affine open formal subschemes $X_i = \mathrm{Spf}\, A_i$, $i \in J$. On each of these X_i, we can construct an open ideal sheaf $\mathcal{A}_i''$ as above that is associated to some open ideal $\mathfrak{a}_i'' \subset A_i$. The construction of $\mathcal{A}_i''$ depends on the choice of a sufficiently big integer n. However, as the index set J is finite, we may pick some n working uniformly on all X_i. Then, since the construction of the $\mathcal{A}_i$ is compatible with flat base change, we can apply 7.3/12 and thereby see that the ideals $\mathcal{A}_i''$ can be glued to produce an open ideal sheaf $\mathcal{A}'' \subset \mathcal{O}_X$ satisfying $\mathcal{A}''\mathcal{O}_{X'} = \mathcal{A}'\mathcal{A}^n\mathcal{O}_{X'}$. We may call $\mathcal{A}''$ a "quasi-coherent" ideal of $\mathcal{O}_X$ since, on each X_i, it is associated to some ideal $\mathfrak{a}_i'' \subset A_i$. By standard methods we can now find an open ideal $\mathcal{A}''' \subset \mathcal{A}'' \subset \mathcal{O}_X$ of *finite type* and, hence, coherent, such that $\mathcal{A}'''\mathcal{O}_{X'} = \mathcal{A}'\mathcal{A}^n\mathcal{O}_{X'}$. Indeed, we can reduce the problem to the scheme X_ℓ obtained from X by dividing out a sufficiently high power $I^{\ell+1}$. By our assumption on X, any such X_ℓ is quasi-compact and quasi-separated. Therefore we know from [EGA I], 6.9.9, that the quasi-coherent ideal $\mathcal{A}'''\mathcal{O}_{X_\ell} \subset \mathcal{O}_{X_\ell}$ is the direct limit of its subideals of finite type. Any such ideal will be coherent, due to 7.3/6. Taking it big enough, its inverse image in $\mathcal{O}_X$, which is of finite type and, hence, coherent, will generate the ideal $\mathcal{A}'\mathcal{A}^n\mathcal{O}_{X'}$ in $\mathcal{O}_{X'}$. For this to work, we can divide again by some power of I and thereby reduce the problem to the scheme case. Since X' is quasi-compact and $\mathcal{A}'\mathcal{A}^n\mathcal{O}_{X'}$ is an ideal of finite type, we may apply [EGA I], Chap. 0, 5.2.3.   □

The assumption in Proposition 11 that X is quasi-compact is quite restrictive. For example, if we are in the situation of 8.4/7 where we work over a complete non-Archimedean field K and consider a separated K-scheme X of finite type, then its rigid analytification X^{rig} in the sense of 5.4/3 is not necessarily quasi-compact any more. Just look at the affine n-space $X = \mathbb{A}_K^n$. Consequently, formal R-models X of X^{rig} over the valuation ring R of K, as defined in 7.4/4, will not automatically be quasi-compact. On the other hand, we will see in 8.4/7 that X^{rig} is quasi-*para*compact and, as a consequence, admits formal R-models X that are quasi-paracompact as well. In the following we show that quasi-paracompactness interacts quite well with admissible formal blowing-up. Thereby we are able to adapt the assertion of Proposition 11 to the quasi-paracompact case, as will be seen in Proposition 15 below.

Definition 12. *A topological (resp. G-topological) space X is called* quasi-para-compact *if there exists an open (resp. admissible open) covering $X = \bigcup_{i \in J} X_i$ such that*:

(i) *X_i is quasi-compact for all $i \in J$ in the sense that each open (resp. admissible open) covering of X_i admits a finite (resp. finite admissible) refinement.*

(ii) *The covering $(X_i)_{i \in J}$ is of finite type, i.e. for each index $i \in J$ the intersection $X_i \cap X_j$ is non-empty for at most finitely many indices $j \in J$.*

Proposition 13. *Let X be an admissible formal R-scheme that is quasi-paracompact and quasi-separated, and let $U \subset X$ be an open formal subscheme that is quasi-compact. Then any coherent open ideal $\mathcal{A}_U \subset \mathcal{O}_U$ extends to a coherent open ideal $\mathcal{A} \subset \mathcal{O}_X$. Furthermore, we can construct $\mathcal{A}$ in such a way that $\mathcal{A}|_V$ coincides with $\mathcal{O}_X|_V$ for any formal open subscheme $V \subset X$ disjoint from U.*

In particular, any admissible formal blowing-up on U admits an extension on X.

Proof. Since $\mathcal{A}_U$ is open and U is quasi-compact, there is an integer $\ell \in \mathbb{N}$ such that $I^\ell \mathcal{O}_U \subset \mathcal{A}_U$. Then we can consider $\mathcal{A}_U / I^\ell \mathcal{O}_U$ as a coherent ideal in $\mathcal{O}_U / I^\ell \mathcal{O}_U$, and it is enough to extend it to an ideal of finite type in $\mathcal{O}_X / I^\ell \mathcal{O}_X$. The latter is possible by Grothendieck and Dieudonné [EGA I], 6.9.6. The construction shows that the extended ideal coincides with $\mathcal{O}_X / I^\ell \mathcal{O}_X$ on all formal open subschemes $V \subset X$ such that $U \cap V = \emptyset$. $\square$

Proposition 14. *Let X be an admissible formal R-scheme, that is quasi-paracompact and quasi-separated. Consider a covering $X = \bigcup_{i \in J} X_i$ of finite type by quasi-compact formal open subschemes $X_i \subset X$, together with admissible formal blowing-ups $\varphi_i \colon X_i' \longrightarrow X_i, i \in J$. Then there is an admissible formal blowing-up $\varphi \colon X' \longrightarrow X$ dominating all φ_i in the sense that for each $i \in J$ there is a unique morphism $\varphi^{-1}(X_i) \longrightarrow X_i'$ such that the diagram*

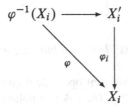

commutes for all $i \in J$.

Proof. Let $\mathcal{A}_i \subset \mathcal{O}_{X_i}$ be a coherent open ideal giving rise to the formal blowing-up $\varphi_i \colon X_i' \longrightarrow X_i$. As explained in Proposition 13, we can extend $\mathcal{A}_i$ to a coherent open ideal $\overline{\mathcal{A}_i} \subset \mathcal{O}_X$, and we may assume $\overline{\mathcal{A}_i}|_U = \mathcal{O}_X|_U$ for each open formal subscheme $U \subset X$ such that $X_i \cap U = \emptyset$. In particular, $\overline{\mathcal{A}_i}$ coincides with $\mathcal{O}_X$ on X_j for almost all indices $j \in J$. Therefore, $\mathcal{A} = \prod_{i \in J} \overline{\mathcal{A}_i}$ is a well-defined coherent open ideal in $\mathcal{O}_X$, and we can consider the associated formal blowing-up $\varphi \colon X' \longrightarrow X$. Since $\overline{\mathcal{A}_i}$ induces an invertible ideal on X' for each i, the universal mapping property of formal blowing-up implies the stated mapping property for φ. $\square$

For later use, we need two consequences of the above results.

Proposition 15. *Let $\varphi'' \colon X'' \longrightarrow X'$ and $\varphi' \colon X' \longrightarrow X$ be admissible formal blowing-ups of admissible formal R-schemes where X is quasi-separated and quasi-paracompact. Then there is an admissible formal blowing-up $\varphi''' \colon X''' \longrightarrow X$ dominating the composition $\varphi' \circ \varphi'' \colon X'' \longrightarrow X$, i.e. such that there is a morphism $\sigma \colon X''' \longrightarrow X''$ satisfying $\varphi''' = \varphi' \circ \varphi'' \circ \sigma$.*

Proof. We choose a covering $X = \bigcup_{i \in J} X_i$ of finite type by quasi-compact open formal subschemes $X_i \subset X$. Then, over each X_i, the composition $\varphi' \circ \varphi''$ is an admissible formal blowing-up of X_i by Proposition 11. Let $\mathcal{A}_i \subset \mathcal{O}_{X_i}$ be the corresponding coherent open ideal. Due to Proposition 13, we can extend $\mathcal{A}_i$ to a coherent open ideal $\overline{\mathcal{A}_i} \subset \mathcal{O}_X$ coinciding with $\mathcal{O}_X$ on each open formal subscheme $V \subset X$ where $V \cap X_i = \emptyset$. Then $\mathcal{A} = \prod_{i \in J} \overline{\mathcal{A}_i}$ is a well-defined coherent open ideal in $\mathcal{O}_X$ and, blowing up $\mathcal{A}$ on X, we obtain an admissible formal blowing-up of X which, due to the universal property of blowing-up, dominates $\varphi' \circ \varphi''$. □

Proposition 16. *Let* $X \longrightarrow Y$ *be a morphism of admissible formal R-schemes and* $Y' \longrightarrow Y$ *an admissible formal blowing-up. Then these morphisms are part of a commutative diagram*

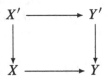

where $X' \longrightarrow X$ *is an admissible formal blowing-up.*

Proof. Let $\mathcal{B} \subset \mathcal{O}_Y$ be the coherent open ideal corresponding to the blowing-up $Y' \longrightarrow Y$ and set $\mathcal{A} = \mathcal{B}\mathcal{O}_X$. Then $\mathcal{A}$ is a coherent open ideal in $\mathcal{O}_X$, and the corresponding formal blowing-up $X' \longrightarrow X$, composed with $X \longrightarrow Y$ will factor through Y', due to the universal property of admissible formal blowing-up. □

8.3   Rig-Points in the Classical Rigid Setting

In this section, we want to deal with admissible formal R-schemes in the classical rigid case. So we assume in the following that R is a complete valuation ring of height 1 with field of fractions K and with $|\cdot|: K \longrightarrow \mathbb{R}_{\geq 0}$ a corresponding absolute value. Then, as in Sect. 7.4, we can consider the functor

$$\text{rig: (admissible formal } R\text{-schemes)} \longrightarrow \text{(rigid } K\text{-spaces)}$$

that is constructed by associating to an affine admissible formal R-scheme Spf A the affinoid K-space Sp $A \otimes_R K$. Any point $x \in$ Sp $A \otimes_R K$ is given by a maximal ideal in $A \otimes_R K$ and, since $A \otimes_R K$ is a localization of A, is induced from a well-defined prime ideal $\mathfrak{p} \subset A$. Of course, $\mathfrak{p}$ cannot be an open ideal in A. However, we will see that there is a unique maximal ideal $\mathfrak{m} \subset A$, which is open and contains $\mathfrak{p}$. Thereby we get a specialization map from the points of Sp $A \otimes_R K$ to the (closed) points of Spf A. To describe this map in convenient terms, we introduce the notion of *rig-points*.

Definition 1. *Let* X *be an admissible formal R-scheme. A* rig-point *of* X, *also called a* locally closed rig-point, *is a morphism* $u: T \longrightarrow X$ *of admissible formal R-schemes such that*

(i) u *is a locally closed immersion, and*
(ii) T *is affine,* $T = \operatorname{Spf} B$, *with* B *a local integral domain of dimension 1. The field of fractions of* B *is called the* residue field *of* u.

A rig-point $u: T \longrightarrow X$ *is called* closed *if* u *is a closed immersion.*[1]

Similarly as in the scheme case, a morphism of admissible formal R-schemes $\varphi: Y \longrightarrow X$ is called a *closed immersion* if there exists an affine open cover $(X_i)_{i \in J}$ of X such that $(\varphi^{-1}(X_i))_{i \in J}$ defines an affine open cover of Y and the induced morphisms $\varphi^{-1}(X_i) \longrightarrow X_i$, $i \in J$, correspond to epimorphisms $\varphi_i^\#: \mathcal{O}_X(X_i) \longrightarrow \mathcal{O}_Y(\varphi^{-1}(X_i))$. Note that then the kernel $\ker \varphi_i^\#$ is saturated in $\mathcal{O}_X(X_i)$ in the sense of Lemma 7.3/7 and, hence, is finitely generated. In particular, $\varphi_*(\mathcal{O}_Y)$ is a coherent $\mathcal{O}_X$-module via the canonical morphism $\varphi^\#: \mathcal{O}_X \longrightarrow \varphi_* \mathcal{O}_Y$, and the kernel $\mathcal{J} = \ker \varphi^\#$ is a coherent ideal in $\mathcal{O}_X$. It follows for any affine open formal subscheme $U \subset X$ that the inverse image $\varphi^{-1}(U)$ is affine open in Y. More generally, a morphism of admissible formal R-schemes $\varphi: Y \longrightarrow X$ is called a *locally closed immersion* if it factors through a closed immersion $Y \longrightarrow U \subset X$ where U is an open formal subscheme of X.

First we want to check, which type of rings B can occur within the context of the above definition. As usual, let $I \subset R$ be an ideal of definition.

Lemma 2. *Let* $T = \operatorname{Spf} B$ *be an admissible formal R-scheme where* B *is a local integral domain of dimension 1. Then* B *is finite over* R *and the integral closure of* B *in its field of fractions* $Q(B)$ *is a valuation ring.*

Proof. First, let us note that the maximal ideal of B is open, since B is I-adically separated. In particular, $B \otimes_R k$ is a local ring where k is the residue field of R. Due to the fact that B is of topologically finite type over R, it follows that $B \otimes_R k$ is of finite type over k. Hence, by Noether normalization, it must be module-finite over k. Now choose an epimorphism of R-algebras $\sigma: R\langle \zeta_1, \ldots, \zeta_r \rangle \longrightarrow B$ such that the residue classes of the elements $x_i = \sigma(\zeta_i)$, $i = 1, \ldots, r$, form a k-basis in $B \otimes_R k$. There is an element $\pi \in R$, $0 < |\pi| < 1$, such that

$$\sigma(\zeta_i \zeta_j) = x_i x_j \in \sum_{i=1}^r R x_i + \pi B, \qquad i, j = 1, \ldots, r.$$

In addition, we may assume $1 \in \sum_{i=1}^r R x_i + \pi B$. Then it follows by iteration that

---

[1] Beyond the classical rigid case, the notion of rig-points is useful when R is a general adic ring of type (V) or (N). Such rig-points will not necessarily be closed, as is the case in classical rigid geometry; cf. Lemma 3 below.

$$B = \sum_{i=1}^{r} Rx_i + \pi B,$$

and by π-adic approximation, that $B = \sum_{i=1}^{r} Rx_i$.

Hence, B is finite over R and its field of fractions $Q(B)$ is finite over the field of fractions $K = Q(R)$. Let $\overline{B} \subset Q(B)$ be the valuation ring corresponding to the unique extension of the absolute value from K to $Q(B)$. By construction of the latter, $\overline{B}$ is integral over R and, being normal, it equals the integral closure of R in $Q(B)$. Then $\overline{B}$ must contain B and, hence, equals the integral closure of B in $Q(B)$.   □

As a consequence, we can observe:

Lemma 3. *Every rig-point of an admissible formal R-scheme X is closed.*

Proof. It is enough to look at the case where X is affine. Thus, consider an affine admissible formal R-scheme $X = \mathrm{Spf}\, A$, and an open formal subscheme $\mathrm{Spf}\, A\langle f^{-1}\rangle \subset X$ induced by some element $f \in A$, as well as a closed rig-point $u\colon \mathrm{Spf}\, B \longrightarrow \mathrm{Spf}\, A\langle f^{-1}\rangle$. Then there is a canonical commutative diagram

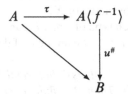

where $u^{\#}$ is surjective. We have to show that the composition $u^{\#} \circ \tau\colon A \longrightarrow B$ is surjective as well. To do this, write B' for the image of A in B. Then B' is an R-algebra of topologically finite type and, hence, by 7.3/8, I-adically complete and separated. Since B is finite over R by Lemma 2, it is finite over B'. In particular, if $\overline{f}$ is the residue class in B of $f \in A\langle f^{-1}\rangle$, we know that its inverse $\overline{f}^{-1} \in B$ is integral over B'. Considering an integral equation of $\overline{f}^{-1}$ over B', the usual trick shows $\overline{f}^{-1} \in B'$, since we know $\overline{f} \in B'$. But then, as the I-adic topology of B restricts to the I-adic topology of B' (use Lemma 7.3/7, or a direct argument involving absolute values), B' is dense in B and, thus, must coincide with B. It follows that the composition $u^{\#} \circ \tau\colon A \longrightarrow B$ is surjective.   □

In the following, we will consider rig-points only up to canonical isomorphism. To be more precise, call two rig-points $u\colon T \longrightarrow X$ and $u'\colon T' \longrightarrow X$ of an admissible formal R-scheme X equivalent if there is an R-isomorphism $\sigma\colon T \xrightarrow{\sim} T'$ such that $u = u' \circ \sigma$. The set of equivalence classes of rig-points of X will be denoted by $\mathrm{rig\text{-}pts}(X)$.

Proposition 4. *Let $\varphi\colon X' \longrightarrow X$ be a morphism of admissible formal R-schemes and $u\colon T' \longrightarrow X'$ a rig-point of X'. Then the composition*

$$\varphi \circ u'\colon T' \longrightarrow X' \longrightarrow X$$

factors uniquely through a rig-point $u\colon T \longrightarrow X$ in the sense that we have a commutative diagram

$$
\begin{array}{ccc}
T' & \longrightarrow & T \\
{\scriptstyle u'}\downarrow & & \downarrow{\scriptstyle u} \\
X' & \xrightarrow{\ \varphi\ } & X .
\end{array}
$$

In particular, φ gives rise to a well-defined map

$$\text{rig-}\varphi\colon \text{rig-pts}(X') \longrightarrow \text{rig-pts}(X), \qquad u' \longmapsto u,$$

between the rig-points of X' and X.

Proof. We may assume that X and X' are affine, say $X = \operatorname{Spf} A$, $X' = \operatorname{Spf} A'$. Furthermore, let $T' = \operatorname{Spf} B'$. If $u\colon \operatorname{Spf} B \longrightarrow \operatorname{Spf} A$ is a rig-point through which the composition $\varphi \circ u'$ factors, then there is a commutative diagram

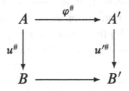

where the vertical maps are surjective. Since B and B' are local integral domains of dimension 1, which are finite over R by Lemma 2, we see that $B \longrightarrow B'$ must be injective. Hence, we can identify B with the subring of B' that equals the image of $u'^{\#} \circ \varphi^{\#}$, and we see that the rig-point $u\colon \operatorname{Spf} B \longrightarrow X$ through which $\varphi \circ u'$ factors, is unique.

To show the existence part of the assertion, set $B = u'^{\#} \circ \varphi^{\#}(A)$. Then, by its definition, B is of topologically finite type and, hence, by 7.3/5, an admissible R-algebra. Furthermore, since B' is a local integral domain of dimension 1, the same must hold for B, as B' is integral over B by Lemma 2. Thus, $A \longrightarrow B$ gives rise to a rig-point $u\colon T \longrightarrow X$, through which the composition $\varphi \circ u'$ factors. $\quad\square$

The construction of the map $\text{rig-}\varphi$ in Proposition 4 shows that the residue field of a rig-point $u\colon T \longrightarrow X$ can shrink under this map. The latter is not possible if φ is an admissible formal blowing-up.

Proposition 5. *Let* $\varphi: X' \longrightarrow X$ *be an admissible blowing-up on an admissible formal R-scheme* X. *Then the associated map*

$$\text{rig-}\varphi: \text{rig-pts}(X') \longrightarrow \text{rig-pts}(X)$$

is bijective and respects residue fields.

Proof. We may assume that X is affine, say $X = \text{Spf } A$. Let $\mathfrak{a} = (f_0, \ldots, f_r) \subset A$ be a coherent open ideal such that φ is the formal blowing-up of $\mathfrak{a}$ on X. In order to exhibit an inverse of rig-φ, consider a rig-point $u: \text{Spf } B \longrightarrow X$. Then $\mathfrak{a}$ becomes invertible over the integral closure $\overline{B}$ of B in $Q(B)$, since $\overline{B}$ is a valuation ring by Lemma 3 and since any finitely generated ideal of a valuation ring is principal. Interpreting $\overline{B}$ as a direct limit of finite extensions of B and using the fact that $\mathfrak{a}$ is finitely generated, we can find a finite subextension $B' \subset \overline{B}$ over B such that the ideal $\mathfrak{a}B' \subset B'$ is invertible. Clearly, B' is a local ring of dimension 1, just as R and $\overline{B}$ are. Furthermore, it is I-adically complete and separated by 7.3/8. Thus, using the universal property of the formal blowing-up φ, the morphism

$$\text{Spf } B' \longrightarrow \text{Spf } B \longrightarrow X$$

factors through a unique morphism $u': \text{Spf } B' \longrightarrow X'$. More precisely, if $\mathfrak{a}B'$ is generated by f_i and $\text{Spf } A_i$ is the open formal subscheme of X' where the invertible sheaf $\mathfrak{a}\mathcal{O}_{X'} \subset \mathcal{O}_{X'}$ is generated by f_i, then u' maps $\text{Spf } B'$ into $\text{Spf } A_i$. Replacing B' by the image of the attached map $A_i \longrightarrow B'$, we may even assume that $u': \text{Spf } B' \longrightarrow X'$ is a closed immersion and, therefore, is a rig-point of X'. Thus, associating to any $u \in \text{rig-pts}(X)$ the rig-point $u' \in \text{rig-pts}(X')$, as just constructed, we obtain a map $\text{rig-pts}(X) \longrightarrow \text{rig-pts}(X')$, which clearly is an inverse of rig-φ.
                                                                                        □

We want to show that the rig-points of an admissible formal R-scheme X correspond bijectively to the points of the associated rigid K-space X_{rig}.

Lemma 6. *Let* $X = \text{Spf } A$ *be an affine admissible formal R-scheme. Then there are canonical bijections between the following sets of points:*

  (i)   *Rig-points of* $\text{Spf } A$, *up to identification via natural isomorphism.*
  (ii) *Non-open prime ideals* $\mathfrak{p} \subset A$ *with* $\dim A/\mathfrak{p} = 1$.
  (iii) *Maximal ideals in* $A \otimes_R K$.

  In more detail, the stated bijections can be described as follows:

  (a) *Given a point of type* (i), *i.e. a rig-point* $u: \text{Spf } B \longrightarrow \text{Spf } A$ *defined by an epimorphism* $u^{\#}: A \longrightarrow B$, *associate to it the prime ideal* $\mathfrak{p} = \ker u^{\#} \subset A$ *as point of type* (ii).
  (b) *Given a point of type* (ii) *represented by a prime ideal* $\mathfrak{p} \subset A$, *associate to it the ideal generated by* $\mathfrak{p}$ *in* $A \otimes_R K$ *as point of type* (iii).

(c) *Given a point of type* (iii), *represented by a maximal ideal* $\mathfrak{m} \subset A \otimes_R K$, *let* $\mathfrak{p} = \mathfrak{m} \cap A$ *and associate to* $\mathfrak{m}$ *the canonical morphism* $\operatorname{Spf} A/\mathfrak{p} \longrightarrow \operatorname{Spf} A$ *as point of type* (i).

Proof. First we show that the maps described above are well-defined in the sense that they produce points of the stated types. Starting with points of type (i), let $u \colon \operatorname{Spf} B \longrightarrow \operatorname{Spf} A$ be a rig-point of X, and let $u^{\#} \colon A \longrightarrow B$ be the associated epimorphism of R-algebras. Then $\mathfrak{p} = \ker u^{\#}$ is a prime ideal that satisfies $\dim A/\mathfrak{p} = \dim B = 1$. Furthermore, since $A/\mathfrak{p}$ contains R as a subring, $\mathfrak{p}$ cannot be open. It follows that $\mathfrak{p}$ is a point of type (ii). Next, consider a prime ideal $\mathfrak{p} \subset A$ giving rise to a point of type (ii), and assume that there is a prime ideal $\mathfrak{q} \subset A$ with $\mathfrak{p} \subsetneq \mathfrak{q}$. Then $\mathfrak{q}$ is a maximal ideal, due to $\dim A/\mathfrak{p} = 1$. Furthermore, such a maximal ideal must be open in A, since otherwise we would have $\pi A + \mathfrak{q} = A$ for $\pi \in R$, $0 < |\pi| < 1$, thus, implying an equation of type $1 - a\pi = q$ for some elements $a \in A$ and $q \in \mathfrak{q}$. But then, due to the geometric series, q would be invertible which, however, is impossible. It follows that $\mathfrak{p} \cdot (A \otimes_R K)$ is a maximal ideal in $A \otimes_R K$ and, thus, a point of type (iii).

Finally, consider a point of type (iii), i.e. a maximal ideal $\mathfrak{m} \subset A \otimes_R K$. Then $K' = (A \otimes_R K)/\mathfrak{m}$ is a field that is finite over K by 2.2/12 and, using 7.3/5, the image of A in K' is an admissible R-algebra, which we denote by B. Extending the absolute value of K to K', which is possible in a unique way, let $R' \subset K'$ be the corresponding valuation ring. Then R' equals the integral closure of R in K', and we claim that $B \subset R'$. In fact, choose an epimorphism $R\langle \zeta \rangle \longrightarrow A$ where ζ is a finite system of variables, and consider on the affinoid K-algebra $A \otimes_R K$ the residue norm derived from the induced epimorphism $K\langle \zeta \rangle \longrightarrow A \otimes_R K$. Fixing some $\pi \in R$, $0 < |\pi| < 1$, the topology of $A \otimes_R K$ restricts to the π-adic topology of A; the latter is true, since A, as an admissible R-algebra, does not admit π-torsion and, thus, embeds into $A \otimes_R K$. But then, by continuity, any bounded part of $A \otimes_R K$, such as A, must be mapped into a bounded part of K', and it follows that $B \subset R'$. Since the extensions $R \subset B \subset R'$ are integral and R, R' are local rings of dimension 1, the same must be true for B. It follows that $A \longrightarrow B$ gives rise to a rig-point $\operatorname{Spf} B \longrightarrow \operatorname{Spf} A$. Furthermore, writing $\mathfrak{p} = \mathfrak{m} \cap A$, we see that the quotient $A/\mathfrak{p}$ is isomorphic to B.

To show that the above described canonical maps are, indeed, bijections, note that these maps are all injective by definition. Furthermore, going from points of type (i) to points of type (ii), then of type (iii) and, finally, of type (i) again, we get the identity map on points of type (i). This is enough to conclude that all three maps are bijective. □

Using the map from points of type (i) in Lemma 6 to those of type (iii), we obtain the following statement:

Proposition 7. *Let X be an admissible formal R-scheme and let X_{rig} be the associated rigid K-space. Then there is a canonical bijection*

$$\text{rig-pts}(X) \xrightarrow{\ \sim\ } X_{\text{rig}}$$

between sets of points, which is functorial in X and associates to a rig-point
$T \longrightarrow X$ the image of the corresponding closed immersion $T_{\text{rig}} \longrightarrow X_{\text{rig}}$.

Choosing an element $\pi \in R$, $0 < |\pi| < 1$, we may use $I = (\pi)$ as an ideal of definition of R. As usual, we set $A_\ell = A/I^{\ell+1}A$ for any R-algebra A and let $X_\ell = X \otimes_R R/I^{\ell+1}$ for any formal R-scheme X. Then the underlying topological spaces of the schemes X_ℓ are canonically identified via the closed immersions $X_\ell \longrightarrow X_{\ell+1}$. If k is the residue field of R, we may, in terms of underlying topological spaces, identify each X_ℓ even with $X_\ell \otimes_{R_\ell} k = X \otimes_R k$. The latter is a k-scheme of finite type if X is a formal R-scheme of topologically finite type; it will be denoted by X_k. Let us call X_k the *special fiber* of X. Since any rig-point $u: T \longrightarrow X$ of an admissible formal R-scheme X induces a closed immersion $u_k: T_k \longrightarrow X_k$, we see that u determines a closed point of the special fiber X_k. Thus, using Proposition 7, we get a canonical specialization map

$$\text{sp:}\ X_{\text{rig}} \longrightarrow X_k$$

that is characterized as follows. Consider a point $x \in X_{\text{rig}}$. To determine its image $\text{sp}(x) \in X_k$, choose an affine open subscheme $U = \text{Spf}\,A$ in X such that x belongs to $U_{\text{rig}} = \text{Sp}\,A \otimes_R K$, and let $\mathfrak{m} \subset A \otimes_R K$ be the corresponding maximal ideal. Then consider the projection $\tau_K: A \otimes_R K \longrightarrow (A \otimes_R K)/\mathfrak{m} = K'$ where K' is finite over K by 3.1/4. Let $\tau: A \longrightarrow B$ for $B = \tau_K(A)$ be the restriction of τ_K. As we have seen, B is a local integral domain of dimension 1 lying between R and the valuation ring of K'. In fact, τ gives rise to the rig-point of X corresponding to x, and the surjections $A \otimes_R k \longrightarrow B \otimes_R k \longrightarrow B \otimes_R k/\text{rad}(B \otimes_R k)$ determine the closed point of the special fiber $U_k = \text{Spec}\,A \otimes_R k \subset X_k$ that equals the image of x under the specialization map sp. Note that the construction of sp is similar to the one considered in [BGR], 7.1.5, although we never use "canonical reductions" of affinoid algebras in the style of [BGR], 6.3.

We want to show:

Proposition 8. *For any admissible formal R-scheme X, the specialization map*
$\text{sp:}\ X_{\text{rig}} \longrightarrow X_k$ is surjective onto the set of closed points of X_k.

Proof. We may assume that X is affine, say $X = \text{Spf}\,A$, and we first look at the special case where A is an algebra of restricted power series, say $A = R\langle \zeta \rangle$ for a finite set of variables $\zeta = (\zeta_1, \ldots, \zeta_n)$. Then consider a closed point $x \in X_k$ and let $\mathfrak{m} \subset A_k = A \otimes_R k$ be the associated maximal ideal. We set $k' = A_k/\mathfrak{m}$ and choose a finite field extension K'/K lifting the extension k'/k. It follows that K' is endowed with an absolute value extending the one given on K, and we denote by $R' \subset K'$ the corresponding valuation ring; it equals the integral closure of R in K'. In order to show that x belongs to the image of the specialization map $\text{sp:}\ X_{\text{rig}} \longrightarrow X_k$, it is enough to show that the canonical projection $p_k: A \longrightarrow A_k \longrightarrow k'$ can be lifted to an R-homomorphism $p: A \longrightarrow R'$. Then, since R' is integral over R and, hence, over $p(A)$, we see that $p(A)$ is a local ring of dimension 1. Therefore the epimorphism $A \longrightarrow p(A)$ gives rise to a rig-point of X specializing into the closed point $x \in X_k$.

To construct the desired lifting p of p_k is easy. Choose a representative $a_i \in R'$ of $p_k(\zeta_i)$ for each i and define a lifting of p_k by

$$p: R\langle \zeta_1, \ldots, \zeta_n \rangle \longrightarrow R', \qquad \zeta_i \longmapsto a_i, \qquad i = 1, \ldots, n.$$

In the general case, we choose an epimorphism $\tau: R\langle \zeta \rangle \longrightarrow A$ where $R\langle \zeta \rangle$ is an R-algebra of restricted power series as before. We can extend τ to an epimorphism $\tau_K: K\langle \zeta \rangle \longrightarrow A \otimes_R K$ between associated affinoid K-algebras. Then A may be interpreted as the subring in $A \otimes_R K$ consisting of all elements $a \in A \otimes_R K$ such that a has residue norm ≤ 1 with respect to the projection τ_K; use 2.3/9. Due to Noether Normalization 2.2/11, there is a K-morphism

$$\iota_K: K\langle \eta \rangle \longrightarrow K\langle \zeta \rangle$$

with a finite set of variables $\eta = (\eta_1, \ldots, \eta_d)$ such that the composition

$$\tau_K \circ \iota_K: K\langle \eta \rangle \longrightarrow A \otimes_R K$$

is a finite monomorphism. Since ι_K is contractive with respect to the Gauß norm, ι_K restricts to an R-morphism $\iota: R\langle \eta \rangle \longrightarrow R\langle \zeta \rangle$. We claim that the resulting R-morphism $\tau \circ \iota: R\langle \eta \rangle \longrightarrow A$ is finite. This fact is readily checked by redoing the proof of Noether normalization in 2.2/11, using coefficients in R instead of K. The important fact is the estimate provided by the Weierstrass division formula in 2.2/8: given any ζ_n-distinguished element $g \in K\langle \zeta \rangle$ of Gauß norm 1 and of order s, any element $f \in K\langle \zeta \rangle$ can uniquely be written as

$$f = qg + r \qquad \text{with} \qquad q \in K\langle \zeta \rangle, \qquad r \in K\langle \zeta_1, \ldots, \zeta_{n-1} \rangle [\zeta_n],$$

where $\deg_{\zeta_n} r < s$ and $|q|, |r| \leq |f|$.

Knowing that $\tau \circ \iota: R\langle \eta \rangle \longrightarrow A$ is a finite monomorphism, we want to apply the Going-down Theorem to it. In order to do this, assume that A is an integral domain. Furthermore, we need to know that $R\langle \eta \rangle$ is a normal ring. The latter follows from the fact that $K\langle \eta \rangle$ is a normal ring; see 2.2/15. In fact, observe that the fraction field Q of $R\langle \eta \rangle$ coincides with the one of $K\langle \eta \rangle$ and consider an element $q \in Q$ that is integral over $R\langle \eta \rangle$. Then $q \in K\langle \eta \rangle$ by 2.2/15, and if we look at an integral equation

$$q^s + c_1 q^{s-1} + \ldots + c_s = 0$$

with coefficients $c_i \in R\langle \eta \rangle$, it follows that q has necessarily Gauß norm ≤ 1 and, thus, belongs to $R\langle \eta \rangle$. Therefore $R\langle \eta \rangle$ is a normal ring.

Now, as in the beginning, let $\mathfrak{m}$ be the maximal ideal in A corresponding to the given closed point $x \in X_k$. Set $\mathfrak{n} = \mathfrak{m} \cap R\langle\eta\rangle$. Using the characterization of rig-points in Lemma 6, we know from the above considered special case that there is a non-open prime ideal $\mathfrak{q} \subset \mathfrak{n}$ satisfying $\dim R\langle\eta\rangle/\mathfrak{q} \overset{.}{=} 1$. Furthermore, by the Going-down Theorem, there is a prime ideal $\mathfrak{p} \subset \mathfrak{m} \subset A$ such that $\mathfrak{p} \cap R\langle\eta\rangle = \mathfrak{q}$. Clearly, $\mathfrak{p}$ is non-open and satisfies $\dim A/\mathfrak{p} = 1$. Thus, by Lemma 6, $\mathfrak{p}$ gives rise to a rig-point of $\operatorname{Spf} A$ that specializes into the point corresponding to the given maximal ideal $\mathfrak{m} \subset A$.

In the general case, consider the injection $A \hookrightarrow A_K = A \otimes_R K$, and let $\mathfrak{p}_1, \ldots, \mathfrak{p}_s \subset A_K$ be the minimal prime ideals in A_K. Set $\mathfrak{p}_i' = \mathfrak{p}_i \cap A$. Then $\operatorname{rad} A_K = \bigcap_{i=1}^s \mathfrak{p}_i$ and, hence, $\operatorname{rad} A = \bigcap_{i=1}^s \mathfrak{p}_i'$. In particular, for any given maximal ideal $\mathfrak{m} \subset A$, there is an index i_0 such that $\mathfrak{p}_{i_0}' \subset \mathfrak{m}$. Write $\mathfrak{p}' = \mathfrak{p}_{i_0}'$ and $\mathfrak{p} = \mathfrak{p}_{i_0}$ and consider the commutative diagram

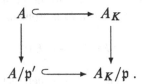

In particular, $A/\mathfrak{p}'$ is an R-algebra of topologically finite type. Since it does not have I-torsion, it is an admissible R-algebra by 7.3/5. Therefore we know from the above considered special case that the projection $A/\mathfrak{p}' \longrightarrow A/\mathfrak{m}$ lifts to an epimorphism $A/\mathfrak{p}' \longrightarrow B$ giving rise to a rig-point $T \longrightarrow \operatorname{Spf} A/\mathfrak{p}'$. Then, as desired, $T \longrightarrow \operatorname{Spf} A/\mathfrak{p}' \longrightarrow \operatorname{Spf} A$ is a rig-point specializing into x. $\square$

8.4 Rigid Spaces in Terms of Formal Models

We consider again the classical rigid situation where R is a complete valuation ring of height 1 with field of fractions K. As usual, let k be the residue field of R and choose a non-unit $\pi \in R - \{0\}$ so that the topology of R coincides with the π-adic one. If X is an admissible formal R-scheme and X_{rig} its associated rigid K-space, we call X a *formal R-model* of X_{rig}; cf. 7.4/4. Given any rigid K-space X_K, one may ask if there will always exist a formal R-model X of X_K, and if yes, in which way such formal models will differ. Assuming some mild finiteness conditions, we will work out satisfying answers to these questions. We thereby obtain a characterization of the category of rigid K-spaces (with certain finiteness conditions) as a localization of an appropriate category of admissible formal R-schemes.

To begin with, let us explain the process of localization of categories.

Definition 1. *Let $\mathfrak{C}$ be a category and S a class of morphisms in $\mathfrak{C}$. Then a localization of $\mathfrak{C}$ by S is a category $\mathfrak{C}_S$ together with a functor $Q: \mathfrak{C} \longrightarrow \mathfrak{C}_S$ such that:*

(i) *$Q(s)$ is an isomorphism in $\mathfrak{C}_S$ for every $s \in S$.*

(ii) *If $F: \mathfrak{C} \longrightarrow \mathfrak{D}$ is a functor such that $F(s)$ is an isomorphism for every $s \in S$, then F admits a unique factorization as follows:*

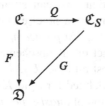

To be more precise, the commutativity of the diagram, as well as the uniqueness of G are meant up to natural equivalence of functors. Without any further assumption, one can show that localizations of categories do always exist.

Proposition 2. *The functor*

$$\mathrm{rig}: (\mathfrak{F}/R) \longrightarrow (\mathfrak{R}/K), \qquad \mathrm{rig}: X \longmapsto X_{\mathrm{rig}},$$

from the category $(\mathfrak{F}/R)$ of admissible formal R-schemes to the category $(\mathfrak{R}/K)$ of rigid K-spaces, as defined in 7.4/3, factors through the localization of $(\mathfrak{F}/R)$ by admissible formal blowing-ups.

Proof. We just have to show that the functor rig transforms an admissible formal blowing-up $X_{\mathcal{A}} \longrightarrow X$ of some admissible formal R-scheme into an isomorphism $X_{\mathcal{A},\mathrm{rig}} \overset{\sim}{\longrightarrow} X_{\mathrm{rig}}$. To do this, we may assume that X is affine, say $X = \mathrm{Spf}\, A$. Then the coherent open ideal $\mathcal{A} \subset \mathcal{O}_X$ is associated to a finitely generated open ideal $\mathfrak{a} = (f_0, \ldots, f_r) \subset A$. Choosing a non-zero non-invertible element $\pi \in R$, we may assume $I = (\pi)$ and we see from 8.2/7 that $X_{\mathcal{A}}$ is covered by the affinoid K-spaces associated to the admissible R-algebras

$$A_i = A\left\langle \frac{f_0}{f_i}, \ldots, \frac{f_r}{f_i} \right\rangle / (\pi\text{-torsion}), \qquad i = 0, \ldots, r.$$

Thus, applying the functor rig to the projection $\mathrm{Spf}\, A_i \longrightarrow \mathrm{Spf}\, A$ and writing $A_K = A \otimes_R K$, we obtain the canonical map

$$\mathrm{Sp}\, A_K\left\langle \frac{f_0}{f_i}, \ldots, \frac{f_r}{f_i} \right\rangle \longrightarrow \mathrm{Sp}\, A_K$$

defining $X_{\mathrm{rig}}(\frac{f_0}{f_i}, \ldots, \frac{f_r}{f_i})$ as a rational subdomain of $X_{\mathrm{rig}} = \mathrm{Sp}\, A_K$. More specifically, one checks that rig transforms the covering $(\mathrm{Spf}\, A_i)_{i=0\ldots r}$ of $X_{\mathcal{A}}$ into the rational covering generated by $f_0, \ldots, f_r$ on X_{rig}, preserving intersections. Of course, one has to realize that, $\mathfrak{a}$ being open in A, it contains a power of π so that the functions $f_0, \ldots, f_r$ will generate the unit ideal in A_K. Thus, we see that rig transforms the morphism $X_{\mathcal{A}} \longrightarrow X$ into an isomorphism. $\square$

Under certain mild conditions we can strengthen Proposition 2 to yield an equivalence of categories. To give a precise statement, recall from 8.2/12 that a formal R-scheme X is called *quasi-paracompact* if it admits an open covering of finite type by quasi-compact open subschemes $U_i \subset X$, $i \in J$, i.e. such that each U_i is disjoint from almost all other U_j, $j \in J$. In a similar way the notion of *quasi-paracompactness* is defined for rigid K-spaces. Furthermore, recall 6.3/2 and 6.3/4 for the characterization of *quasi-separated* rigid K-spaces.

Theorem 3 (Raynaud). *Let R be a complete valuation ring of height 1 with field of fractions K. Then the functor* rig *induces an equivalence between*

(i) $(\mathrm{FSch}/R)_S$, *the category of all admissible formal R-schemes that are quasi-paracompact, localized by the class S of admissible formal blowing-ups, and*

(ii) (Rig/K), *the category of all quasi-separated rigid K-spaces that are quasi-paracompact.*

The *proof* will consist in establishing the following steps:

Lemma 4.

(a) *The functor* rig *transforms admissible formal blowing-ups into isomorphisms.*

(b) *Two morphisms $\varphi, \psi \colon X \longrightarrow Y$ of admissible formal R-schemes coincide if the associated rigid morphisms $\varphi_{\mathrm{rig}}, \psi_{\mathrm{rig}}$ coincide.*

(c) *Let X, Y be admissible formal R-schemes that are quasi-paracompact, and let $\varphi_K \colon X_{\mathrm{rig}} \longrightarrow Y_{\mathrm{rig}}$ be a morphism between the associated rigid K-spaces. Then there exist an admissible formal blowing-up $\tau' \colon X' \longrightarrow X$ and a morphism of formal R-schemes $\varphi' \colon X' \longrightarrow Y$ such that $\varphi'_{\mathrm{rig}} = \varphi_K \circ \tau'_{\mathrm{rig}}$.*

(d) *Assume that $\varphi_K \colon X_{\mathrm{rig}} \longrightarrow Y_{\mathrm{rig}}$ as in (c) is an isomorphism and that X, Y are quasi-compact. Then we can choose $\varphi' \colon X' \longrightarrow Y$ satisfying $\varphi'_{\mathrm{rig}} = \varphi_K \circ \tau'_{\mathrm{rig}}$ with the additional property that it is an admissible formal blowing-up of Y.*

(e) *Each rigid K-space X_K that is quasi-separated and quasi-paracompact, admits a quasi-paracompact admissible formal R-scheme X as a formal model, i.e. with X satisfying $X_{\mathrm{rig}} \simeq X_K$.*

First note that, in the classical rigid case, an admissible formal R-scheme X is automatically quasi-separated, since its special fiber is a scheme of locally finite type over the residue field k of R; see Sect. 8.3. Therefore it is clear that the functor of associating to an admissible formal R-scheme X its corresponding rigid K-space X_{rig} restricts to a functor

$$\text{rig: (FSch}/R) \longrightarrow (\text{Rig}/K).$$

We begin by showing that the assertions (a)–(c) and (e) of Lemma 4 imply that the functor rig satisfies the conditions of a localization of (FSch/R) by the class S of all admissible formal blowing-ups; note that assertion (d) of Lemma 4 will be necessary for the proof of assertion (e).

Of course, we realize from (a) that rig(s) is an isomorphism for every $s \in S$. Next, consider a functor $F : (\text{FSch}/R) \longrightarrow \mathfrak{D}$ to some category $\mathfrak{D}$ where $F(s)$ is an isomorphism for all $s \in S$. In order to define a functor $G : (\text{Rig}/K) \longrightarrow \mathfrak{D}$ with $F = G \circ \text{rig}$, we proceed as follows. For any object X_K in (Rig/K) we pick an R-model X in (FSch/R) (with $X_{\text{rig}} \simeq X_K$) and set $G(X_K) = F(X)$. The latter is possible due to assertion (e). Furthermore, let $\varphi_K : X_K \longrightarrow Y_K$ be a morphism in (Rig/K). Then, if X and Y are the R-models we have picked for X_K and Y_K, we use (c) and choose an admissible formal blowing-up $\tau' : X' \longrightarrow X$ such that there is an R-morphism $\varphi' : X' \longrightarrow Y$ satisfying $\varphi'_{\text{rig}} = \varphi_K \circ \tau'_{\text{rig}}$. Then we define the composition

$$G(\varphi_K) : F(X) \xrightarrow{F(\tau')^{-1}} F(X') \xrightarrow{F(\varphi')} F(Y)$$

as the image of the morphism φ_K under G. To show that $G(\varphi_K)$ is well-defined, consider a second admissible blowing-up $\tau'' : X'' \longrightarrow X$ and an R-morphism $\varphi'' : X'' \longrightarrow Y$ such that $\varphi''_{\text{rig}} = \varphi_K \circ \tau''_{\text{rig}}$. Let $\mathcal{A}', \mathcal{A}'' \subset \mathcal{O}_X$ be the coherent open ideals corresponding to τ', τ'', and let $\tau''' : X''' \longrightarrow X$ be the formal blowing-up of the product $\mathcal{A}'\mathcal{A}''$ on X. Then τ''' dominates τ', τ'', and we thereby get a diagram

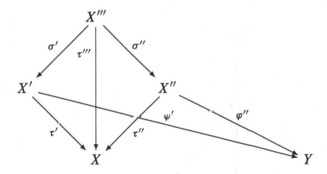

where the square with the diagonal τ''' is commutative. Furthermore, the compositions $\varphi' \circ \sigma'$ and $\varphi'' \circ \sigma''$ coincide by (b), since they coincide when applying the functor rig. Therefore the whole diagram is commutative. Since $\tau', \tau'', \tau''' \in S$, it follows that the compositions

$$F(X) \xrightarrow{F(\tau')^{-1}} F(X') \xrightarrow{F(\varphi')} F(Y)$$

$$F(X) \xrightarrow{F(\tau'')^{-1}} F(X'') \xrightarrow{F(\varphi'')} F(Y)$$

coincide and, hence that $G(\varphi_K)$ is well-defined. For G being a functor, it remains to show that G respects the composition of morphisms. Thus, consider a composition of morphisms $X_K \xrightarrow{\varphi_K} Y_K \xrightarrow{\psi_K} Z_K$ in (Rig/K). Then the corresponding composition $G(\psi_K) \circ G(\varphi_K)$ in $\mathfrak{D}$ is constructed via a diagram of type

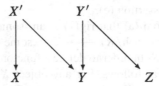

in (FSch/R) where the vertical arrows are admissible formal blowing-ups. By 8.2/16, there is a commutative diagram

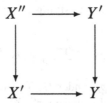

with an admissible blowing-up $X'' \longrightarrow X'$ and furthermore, by 8.2/15, we can dominate the composition of admissible blowing-ups $X'' \longrightarrow X' \longrightarrow X$ by an admissible blowing-up $X''' \longrightarrow X$, thereby getting the following commutative diagram:

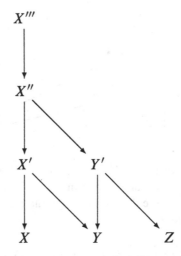

Since the composition $G(\psi_K \circ \varphi_K)$ can be thought to be constructed via the resulting diagram we see that $G(\psi_K \circ \varphi_K) = G(\psi_K) \circ G(\varphi_K)$. That G is unique, as required, is clear from the construction.

Next, in order to prove the assertions of Lemma 4, we gather some general facts that will be needed.

Lemma 5. *Let X be an admissible formal R-scheme that is quasi-paracompact, and let $\mathfrak{U}_K$ be an admissible covering of finite type of the associated rigid K-space X_{rig}, consisting of quasi-compact open subspaces of X_{rig}. Then there is an admissible formal blowing-up $\tau \colon X' \longrightarrow X$ together with an open covering $\mathfrak{U}'$ of X' such that the associated family $\mathfrak{U}'_{\mathrm{rig}}$ of rigid K-subspaces of X_{rig} coincides with $\mathfrak{U}_K$.*

Proof. We start with the case where X is affine. Then $\mathfrak{U}_K$ is a finite covering. By the Theorem of Gerritzen–Grauert 3.3/20, each $U_K \in \mathfrak{U}_K$ is a finite union of rational subdomains of X_{rig}, and we may assume that U_K itself is a rational subdomain in X_{rig}. Then U_K is of type

$$X_{\mathrm{rig}}\left(\frac{f_1}{f_0}, \ldots, \frac{f_n}{f_0}\right)$$

for some global sections $f_0, \ldots, f_n$ generating the unit section in $\mathcal{O}_X$. Multiplying the f_i with a suitable constant in R, we may even assume $f_i \in \mathcal{O}_X$ for all i. So we can consider the coherent open ideal $\mathcal{A} \subset \mathcal{O}_X$ generated by the f_i, as well as the associated formal blowing-up $X' \longrightarrow X$. Then the part of X' where f_0 generates the ideal $\mathcal{A}\mathcal{O}_{X'} \subset \mathcal{O}_{X'}$ constitutes an open formal subscheme $U \subset X'$ inducing the admissible open subspace $U_K \subset X_{\mathrm{rig}}$. Working with all $U_K \in \mathfrak{U}_K$ this way, we can blow up the product of the corresponding coherent open ideals in $\mathcal{O}_X$. Thereby we obtain an admissible formal R-scheme X' admitting a system $\mathfrak{U}'$ of open formal subschemes that induce the system $\mathfrak{U}_K$ on X_{rig}. That $\mathfrak{U}'$ covers X' will be shown below.

In the general case we work locally on X with respect to an affine open covering $(X_j)_{j \in J}$ of finite type. Restricting $\mathfrak{U}_K$ to any $X_{j,\mathrm{rig}}$ and using the fact that X_{rig} is quasi-separated, we obtain an admissible covering of finite type of $X_{j,\mathrm{rig}}$, consisting of quasi-compact admissible open subspaces of $X_{j,\mathrm{rig}}$. As shown above, one can construct a coherent open ideal $\mathcal{A}_j \subset \mathcal{O}_{X_j}$ such that, after blowing up $\mathcal{A}_j$ on X_j, there exist open formal subschemes of the blowing-up $X_{j,\mathcal{A}_j}$ of $\mathcal{A}_j$ on X_j, giving rise to formal R-models of the members of $\mathfrak{U}_K|_{X_j}$. Extending each $\mathcal{A}_j$ to a coherent open ideal $\overline{\mathcal{A}_j} \subset \mathcal{O}_X$ as in 8.2/13 and setting $\mathcal{A} = \prod_{j \in J} \overline{\mathcal{A}_j}$, we can represent all members of $\mathfrak{U}_K$ as open formal subschemes of the blowing-up $X' = X_{\mathcal{A}}$ of $\mathcal{A}$ on X, thereby obtaining a family $\mathfrak{U}'$ as required.

Finally, to show that $\mathfrak{U}'$ covers X', we can use the surjectivity of the specialization map from X'_{rig} to the closed points of X'_k, for k the residue field of R. In fact, if x is a closed point of X', we may view it as a closed point of X'_k. Then we know from 8.3/8 that x is induced by a rig-point $u: T \longrightarrow X'$ or, using 8.3/7, by the corresponding closed point $u_{\text{rig}}: T_{\text{rig}} \longrightarrow X'_{\text{rig}}$. Consequently, u_{rig} factors through a member of $\mathfrak{U}_K$, and it follows that u factors through a member of $\mathfrak{U}'$. In particular, the open formal subscheme $V = \bigcup_{U' \in \mathfrak{U}'} U' \subset X'$ contains all closed points of X'. But then the closed part $X' - V$ does not contain any closed point of X' and, therefore, must be empty; just look at the special fiber X'_k of X', which is locally of finite type over k, and consult [Bo], 8.3/6, for example. Thus, $\mathfrak{U}'$ covers X', as claimed.   □

Lemma 6. *Let A be an admissible R-algebra. Consider A as a subring of the associated affinoid K-algebra $A_{\text{rig}} = A \otimes_R K$, and let $f_1, \ldots, f_n \in A_{\text{rig}}$ be elements satisfying $|f_i|_{\sup} \leq 1$ for $i = 1, \ldots, n$. Then $A' = A[f_1, \ldots, f_n]$ is an admissible R-algebra that is finite over A. Furthermore, if $c \in R - \{0\}$ is chosen in such a way that $cf_1, \ldots, cf_n$ belong to A, the canonical morphism $\tau: \operatorname{Spf} A' \longrightarrow \operatorname{Spf} A$ can be viewed as the formal blowing-up of the coherent open ideal $\mathfrak{a} = (c, cf_1, \ldots, cf_n)$ of A on $\operatorname{Spf} A$.*

Proof. We choose an epimorphism $R\langle \zeta \rangle \longrightarrow A$ for a finite system of variables ζ and consider on A_{rig} the residue norm with respect to the induced epimorphism $K\langle \zeta \rangle \longrightarrow A_{\text{rig}}$. Then A consists of all elements $a \in A_{\text{rig}}$ with $|a| \leq 1$, use 3.1/5 (iii), and we see from 3.1/17 that A' is integral and, hence, finite over A, since it is of finite type over A. Furthermore, as $cA' \subset A$, it is easily seen that A' is an R-algebra of topologically finite type. Then A' is an admissible R-algebra by 7.3/5, since it does not admit π-torsion.

In order to show that $\tau: \operatorname{Spf} A' \longrightarrow \operatorname{Spf} A$ is the formal blowing-up of the ideal $\mathfrak{a} \subset A$, it is enough to show that τ satisfies the universal property of admissible formal blowing-up. To do this note that the ideal $\mathfrak{a} A' \subset A'$ is generated by c and, hence, is invertible, since c is not a zero divisor in $A' \subset A_{\text{rig}}$. Furthermore, consider a homomorphism of admissible R-algebras $A \longrightarrow D$ such that the ideal $\mathfrak{a} D \subset D$ is invertible. Let us write f_i for the image of f_i in D again. If $\mathfrak{a} D$ is generated by c, then $cf_i \in cD$ and, hence, $f_i \in D$ for all i, since c is not a zero divisor in D. But then $A \longrightarrow D$ admits a unique extension $A' \longrightarrow D$. If, on the other hand, $\mathfrak{a} D$ is generated by cf_i for some i, then look at the inclusions $D \hookrightarrow D[f_i] \hookrightarrow D \otimes_R K$. Since $c \in (cf_i)D$ and c is not a zero divisor in D, we see that f_i is invertible in $D[f_i]$ with an inverse $f_i^{-1} \in D$. Using the fact that f_i is integral over D, there is an integral equation

$$f_i^s + d_1 f_i^{s-1} + \ldots + d_s = 0$$

with coefficients $d_j \in D$. Multiplication with $f_i^{-s+1} \in D$ yields

$$f_i + d_1 + \ldots + d_s f_i^{-s+1} = 0$$

and, hence, $f_i \in D$. Thus, f_i is a unit in D and, as before, $\mathfrak{a}D$ is generated by c. Again, $A \longrightarrow D$ admits a unique extension $A' \longrightarrow D$, and this is enough for showing that $\tau \colon \mathrm{Spf}\, A' \longrightarrow \mathrm{Spf}\, A$ satisfies the universal property of blowing up $\mathfrak{a}$ on $\mathrm{Spf}\, A$.   □

Now we are able to establish the assertions (a)–(e) of Lemma 4.

(a) This is a consequence of Proposition 2.
(b) Consider two morphisms $\varphi, \psi \colon X \longrightarrow Y$ in (FSch/R) such that φ_{rig} coincides with ψ_{rig}. It follows from 8.3/7 that φ and ψ coincide on the level of rig-points as maps $\mathrm{rig\text{-}pts}(X) \longrightarrow \mathrm{rig\text{-}pts}(Y)$. Since this map is compatible with the specialization maps $\mathrm{sp} \colon \mathrm{rig\text{-}pts}(X) \longrightarrow X_k$ and $\mathrm{sp} \colon \mathrm{rig\text{-}pts}(Y) \longrightarrow Y_k$, we see from 8.3/8 that φ and ψ coincide as maps from closed points of X to the closed points of Y. But then, since X_k and Y_k are of locally finite type over k, it is clear that φ and ψ must coincide as maps between the underlying point sets of X and Y. Therefore, in order to show $\varphi = \psi$, we can assume that X and Y are affine, say $X = \mathrm{Spf}\, A$ and $Y = \mathrm{Spf}\, B$. But then, since the canonical maps $A \longrightarrow A \otimes_R K$ and $B \longrightarrow B \otimes_R K$ are injective, due to the fact that X and Y are admissible, it is obvious that $\varphi_{\mathrm{rig}} = \psi_{\mathrm{rig}}$ implies $\varphi = \psi$, thereby finishing the proof of Lemma 4 (b).
(c) Consider two admissible formal R-schemes X, Y that are quasi-paracompact, and a morphism $\varphi_K \colon X_{\mathrm{rig}} \longrightarrow Y_{\mathrm{rig}}$ between associated rigid K-spaces. We have to look for an admissible blowing-up $\tau \colon X' \longrightarrow X$ together with a morphism of formal R-schemes $\varphi \colon X' \longrightarrow Y$ satisfying $\varphi_{\mathrm{rig}} = \varphi_K \circ \tau_{\mathrm{rig}}$. To do this, let us start with the case where X and Y are affine, say $X = \mathrm{Spf}\, A$ and $Y = \mathrm{Spf}\, B$. Then $\varphi_K \colon X_{\mathrm{rig}} \longrightarrow Y_{\mathrm{rig}}$ is given by a morphism $\varphi_K^\# \colon B_{\mathrm{rig}} \longrightarrow A_{\mathrm{rig}}$ between associated affinoid K-algebras $B_{\mathrm{rig}} = B \otimes_R K$ and $A_{\mathrm{rig}} = A \otimes_R K$. Since B is an admissible R-algebra, we can view it as a subalgebra of B_{rig}, and we claim that

$$B \subset \{g \in B_{\mathrm{rig}} \,;\, |g|_{\sup} \leq 1\}.$$

Indeed, choose an epimorphism $\alpha \colon R\langle \zeta_1, \ldots, \zeta_r \rangle \longrightarrow B$, for a finite system of variables ζ_i, and look at the resulting epimorphism $\alpha_K \colon K\langle \zeta_1, \ldots, \zeta_r \rangle \longrightarrow B_{\mathrm{rig}}$ obtained via tensoring with K over R. Then all elements $g \in B$ have residue norm $|g| \leq 1$ with respect to α_K and, hence, satisfy $|g|_{\sup} \leq 1$ by 3.1/9. Let $g_i = \alpha(\zeta_i)$ and set $f_i = \varphi_K^\#(g_i)$, $i = 1, \ldots, r$.

Since $\varphi_K^\#$ is contractive with respect to the supremum norm by 3.1/7, we see that $|f_i|_{\sup} \leq 1$ for all i. Furthermore, $A' = A[f_1, \ldots, f_r]$ is an admissible R-algebra according to Lemma 6, and it follows from the completeness of A' that the map

$$R\langle \zeta_1, \ldots, \zeta_r \rangle \xrightarrow{\alpha} B \hookrightarrow B_{\mathrm{rig}} \xrightarrow{\varphi_K^{\#}} A_{\mathrm{rig}}$$

will factor through A'. Thus, $\varphi_K^{\#} \colon B_{\mathrm{rig}} \longrightarrow A_{\mathrm{rig}}$ restricts to a well-defined R-morphism $\varphi^{\#} \colon B \longrightarrow A'$ giving rise to a morphism of admissible formal R-schemes $\varphi \colon X' \longrightarrow Y$ with $X' = \mathrm{Spf}\, A'$. But then the inclusion $A \hookrightarrow A'$ induces an admissible formal blowing-up $\tau \colon X' \longrightarrow X$ by Lemma 6 satisfying $\varphi_{\mathrm{rig}} = \varphi_K \circ \tau_{\mathrm{rig}}$, as required.

Now let us consider the general case where X and Y are quasi-paracompact. We fix affine open coverings of finite type $\mathfrak{U}$ of X and $\mathfrak{V}$ of Y, and consider the induced admissible coverings $\mathfrak{U}_{\mathrm{rig}}$ and $\mathfrak{V}_{\mathrm{rig}}$ of the associated rigid K-spaces X_{rig} and Y_{rig}. Then $\mathfrak{U}_{\mathrm{rig}}$ and $\mathfrak{V}_{\mathrm{rig}}$ are of finite type. Restricting the pull-back $\varphi_K^{-1}(\mathfrak{V}_{\mathrm{rig}})$, which is an admissible covering of X_{rig}, to each member $U_{\mathrm{rig}} \in \mathfrak{U}_{\mathrm{rig}}$, we can find a refinement $\mathfrak{U}_K$ of $\mathfrak{U}_{\mathrm{rig}}$ that is an admissible affinoid covering of finite type again, but where, in addition, any member $U_K \in \mathfrak{U}_K$ is mapped by φ_K into some member $V_{\mathrm{rig}} \in \mathfrak{V}_{\mathrm{rig}}$. Furthermore, using Lemma 5 in conjunction with 8.2/15, we may even assume that the covering $\mathfrak{U}_K$ is induced from an affine open covering of X, which we will denote by $\mathfrak{U}$ again. Now, for any $U \in \mathfrak{U}$, there is a member $V_{\mathrm{rig}} \in \mathfrak{V}_{\mathrm{rig}}$ such that U_{rig}, the admissible open subspace of X_{rig} induced from U, is mapped into V_{rig}. From the affine case we know that there is an admissible formal blowing-up $\tau_U \colon U' \longrightarrow U$ together with a morphism of formal R-schemes $\varphi_U \colon U' \longrightarrow V \hookrightarrow Y$ such that $\varphi_{U,\mathrm{rig}} \colon U'_{\mathrm{rig}} \longrightarrow Y_{\mathrm{rig}}$ coincides with the composition $\varphi_K|_{U_{\mathrm{rig}}} \circ \tau_{U,\mathrm{rig}}$. Using Proposition 8.2/14, we can dominate all blowing-ups τ_U by an admissible formal blowing-up $\tau \colon X' \longrightarrow X$ that, restricted to $\tau^{-1}(U)$ for each $U \in \mathfrak{U}$, factors through $\tau_U^{-1}(U)$ via some morphism $\sigma_U \colon \tau^{-1}(U) \longrightarrow \tau_U^{-1}(U)$. It follows from assertion (b) that the compositions $\varphi_U \circ \sigma_U$ can be glued to yield a well-defined morphism of admissible formal R-schemes $\varphi \colon X' \longrightarrow Y$ satisfying $\varphi_{\mathrm{rig}} = \varphi_K \circ \tau_{\mathrm{rig}}$. This settles assertion (c) of Lemma 4.

(d) Assume that X, Y are quasi-compact and that we have an isomorphism $\varphi_K \colon X_{\mathrm{rig}} \xrightarrow{\sim} Y_{\mathrm{rig}}$. Then, using (c), there is a diagram

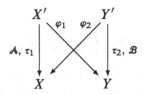

with admissible formal blowing-ups τ_1, τ_2, say given by the coherent open ideals $\mathcal{A} \subset \mathcal{O}_X$ and $\mathcal{B} \subset \mathcal{O}_Y$, such that

$$\varphi_{1,\mathrm{rig}} = \varphi_K \circ \tau_{1,\mathrm{rig}}, \qquad \varphi_{2,\mathrm{rig}} = \varphi_K^{-1} \circ \tau_{2,\mathrm{rig}}.$$

Furthermore, if σ_1 is the formal blowing-up of $\mathcal{B}\mathcal{O}_{X'}$ on X' and σ_2 the formal blowing-up of $\mathcal{A}\mathcal{O}_{Y'}$ on Y', the morphism $\varphi_1 \circ \sigma_1$ factors uniquely through

Y', and $\varphi_2 \circ \sigma_2$ factors uniquely through X', due to the universal property of admissible blowing-up. Thus, we can enlarge the diagram as follows:

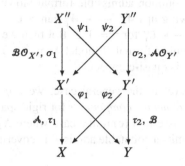

Since all vertical maps of the diagram induce isomorphisms on the level of associated rigid spaces and since all diagonal maps give rise to φ_K or its inverse on the level of X_{rig} and Y_{rig}, we can conclude from (b) that the diagram is commutative. Furthermore, using the universal property of the formal blowing-up σ_2 and the fact that the ideal $\mathcal{A}\mathcal{O}_{X''} \subset \mathcal{O}_{X''}$ is invertible, we see that the morphism ψ_1 factors uniquely through a morphism $\alpha_1 \colon X'' \longrightarrow Y''$. Likewise, ψ_2 will factor through a unique morphism $\alpha_2 \colon Y'' \longrightarrow X''$. Since, on the level of X_{rig} and Y_{rig}, the morphisms α_1, α_2 coincide with φ_K and its inverse, it follows from (b) again that α_1 and α_2 are inverse to each other. But then, using 8.2/11, namely that in the quasi-compact case the composition of two admissible formal blowing-ups yields an admissible formal blowing-up again, the assertion (d) of Lemma 4 follows.

(e) Consider a quasi-separated and quasi-paracompact rigid K-space X_K and an admissible covering of finite type $(X_{i,K})_{i \in J}$ of X_K by quasi-compact open subspaces $X_{i,K} \subset X_K$. We may even assume that $X_{i,K}$ is affinoid for all $i \in J$. Any finite union of affinoid open subspaces of X_K yields a quasi-compact open subspace of X_K and we will start by showing that quasi-compact open subspaces of X_K admit formal R-models. Thus, assuming that J is finite, we can proceed by induction on the cardinality of J. If J consists of just one element, X_K is affinoid, say $X_K = \operatorname{Sp} A_K$. Fixing an epimorphism of type $\alpha \colon K\langle \zeta_1, \ldots, \zeta_r \rangle \longrightarrow A_K$, let $A = \alpha(R\langle \zeta_1, \ldots, \zeta_r \rangle)$. Then $X = \operatorname{Spf} A$ is a formal R-model of X_K. Next, taking care of the induction step, assume $X_K = U_{1,K} \cup U_{2,K}$ with quasi-compact admissible open subspaces $U_{i,K} \subset X_K$ that admit formal R-models U_i for $i = 1, 2$. Let $W_K = U_{1,K} \cap U_{2,K}$. Since X_K is quasi-separated, W_K is quasi-compact. Thus, there is a finite admissible affinoid covering of W_K, and the latter can be enlarged to yield a finite admissible affinoid covering of $U_{1,K}$. Then, applying Lemma 5, there is an admissible formal blowing-up $U_1' \longrightarrow U_1$ such that the open immersion $W_K \hookrightarrow U_{1,K}$ is represented by an open immersion of admissible formal R-schemes $W_1' \hookrightarrow U_1'$. In the same way, we can find an admissible formal blowing-up $U_2' \longrightarrow U_2$ such that the open immersion $W_K \hookrightarrow U_{2,K}$ is represented by an open immersion of admissible formal R-schemes $W_2' \hookrightarrow U_2'$. In particular, W_1' and W_2' are two formal R-models of W_K and, by applying (d), there exists a common admissible formal blowing-up W'' of W_1' and W_2'. By 8.2/13, the blowing-ups $W'' \longrightarrow W_i'$ can be extended to admissible formal blowing-ups $U_i'' \longrightarrow U_i'$ for $i = 1, 2$. But then we can glue U_1'' to U_2'' along W'', thereby obtaining a formal R-model X of X_K. This settles Lemma 4 (e) in the case where X_K is quasi-compact.

To prove Lemma 4 (e) in the general case, we need to recall the concept of *connectedness* and of *connected components* for rigid spaces from 5.3/9 and 5.3/10. In fact, we will reduce assertion (e) to the case where X_K is connected and, being quasi-paracompact, admits a countable admissible covering by quasi-compact open subspaces.

Decomposing X_K into its connected components in the sense of 5.3/10, we may assume that X_K is connected. Then we will construct a countable admissible covering $(U_{n,K})_{n \in \mathbb{N}}$ of X_K consisting of quasi-compact open subspaces $U_{n,K} \subset X_K$ with the additional property that

$$U_{n,K} \cap U_{m,K} = \emptyset \quad \text{for} \quad m < n - 1.$$

To do this, fix an admissible covering of finite type $(X_{i,K})_{i \in J}$ of X_K where all $X_{i,K}$ are connected. We start with $U_{0,K} = X_{i_0,K}$ for some $i_0 \in J$ and define $U_{n+1,K}$ for $n \in \mathbb{N}$ inductively as the union of all $X_{i,K}$ that are not yet contained in the union

$$V_{n,K} = U_{0,K} \cup \ldots \cup U_{n,K},$$

but meet $U_{n,K}$. As $(X_{i,K})_{i \in J}$ is a covering of finite type, U_{n+1} consists of only finitely many sets $X_{i,K}$ and, hence, is quasi-compact. Furthermore, it is easily seen that $\bigcup_{n=0}^{\infty} U_{n,K}$ equals the connected component of X_K containing $X_{i_0,K}$. However, as X_K was supposed to be connected, this component coincides with X_K. Therefore $(U_{n,K})_{n \in \mathbb{N}}$ is an admissible covering of X_K consisting of admissible open subsets that are quasi-compact.

To construct a formal R-model X of X_K, we proceed by induction on n via the procedure we have used above in the quasi-compact case. Thus, assume that we have already obtained a formal R-model V_n of the union $V_{n,K}$, together with open immersions $U_i \hookrightarrow V_n$ for $i \le n$ representing the open immersions $U_{i,K} \hookrightarrow V_{n,K}$. To obtain a formal R-model of $V_{n+1,K} = V_{n,K} \cup U_{n+1,K}$, we start out from the formal R-model U_n of $U_{n,K}$ and a certain formal R-model U_{n+1} of $U_{n+1,K}$, which exists, since $U_{n+1,K}$ is quasi-compact. Then, in order to glue U_n to U_{n+1}, we need to perform suitable admissible formal blowing-ups on U_n and U_{n+1} first. Since $U_{n+1,K}$ does not meet any $U_{i,K}$ for $i < n$, we can glue U_{n+1} to V_n after extending the blowing-up on the side of U_n to all of V_n. Now, due to 8.2/13, such an extension exists and can be chosen in such a way that it is an isomorphism over any open $V \subset V_n$ disjoint from U_n. In particular, the extension of the blowing-up leaves all U_i with $i < n - 1$ unchanged. From this it follows that the R-models

V_n of $V_{n,K}$ "converge" towards a well-defined R-model X of X_K, as n progresses towards infinity. This settles assertion (e) of Lemma 4 and thereby also the proof of Theorem 3.   □

In Sect. 5.4 we have associated to any K-scheme of locally finite type X a rigid K-space X^{rig}, called the *rigid analytification* of X. We want to show:

Proposition 7. *Let X be a separated K-scheme of finite type. Then the associated rigid analytification X^{rig} is a separated and quasi-paracompact rigid K-space and, hence, admits a formal R-model X.*

Proof. The explicit construction of rigid analytifications in Sect. 5.4 shows that for any closed immersion $f: X \longrightarrow Y$ of K-schemes (of locally finite type) the associated rigid analytification $f^{\mathrm{rig}}: X^{\mathrm{rig}} \longrightarrow Y^{\mathrm{rig}}$ is a closed immersion. Furthermore, it is seen in the same way that rigid analytification respects cartesian products (in fact, more generally, fiber products). Therefore, if X is separated, its rigid analytification will be separated as well.

To show that X^{rig} is quasi-paracompact, we use the fact that X, as a separated scheme of finite type, admits a so-called Nagata compactification $\overline{X}$; see Conrad [C]. This is a proper K-scheme containing X as a dense open subscheme. Then, by Chow's Lemma [EGA II], 5.6, there is an epimorphism $\omega: P \longrightarrow \overline{X}$ from a projective scheme P onto $\overline{X}$. Now consider the associated morphism of rigid K-spaces $\omega^{\mathrm{rig}}: P^{\mathrm{rig}} \longrightarrow \overline{X}^{\mathrm{rig}}$, which is surjective as well. As we have seen in Sect. 5.4, the rigid analytification $\mathbb{P}_K^{n,\mathrm{rig}}$ of the projective n-space $\mathbb{P}_K^n$ admits an admissible affinoid covering that is finite. Likewise, the same is true for P^{rig}, and it follows from the surjectivity of ω^{rig} that any admissible affinoid covering of $\overline{X}^{\mathrm{rig}}$ admits a finite refinement. In particular, there is a finite admissible affinoid covering $(U_\nu)_{\nu \in N}$ of $\overline{X}^{\mathrm{rig}}$, say for $N = \{1, \ldots, n\}$.

Let $Z = \overline{X}^{\mathrm{rig}} - X^{\mathrm{rig}}$. It follows that $U_\nu \cap Z$ is Zariski closed in U_ν for each $\nu \in N$, and we claim that its complement $U_\nu - Z$ is quasi-paracompact. In fact, choose global sections $f_1, \ldots, f_r$ on U_ν whose zero set is $U_\nu \cap Z$. Then, fixing some $\pi \in R, 0 < |\pi| < 1$, the final term of the equation

$$U_\nu - Z = \bigcup_{i=1}^{r} \{x \in U_\nu \, ; \, |f_i(x)| > 0\}$$

$$= \bigcup_{m \in \mathbb{Z}} \{x \in U_\nu \, ; \, |\pi|^{m+1} \leq \max_{i=1,\ldots,r} |f_i(x)| \leq |\pi|^m\}$$

yields an admissible covering of finite type of $U_\nu - Z$ by quasi-compact open subsets. Hence, $U_\nu - Z$ is quasi-paracompact for each $\nu \in N$, and we claim that the same is true for

$$X^{\mathrm{rig}} = (U_1 - Z) \cup \ldots \cup (U_n - Z).$$

To justify this, we can proceed by induction on n. Writing $U' = U_1 \cup \ldots \cup U_{n-1}$, assume that

$$U' - Z = (U_1 - Z) \cup \ldots \cup (U_{n-1} - Z)$$

is quasi-paracompact. Then, choosing admissible affinoid coverings of finite type $\mathfrak{U}'$ on $U' - Z$ and $\mathfrak{U}''$ on $U_n - Z$ and using the fact that $\overline{X}^{\mathrm{rig}}$ is separated, these restrict to admissible coverings of finite type $\mathfrak{U}'_{\mathrm{res}}$ and $\mathfrak{U}''_{\mathrm{res}}$ on the intersection

$$(U' - Z) \cap (U_n - Z) = (U' - Z) \cap U_n = U' \cap (U_n - Z),$$

where the members of $\mathfrak{U}'_{\mathrm{res}}$ are affinoid and the members of $\mathfrak{U}''_{\mathrm{res}}$ are at least quasi-compact, since U' has this property. We claim that the union $\mathfrak{U}'_{\mathrm{res}} \cup \mathfrak{U}''_{\mathrm{res}}$ is an admissible covering of finite type again. To verify this, fix an open affinoid subspace $V \subset (U' \cap U_n) - Z$. It is covered by finitely many members of $\mathfrak{U}'_{\mathrm{res}}$ and, likewise, of $\mathfrak{U}''_{\mathrm{res}}$. Then, since $\mathfrak{U}'_{\mathrm{res}}, \mathfrak{U}''_{\mathrm{res}}$ are coverings of finite type, V will meet only finitely many members of $\mathfrak{U}'_{\mathrm{res}}$ and $\mathfrak{U}''_{\mathrm{res}}$. Therefore it follows that the covering $\mathfrak{U}'_{\mathrm{res}} \cup \mathfrak{U}''_{\mathrm{res}}$ must be of finite type. Hence, the same is true for the covering $\mathfrak{U}' \cup \mathfrak{U}''$. Since the latter is an admissible affinoid covering of X^{rig}, we are done.   □

A typical example of a rigid K-space X_K that is not quasi-paracompact, can be obtained by gluing an infinity of unit discs $\mathbb{B}^1_K = \mathrm{Sp}\, K\langle \zeta \rangle$ along the open unit disk $\mathbb{B}^1_+ = \{x \in \mathbb{B}^1_K \,;\, |\zeta(x)| < 1\}$. Since X_K is not quasi-separated, it cannot admit a formal R-model.

Chapter 9
More Advanced Stuff

9.1 Relative Rigid Spaces

So far we have considered formal schemes over adic base rings R that are part of the following classes mentioned in Sect. 7.3:

(V) R is an adic valuation ring with a finitely generated ideal of definition (which automatically is principal by 7.1/6).

(N) R is a Noetherian adic ring with an ideal of definition I such that R does not have I-torsion.

Instead of $S = \mathrm{Spf}\, R$ we can just as well work over more global bases. The following types of formal base schemes S will be of interest:

(V′) S is an admissible formal R-scheme where R is an adic valuation ring of type (V) as above. Thus, the topology of $\mathcal{O}_S$ is generated by the ideal $\pi \mathcal{O}_S$ where $\pi \in R$ is a suitable element generating the adic topology of R.

(N′) S is a Noetherian formal scheme (of quite general type) such that the topology of its structure sheaf $\mathcal{O}_S$ is generated by a coherent ideal $\mathcal{J} \subset \mathcal{O}_S$ and such that $\mathcal{O}_S$ does not admit $\mathcal{J}$-torsion.

Over base schemes S of this type, it is possible to consider admissible formal S-schemes, or just formal S-schemes that are locally of topologically finite presentation. Then, taking into account the Theorem of Raynaud 8.4/3, we can extend the notion of rigid spaces to such more general situations as follows:

Definition 1 (Raynaud). *Let S be a formal scheme of type* (V′) *or* (N′)*, as defined above, and let* (FSch/S) *be the category of admissible formal S-schemes. Then the category* (Rig/S) *of rigid S-spaces is defined as the localization of* (FSch/S) *by admissible formal blowing-ups.*

Thus, as object, a rigid S-space is the same as an admissible formal S-scheme, whereas on the level of morphisms, admissible formal blowing-ups are viewed

S. Bosch, *Lectures on Formal and Rigid Geometry*, Lecture Notes in Mathematics 2105, DOI 10.1007/978-3-319-04417-0_9, © Springer International Publishing Switzerland 2014

as isomorphisms. To get an intuitive picture of such a rigid space, one may generalize the concept of rig-points as developed in Sect. 8.3 and view any rigid S-space X as a family of classical rigid spaces X_s over the rig-points $s \in S$.

We want to work out in more detail how the category (Rig/S) of rigid S-spaces is obtained from (FSch/S), assuming that we restrict ourselves to formal S-schemes that are *quasi-separated* and *quasi-paracompact*. To define (Rig/S), take as objects the objects of (FSch/S). Furthermore, for two objects X, Y of (Rig/S), a morphism $X \longrightarrow Y$ is given by an equivalence class of diagrams in (FSch/S) of type

where $X' \longrightarrow X$ is an admissible formal blowing-up. Two such diagrams

$$X \longleftarrow X_1' \longrightarrow Y, \qquad X \longleftarrow X_2' \longrightarrow Y,$$

are called equivalent if there is a third diagram $X \longleftarrow X'' \longrightarrow Y$ of this type, together with factorizations $X'' \longrightarrow X_1$ and $X'' \longrightarrow X_2$ making the following diagram commutative:

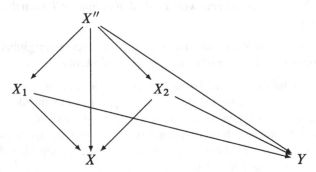

It is not difficult to check directly, using 8.2/10, that the just described relation really is an equivalence relation.

On the other hand, it might be more appropriate to interpret the set of morphisms $\text{Hom}_{(\text{Rig}/S)}(X, Y)$ as the direct limit of the sets $\text{Hom}_{(\text{FSch}/S)}(X', Y)$ where X' varies over all admissible formal blowing-ups of X; here direct limits are meant in the style of Artin [A], Sect. I.1. To do this, consider the full subcategory $\mathfrak{B}$ of the category of all X-objects $X' \longrightarrow X$ in (FSch/S) whose structural morphisms are admissible formal blowing-ups, and consider the contravariant functor $\mathfrak{B} \longrightarrow (\text{Sets})$ associating to any object $X' \longrightarrow X$ of $\mathfrak{B}$ the set $\text{Hom}_{(\text{FSch}/S)}(X', Y)$. Viewing this as a covariant functor $\mathfrak{B}^0 \longrightarrow (\text{Sets})$, we have

$$\text{Hom}_{(\text{Rig}/S)}(X, Y) = \varinjlim_{X' \in \mathfrak{B}^0} \text{Hom}_{(\text{FSch}/S)}(X', Y).$$

To compose two morphisms $X \longrightarrow Y$ and $Y \longrightarrow Z$ in (Rig/S), say given by diagrams

$$X \longleftarrow X' \longrightarrow Y, \qquad Y \longleftarrow Y' \longrightarrow Z$$

in (FSch/S), we use a diagram of type

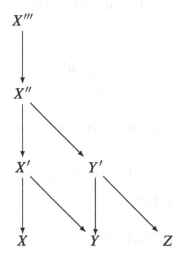

where $X'' \longrightarrow X'$ is the pull-back (in the sense of 8.2/16) of the admissible formal blowing-up $Y' \longrightarrow Y$ and where $X''' \longrightarrow X$ is an admissible formal blowing-up dominating the composition $X'' \longrightarrow X' \longrightarrow X$; see 8.2/15. It is straightforward to show that the objects of (FSch/S), together with the described morphisms, satisfy the universal property of a localization of (FSch/S) by the class of admissible formal blowing-ups.

9.2 An Example: Raynaud's Universal Tate Curve

As a typical example of a rigid space in the sense of 9.1/1, we want to construct Raynaud's universal family of Tate elliptic curves. The latter is defined over the formal base scheme $S = \mathrm{Spf}\,\mathbb{Z}[\![Q]\!]$ where Q is a variable. Note that $\mathbb{Z}[\![Q]\!]$ is not a valuation ring and neither a ring that can be accessed in terms of classical rigid geometry, since it is not of class (V) or (V'). However, it is an adic ring of class (N) with ideal of definition generated by Q. To begin with, we first carry out the construction of Tate curves over a complete valuation ring R of height 1 with field of fractions K. Let $q \in R$ where $0 < |q| < 1$. Then the multiplicative group scheme $\mathbb{G}_{m,K}$ can be viewed as a rigid K-group via rigid analytification. The set of K-valued points $\mathbb{G}_{m,K}(K)$ coincides with K^* and, thus, we may consider the infinite cyclic group $q^{\mathbb{Z}}$ generated by q as a closed analytic subgroup of $\mathbb{G}_{m,K}$. Then we can build the quotient $E_q = \mathbb{G}_{m,K}/q^{\mathbb{Z}}$ in the category of rigid K-spaces, in fact, in the category of rigid K-groups, and this is the Tate elliptic curve over K that is associated to the parameter q. To describe E_q in more concrete terms, consider the unit disk $B = \mathbb{B}^1_K = \operatorname{Sp} K\langle \zeta \rangle$ and look at the affinoid subdomains given by the annuli

$$U_1 = B(q\zeta^{-2}) = \left\{x \in B \,;\, |q|^{\frac{1}{2}} \leq |\zeta(x)| \leq 1\right\}$$
$$= \operatorname{Sp} K\langle \zeta, q\zeta^{-2}\rangle,$$

$$U_2 = B(q\zeta^{-1}, q^{-1}\zeta^2) = \left\{x \in B \,;\, |q| \leq |\zeta(x)| \leq |q|^{\frac{1}{2}}\right\}$$
$$= \operatorname{Sp} K\langle \zeta, q\zeta^{-1}, q^{-1}\zeta^2\rangle.$$

Furthermore, looking at the peripheries of these annuli, note that U_1 contains the affinoid subdomains

$$U_1^+ = B(\zeta^{-1}) = \left\{x \in B \,;\, |\zeta(x)| = 1\right\}$$
$$= \operatorname{Sp} K\langle \zeta, \zeta^{-1}\rangle,$$

$$U_1^- = B(q^{-1}\zeta^2, q\zeta^{-2}) = \left\{x \in B \,;\, |\zeta(x)| = |q|^{\frac{1}{2}}\right\}$$
$$= \operatorname{Sp} K\langle \zeta, q^{-1}\zeta^2, q\zeta^{-2}\rangle,$$

just as U_2 contains the affinoid subdomains

$$U_2^+ = B(q^{-1}\zeta^2, q\zeta^{-2}) = \left\{x \in B \,;\, |\zeta(x)| = |q|^{\frac{1}{2}}\right\}$$
$$= \operatorname{Sp} K\langle \zeta, q^{-1}\zeta^2, q\zeta^{-2}\rangle,$$

$$U_2^- = B(q^{-1}\zeta, q\zeta^{-1}) = \left\{x \in B \,;\, |\zeta(x)| = |q|\right\}$$
$$= \operatorname{Sp} K\langle \zeta, q^{-1}\zeta, q\zeta^{-1}\rangle.$$

Then, clearly, U_1^- coincides with U_2^+ and there is a canonical isomorphism

$$K\langle \zeta, q^{-1}\zeta, q\zeta^{-1}\rangle \overset{\sim}{\longrightarrow} K\langle \zeta, \zeta^{-1}\rangle, \qquad \zeta \longmapsto q\zeta,$$

which corresponds to an isomorphism

$$\tau : U_1^+ \overset{\sim}{\longrightarrow} U_2^-, \qquad x \longmapsto qx.$$

Now observe that

$$\mathbb{G}_{m,K} = \bigcup_{n\in\mathbb{Z}} q^n(U_1 \cup U_2)$$

is an admissible affinoid covering of $\mathbb{G}_{m,K}$ viewed as a rigid K-space, in fact a covering of finite type by annuli. In order to construct the quotient $E_q = \mathbb{G}_{m,K}/q^{\mathbb{Z}}$, one just glues the union $U_1 \cup U_2$ to itself by identifying U_1^+ with U_2^- via the isomorphism τ.

Let us exhibit a formal R-model $\mathcal{E}_q$ of E_q. Looking at the canonical epimorphisms

$$\varphi_1 : K\langle \zeta, \eta \rangle \longrightarrow K\langle \zeta, q\zeta^{-2}\rangle \simeq K\langle \zeta, \eta\rangle/(q^{-1}\zeta^2\eta - 1),$$

$$\varphi_2 : K\langle \zeta, \eta, \xi \rangle \longrightarrow K\langle \zeta, q^{-1}\zeta^2, q\zeta^{-1}\rangle \simeq K\langle \zeta, \eta, \xi\rangle/(q^{-1}\zeta^2 - \eta, q^{-1}\zeta\xi - 1),$$

one can check that

$$\ker \varphi_1 \cap R\langle \zeta, \eta\rangle = (\zeta^2\eta - q)R\langle \zeta, \eta\rangle,$$

$$\ker \varphi_2 \cap R\langle \zeta, \eta, \xi\rangle = (\zeta^2 - q\eta, \zeta\xi - q, \zeta - \eta\xi)R\langle \zeta, \eta, \xi\rangle.$$

This is done using the multiplicativity of the Gauß norm on $K\langle \zeta, \eta\rangle$ and, in the case of the second equation, by dividing out the generator

$$\zeta - \eta\xi = q^{-1}\xi(\zeta^2 - q\eta) - q^{-1}\zeta(\zeta\xi - q)$$

first. The ideals just constructed give rise to flat and, hence, admissible formal R-schemes

$$\mathcal{U}_1 = \operatorname{Spf} R\langle \zeta, q\zeta^{-2}\rangle = \operatorname{Spf} R\langle \zeta, \eta\rangle/(\zeta^2\eta - q),$$

$$\mathcal{U}_2 = \operatorname{Spf} R\langle \zeta, q^{-1}\zeta^2, q\zeta^{-1}\rangle = \operatorname{Spf} R\langle \zeta, \eta, \xi\rangle/(\zeta^2 - q\eta, \zeta\xi - q, \zeta - \eta\xi),$$

which are formal R-models of U_1 and U_2. Then we can consider the open formal subschemes

$$\mathcal{U}_1^+ = \mathcal{U}_1(\zeta^{-1}), \qquad \mathcal{U}_1^- = \mathcal{U}_1(q^{-1}\zeta^2),$$

$$\mathcal{U}_2^+ = \mathcal{U}_2(q\zeta^{-2}), \qquad \mathcal{U}_2^- = \mathcal{U}_2(q^{-1}\zeta)$$

of $\mathcal{U}_1$ and $\mathcal{U}_2$, and one checks that $\mathcal{U}_1^-$ coincides canonically with $\mathcal{U}_2^+$. Furthermore, multiplication with q yields an isomorphism $\mathcal{U}_1^+ \overset{\sim}{\longrightarrow} \mathcal{U}_2^-$. In fact, the open immersions

$$\mathcal{U}_1^+ \hookrightarrow \mathcal{U}_1, \qquad \mathcal{U}_1^- \hookrightarrow \mathcal{U}_1, \qquad \mathcal{U}_2^+ \hookrightarrow \mathcal{U}_2, \qquad \mathcal{U}_2^- \hookrightarrow \mathcal{U}_2,$$

together with the just mentioned canonical isomorphisms, represent the open immersions

$$U_1^+ \hookrightarrow U_1, \qquad U_1^- \hookrightarrow U_1, \qquad U_2^+ \hookrightarrow U_2, \qquad U_2^- \hookrightarrow U_2,$$

and their identifications on the level of formal R-models. In other words, we can identify $\mathcal{U}_1^-$ with $\mathcal{U}_2^+$, as well as $\mathcal{U}_1^+$ with $\mathcal{U}_2^-$, and thereby construct a formal R-model $\mathcal{E}_q$ of E_q.

The latter model can also be obtained as a quotient of a formal R-model $\mathcal{G}$ of the multiplicative group $\mathbb{G}_{m,K}$. Indeed, glue $\mathcal{U}_1$ to $\mathcal{U}_2$ via the canonical identification $\mathcal{U}_1^- \overset{\sim}{\longrightarrow} \mathcal{U}_2^+$ and define an admissible formal R-scheme

$$\mathcal{G} = \bigcup_{n \in \mathbb{Z}} q^n (\mathcal{U}_1 \cup \mathcal{U}_2)$$

by using an infinite number of copies of $\mathcal{U}_1 \cup \mathcal{U}_2$, say denoted by $q^n(\mathcal{U}_1 \cup \mathcal{U}_2)$ for $n \in \mathbb{Z}$, and glue $q^{n+1}(\mathcal{U}_1 \cup \mathcal{U}_2)$ to $q^n(\mathcal{U}_1 \cup \mathcal{U}_2)$ via the canonical isomorphism $\mathcal{U}_1^+ \overset{\sim}{\longrightarrow} \mathcal{U}_2^-$ induced from multiplication by q. The resulting formal R-scheme $\mathcal{G}$ is a formal R-model of $\mathbb{G}_{m,K}$, although $\mathcal{G}$ cannot be viewed as a formal R-group scheme. However, multiplication by q is defined on $\mathcal{G}$ and we see that the formal model $\mathcal{E}_q$ of E_q may be viewed as the quotient $\mathcal{G}/q^{\mathbb{Z}}$.

Now observe that the construction of the admissible formal scheme $\mathcal{E}_q$ is already possible over the base $S = \operatorname{Spf} \mathbb{Z}[\![Q]\!]$, for a variable Q replacing the parameter q. Associated to this object of (FSch/S) is a rigid S-space E_Q in (Rig/S), which may be viewed as the family of all Tate elliptic curves. In fact, if E_q is a Tate elliptic curve over some complete valuation ring R of height 1, we can look at the canonical morphism $\sigma \colon \operatorname{Spf} R \longrightarrow S$ given by

$$\mathbb{Z}[\![Q]\!] \longrightarrow R, \qquad Q \longmapsto q,$$

thereby obtaining E_q as the pull-back of E_Q with respect to σ.

9.3   The Zariski–Riemann Space

In the following, let S be a formal scheme of type (V$'$) or (N$'$), as introduced in Sect. 9.1, and let X be an admissible formal S-scheme where we will always assume that X is *quasi-separated* and *quasi-paracompact*. Then we can consider the family $(X_{\mathcal{A}})_{\mathcal{A} \in \mathfrak{B}(X)}$ of all admissible formal blowing-ups $X_{\mathcal{A}} \longrightarrow X$, parametrized by the set $\mathfrak{B}(X)$ of coherent open ideals $\mathcal{A} \subset \mathcal{O}_X$. For $\mathcal{A}, \mathcal{B} \in \mathfrak{B}(X)$ we write $\mathcal{A} \leq \mathcal{B}$ if the ideal $\mathcal{A}$ becomes invertible on $X_{\mathcal{B}}$. Due to the universal property of admissible formal blowing-up 8.2/9, the latter implies that the blowing-up $X_{\mathcal{B}} \longrightarrow X$ factors through a unique morphism $\Phi_{\mathcal{A}\mathcal{B}} \colon X_{\mathcal{B}} \longrightarrow X_{\mathcal{A}}$. Furthermore, given $\mathcal{A}, \mathcal{B} \in \mathfrak{B}(X)$, we have $\mathcal{A}\mathcal{B} \in \mathfrak{B}(X)$, as well as $\mathcal{A} \leq \mathcal{A}\mathcal{B}$ and $\mathcal{B} \leq \mathcal{A}\mathcal{B}$. It is clear that the $X_{\mathcal{A}}$ together with the morphisms $\Phi_{\mathcal{A}\mathcal{B}}$ define a projective system of S-morphisms so that we can look at the projective limit

$$\langle X \rangle = \varprojlim_{\mathcal{A} \in \mathfrak{B}(X)} X_{\mathcal{A}}.$$

This projective limit is, first of all, meant in terms of topological spaces. Furthermore, for each $\mathcal{A} \in \mathfrak{B}(X)$, there is a canonical projection $\pi_{\mathcal{A}}: \langle X \rangle \longrightarrow X_{\mathcal{A}}$, and we can consider

$$\mathcal{O}_{\langle X \rangle} = \varinjlim_{\mathcal{A} \in \mathfrak{B}(X)} \pi_{\mathcal{A}}^{-1}(\mathcal{O}_{X_{\mathcal{A}}})$$

as a sheaf of rings on $\langle X \rangle$. It is not hard to see that the stalks

$$\mathcal{O}_{\langle X \rangle, x} = \varinjlim_{\mathcal{A} \in \mathfrak{B}(X)} \mathcal{O}_{\mathcal{A}, \pi_{\mathcal{A}}(x)}, \qquad x \in \langle X \rangle,$$

being direct limits of local rings, are local again so that $\langle X \rangle = (\langle X \rangle, \mathcal{O}_{\langle X \rangle})$ is a locally ringed space.

Definition 1. *Let X be an admissible formal S-scheme (by the convention of the present section assumed to be quasi-separated and quasi-paracompact). Then the associated locally ringed space*

$$\langle X \rangle = \varprojlim_{\mathcal{A} \in \mathfrak{B}(X)} X_{\mathcal{A}}$$

is called the Zariski–Riemann space[1] *associated to X.*

Without proof, let us mention a few facts on the topology of $\langle X \rangle$. For more details, consult [FK], Chap. II.3.

Proposition 2. *Let X be an admissible formal S-scheme and $\langle X \rangle$ the associated Zariski–Riemann space. Then:*

(i) *$\langle X \rangle$ is non-empty if X is non-empty.*
(ii) *$\langle X \rangle$ is sober[2] and, in particular a T_0-space, but not necessarily Hausdorff.*
(iii) *$\langle X \rangle$ is quasi-separated and locally quasi-compact, even quasi-paracompact. It is quasi-compact if X is quasi-compact.*

Passing from X to the associated rigid S-space X_{rig}, we see that the Zariski–Riemann space $\langle X \rangle$, in a certain sense, takes into account all formal S-models of X_{rig}, just as X_{rig} itself does. Thus, one can well imagine that there is a certain equivalence between X_{rig} and $\langle X \rangle$, although one must be aware of the fact that, in the classical rigid case, X_{rig} is a locally ringed space with respect to a *Grothendieck*

---

[1] Zariski–Riemann spaces were first introduced by Zariski calling them Riemann manifolds. Later, Nagata preferred the term Zariski–Riemann space when he used these spaces for the compactification of algebraic varieties.

[2] A topological space is called *sober* if every irreducible closed subset admits a unique generic point.

topology, whereas $\langle X \rangle$ is a locally ringed space in the ordinary sense. Of course, this difference is in accordance with the fact that $\langle X \rangle$ includes "much more" points than X_{rig}.

To look a bit closer on the relationship between X_{rig} and $\langle X \rangle$, let us restrict to the classical rigid case where S consists of an adic valuation ring R of height 1 with field of fractions K. As we have shown in Sect. 8.3, there is a well-defined specialization map

$$\text{sp}: X_{\text{rig}} \longrightarrow X,$$

mapping a point of X_{rig} to a closed point of X. Since sp is functorial, the map factors through all formal models $X_{\mathcal{A}}$, as $\mathcal{A}$ varies in $\mathfrak{B}(X)$, thus, giving rise to a specialization map

$$\text{sp}: X_{\text{rig}} \longrightarrow \langle X \rangle.$$

Proposition 3. *In the classical rigid case, let X be an admissible formal R-scheme. Then the specialization map* sp: $X_{\text{rig}} \longrightarrow \langle X \rangle$ *enjoys the following properties*:

(i) sp *is injective.*
(ii) *The image of* sp *is dense in $\langle X \rangle$ with respect to the constructible topology.*[3]

There are examples of abelian sheaves $\mathcal{F}$ on a rigid K-space X_K where all stalks $\mathcal{F}_x$ for $x \in X_K$ are trivial, although $\mathcal{F}$ is not trivial itself; see 5.2/2. This shows that in order to handle general abelian sheaves, rigid K-spaces are not equipped with sufficiently many points that can give rise to stalk functors. On the other hand, the Zariski–Riemann space associated to a formal model of X_K does not suffer from such a problem and, indeed, can serve as an excellent replacement for X_K, due to the following fact:

Proposition 4. *In the classical rigid case, let X_K be a rigid K-space with a formal R-model X. Then the specialization map* sp: $X_K \longrightarrow \langle X \rangle$ *induces a natural equivalence between the category of abelian sheaves on X_K and the category of abelian sheaves on $\langle X \rangle$.*

Without giving a full proof, let us just indicate how to pass back and forth between abelian sheaves on X_K and $\langle X \rangle$. For any admissible open subset $U_K \subset X_K$, consider the open subset

$$\text{sp}_*(U_K) = \bigcup W$$

---

[3]The definition of the constructible topology is based on the notion of constructible sets [EGA I], Chap. 0, 2.3.10, and ind-constructible sets [EGA I], Chap. I, 7.2.2. For a convenient adaptation to our situation see [W], 3.3.

of the Zariski–Riemann space $\langle X \rangle$ where the union runs over all open subsets $W \subset \langle X \rangle$ such that $W \cap \mathrm{sp}(X_K) \subset \mathrm{sp}(U_K)$. For example, if U_K is quasi-compact and, thus, is represented by some open formal subscheme $U' \subset X_{\mathcal{A}}$ for a coherent open ideal $\mathcal{A} \in \mathfrak{B}(X)$, then $\mathrm{sp}_*(U_K) = \pi_{\mathcal{A}}^{-1}(U')$. Furthermore, one shows for any admissible open subset $U_K \subset X_K$ that a given union $U_K = \bigcup_{i \in J} U_{i,K}$, consisting of admissible open subsets $U_{i,K} \subset U_K$, is an admissible covering of U_K if and only if $\mathrm{sp}_*(U_K) = \bigcup_{i \in J} \mathrm{sp}_*(U_{i,K})$. Now start with an abelian sheaf $\mathcal{F}_K$ on X_K and set $\mathcal{F}(\mathrm{sp}_*(U_K)) = \mathcal{F}_K(U_K)$ for any quasi-compact admissible open subset $U_K \subset X_K$. Since the associated sets of type $\mathrm{sp}_*(U_K)$ define a basis of the topology on $\langle X \rangle$, we can view $\mathcal{F}$ as a sheaf on $\langle X \rangle$. Conversely, given any abelian sheaf $\mathcal{F}$ on $\langle X \rangle$, we can define an abelian sheaf $\mathcal{F}_K$ on X_K by setting $\mathcal{F}_K(U_K) = \mathcal{F}(\mathrm{sp}_*(U_K))$ for any quasi-compact admissible open subset $U_K \subset X_K$. It is not hard to see that the described correspondence between abelian sheaves on X_{rig} and $\langle X \rangle$ is an equivalence of categories.

9.4 Further Results on Formal Models

Working with a scheme X_K over the field of fractions K of a discrete valuation ring R, the arithmetic nature of X_K can quite often be uncovered by looking at suitable R-models X of X_K. In fact, one is interested in models where certain properties already present on X_K extend to the level of X. For example if X_K is a proper smooth curve, we can construct the minimal regular model X of X_K, so to say a best possible R-model that is still proper. Or we can consider an abelian variety X_K over K and look at the Néron model X of X_K. This is a best possible R-model that is smooth.

In the same spirit we can start with a classical rigid space X_K, or with a rigid space in the style of 9.1/1, and try to extend certain properties from X_K to the level of suitable formal models. This is the theme we want to discuss in the present section. However, for rigid spaces in the style of 9.1/1, which are given as objects in a localized category, specific properties have still to be introduced in a way that is compatible with the classical rigid case. The whole subject is rather extensive and so we can only highlight some of the main points at this place. For further information we refer to the series of articles on *Formal and rigid geometry* [F I], [F II], [F III], [F IV], as well as to the monograph [EGR].

Let S be a formal base scheme of type (V') or (N'), as in Sect. 9.1, and let X_{rig} be a rigid S-space in the sense of 9.1/1. Without explicitly saying so, we will always assume such rigid spaces, as well as their formal S-models, to be *quasi-separated* and *quasi-paracompact*.

If (P) is a property applicable to schemes or formal schemes, we can basically proceed in two ways in order to extend the notion of (P) to rigid S-spaces like X_{rig}. The first possibility is to say that X_{rig} satisfies (P) if there exists a formal S-model X of X_{rig} satisfying (P). For example, on the level of morphisms, one can proceed like this with open (resp. closed) immersions. Thus, call a morphism of rigid S-spaces $\tau_{\text{rig}}\colon U_{\text{rig}} \longrightarrow X_{\text{rig}}$ an *open immersion* (resp. a *closed immersion*) if τ_{rig} admits an open (resp. closed) immersion of admissible formal S-schemes $\tau\colon U \longrightarrow X$ as a formal S-model. That such a definition coincides with the usual one in the classical rigid case follows from 8.4/5 for open immersions, whereas closed immersions can be handled relying on 8.4/6. Similarly one can proceed with *proper morphisms*; for the compatibility of properness in terms of formal schemes with the definition 6.3/6 in the classical rigid case see Lütkebohmert [L] or Temkin [Te].

Another more direct approach to define certain properties on general rigid spaces consists in looking at the validity of (P) on the "complement" of the special fiber of formal S-models X associated to X_{rig}. To be more precise, let $\mathcal{I} \subset \mathcal{O}_S$ be an ideal of definition. Then, for any formal S-model X, the scheme $X_0 = X \otimes_S \mathcal{O}_S/\mathcal{I}$ is called the *special fiber* of X. If $(U_i)_{i \in J}$ is an affine open covering of X, say $U_i = \text{Spf}\, A_i$, and if, on U_i, the coherent open ideal $\mathcal{I}\mathcal{O}_X \subset \mathcal{O}_X$ is associated to the ideal $\mathfrak{a}_i \subset A_i$, we view the ordinary scheme $\text{Spec}\, A_i - V(\mathfrak{a}_i)$ as the complement of the special fiber on U_i. In general, such a complement is not well-defined globally on X. However, if we restrict ourselves to *closed* points and consider the classical rigid situation, then the complement of the special fiber of X makes sense globally, as it coincides with the point set of the rigid space X_{rig} associated to X in the sense of Sect. 7.4.

Now, if (P) is a scheme property, we can consider an affine open covering $(U_i)_{i \in J}$ of X as before and say that X_{rig} satisfies (P) if all schemes $\text{Spec}\, A_i - V(\mathfrak{a}_i)$ satisfy (P). Of course, in order that (P) defines a reasonable property on the associated rigid S-space X_{rig}, one has to check that the validity of (P) is independent of the chosen covering $(U_i)_{i \in J}$ and invariant under admissible formal blowing-up. Then, in most cases, it is a truly demanding venture, to find out whether or not a rigid S-space satisfying (P) will always admit a formal S-model satisfying (P).

As a first example that can successfully be handled along these lines, let us mention the property (P) of being flat, for morphisms of rigid S-spaces or coherent modules on rigid S-spaces and their formal models. Flatness on the rigid level is defined via flatness on complements of the special fiber, a method that is compatible with the usual notion of flatness in the classical rigid case. The main result on flatness is then the existence of flat formal models, due to Raynaud and Gruson; see [RG], as well as [F II].

Theorem 1 (Flattening Theorem). *Let $\varphi\colon X \longrightarrow Y$ be a quasi-compact morphism of admissible formal S-schemes and assume that the associated morphism $\varphi_{\text{rig}}\colon X_{\text{rig}} \longrightarrow Y_{\text{rig}}$ between rigid S-spaces is flat. Then there exists a commutative diagram of admissible formal S-schemes*

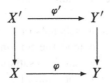

where φ' is flat, $Y' \longrightarrow Y$ is the formal blowing-up of some coherent open ideal $\mathcal{A} \subset \mathcal{O}_Y$, and where $X' \longrightarrow X$ is the formal blowing-up of the ideal $\mathcal{A}\mathcal{O}_X \subset \mathcal{O}_X$ on X.

Let us point out that X' can also be viewed as the *strict transform* of X with respect to the admissible formal blowing-up $Y' \longrightarrow Y$. The latter is constructed from the fiber product $X'' = X \times_Y Y'$ (a formal S-scheme of locally topologically finite presentation, but *not* necessarily admissible) by dividing out all torsion with respect to the ideal generated by the pull-back of $\mathcal{A}$. The existence of flat models has an interesting consequence for classical rigid spaces.

Corollary 2. *In the classical rigid case, let R be an adic valuation ring of height 1 with field of fractions K. Furthermore, let $\varphi_K \colon X_K \longrightarrow Y_K$ be a flat morphism of quasi-compact and quasi-separated rigid K-spaces. Then its image $\varphi_K(X_K)$ is admissible open in Y_K.*

Proof. Due to Theorem 1, there exists a flat formal R-model $\varphi \colon X \longrightarrow Y$ of φ_K. Tensoring φ with the residue field k of R yields a morphism of k-schemes $\varphi_k \colon X_k \longrightarrow Y_k$ that is flat and of finite presentation. It is known that the image of φ_k is a quasi-compact open subscheme $V_k \subset Y_k$; see [EGA IV], 2.4.6. Now, if $V \subset Y$ is the corresponding open formal subscheme of Y, then, clearly, φ factors through V, and the induced morphism $X \longrightarrow V$ is faithfully flat. Finally, a local consideration involving rig-points, as introduced in Sect. 8.3, shows that φ_K must map X_K onto the admissible open subspace $V_{\mathrm{rig}} \subset X_K$ associated to V. $\square$

Another property (P) that can be defined on general rigid S-spaces X_{rig} by requiring (P) to be satisfied on complements of the special fiber, is the notion of smoothness. Also in this case one may ask if any smooth (or even étale) morphism of rigid S-spaces will admit a smooth (resp. étale) formal S-model. However, the answer will be negative in general. To give a simple example, one may look at the classical rigid situation where R is an adic valuation ring of height 1 with a fraction field K that is algebraically closed. Then, for any $q \in R$, $0 < |q| < 1$, the annulus $X_K = \mathrm{Sp}\, K\langle \zeta, q\zeta^{-1} \rangle$ is a smooth rigid K-space, which does not admit a smooth R-model. A canonical R-model of X_K is given by the formal R-scheme $X = \mathrm{Spf}\, R\langle \zeta, \eta \rangle / (\zeta\eta - q)$, which is not smooth. If there were a smooth formal R-model X of X_K, it would be connected, since X_K is connected. In particular, the special fiber X_k over the residue field k of R would be connected and, hence, integral since X_k is smooth. Then, starting out from an affine open covering $(U_i)_{i \in J}$ of X, all special fibers $U_{i,k}$ would be integral and we would get a finite affinoid covering $(U_{i,K})_{i \in J}$ on X_K such that there is a multiplicative residue norm on each of the affinoid K-algebras $A_i = \mathcal{O}_{X_K}(U_{i,K})$. The latter norm would coincide with the supremum norm on A_i, as can be concluded from 3.1/17. However, over an algebraically closed field K, the affinoid subdomains of the unit ball $\mathbb{B}_K^1$ are well-known, [BGR], 9.7.2/2, and it follows that such a covering cannot exist.

Thus, expecting the existence of smooth formal S-models for smooth rigid S-spaces would be too much. Stepping back a bit, one may replace smoothness by the weaker property (P) that the structural morphism $X_{\text{rig}} \longrightarrow S_{\text{rig}}$ has geometrically reduced fibers. Here is an advanced result on the existence of formal S-models with such a property (P)[4]:

Theorem 3 (Reduced Fiber Theorem). *Let X be a quasi-compact admissible formal S-scheme such that X/S is flat and $X_{\text{rig}}/S_{\text{rig}}$ has reduced geometric fibers, equidimensional of dimension d. Then there is a commutative diagram of admissible formal S-schemes*

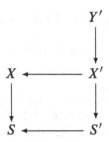

where

(i) $X' = X \times_S S'$,
(ii) $S' \longrightarrow S$ *is surjective and* $S'_{\text{rig}} \longrightarrow S_{\text{rig}}$ *is étale,*
(iii) $Y' \longrightarrow X'$ *is finite and* $Y'_{\text{rig}} \longrightarrow X'_{\text{rig}}$ *is an isomorphism,*
(iv) $Y' \longrightarrow S'$ *is flat and has reduced geometric fibers.*

So in order to transform the formal model X/S of $X_{\text{rig}}/S_{\text{rig}}$ into a flat one that has reduced geometric fibers, one has to apply, first of all, a base change S'/S that is étale on the rigid level. Then, still, the resulting formal S'-scheme $X' = X \times_S S'$ needs a finite extension Y'/X' that is an isomorphism on the rigid level.

Also note that, due to Theorem 1, the assumption of X/S to be flat may be replaced by requiring $X_{\text{rig}}/S_{\text{rig}}$ to be flat. Furthermore, at least in the Noetherian case (N'), the assumption on the equidimensionality of the fibers of $X_{\text{rig}}/S_{\text{rig}}$ can be avoided.

Finally, let us point out that Theorem 3 is, in fact, a relative version of the so-called *Finiteness Theorem of Grauert and Remmert* in [GR], a deep result from the beginnings of classical rigid geometry. To state the theorem, consider an algebraically closed field K occurring as field of fractions of a complete valuation ring R of height 1, as well as a *reduced* affinoid K-algebra A_K. The theorem asserts that, together with its t-adic topology for arbitrary $t \in R, 0 < |t| < 1$, the R-algebra

---

[4]For details see [F IV]. The theorem has been proved in the classical rigid case and in the Noetherian case (N').

$$A = \{f \in A_K \; ; \; |f|_{\sup} \leq 1\},$$

is of topologically finite type. From this one deduces that the special fiber $A \otimes_R k$ for k the residue field of R is *reduced*. There is also a version of this result for fields K that are not necessarily algebraically closed and where we assume that A_K is *geometrically reduced*. Then it might be necessary to apply a finite separable extension K' of K to the situation before one can assert that A is of topologically finite type and the special fiber $A \otimes_R k$ is *geometrically reduced*. The extension K'/K corresponds to the étale base change $S'_{\mathrm{rig}}/S_{\mathrm{rig}}$ in Theorem 3.

Appendix A

Classical Valuation Theory

In the following, let K be a field with a non-Archimedean absolute value denoted by $|\cdot|\colon K \longrightarrow \mathbb{R}_{\geq 0}$; cf. 2.1/1. We will always assume that such an absolute value is *non-trivial*, i.e. that its values in $\mathbb{R}_{\geq 0}$ are not restricted to 0 and 1. Furthermore, let V be a K-vector space. A *vector space norm* on V (cf. 2.3/4) is a map $\|\cdot\|\colon V \longrightarrow \mathbb{R}_{\geq 0}$ satisfying the following conditions for elements $x, y \in V$ and $\alpha \in K$:

(i) $\|x\| = 0 \Longleftrightarrow x = 0$,

(ii) $\|x + y\| \leq \max\{\|x\|, \|y\|\}$.

(iii) $\|\alpha x\| = |\alpha|\,\|x\|$,

When no confusion is possible, we will usually make no notational difference between the absolute value $|\cdot|$ on K and the vector space norm $\|\cdot\|$ on V, thus always writing $|x|$ instead of $\|x\|$ for elements $x \in V$. To give an example of a K-vector space norm, let V be a finite dimensional K-vector space and fix a basis $v_1, \ldots, v_d$ on it. Then we define the corresponding *maximum norm* $|\cdot|_{\max}$ on V as follows. Given an element $x \in V$, write it as a linear combination $x = \sum_{i=1}^{d} \alpha_i v_i$ with coefficients $\alpha_i \in K$ and set

$$|x|_{\max} = \max_{i=1\ldots d} |\alpha_i|.$$

One easily checks that $|\cdot|_{\max}$ defines a vector space norm on V. Furthermore, if K is complete under its absolute value, V is complete under such a maximum norm.

As usual, any vector space norm on a K-vector space V defines a topology on V. Two such norms $|\cdot|_1$ and $|\cdot|_2$ are called *equivalent* if they induce the same topology on V. The latter amounts to the fact that there exist constants $c, c' > 0$ such that $|x|_1 \leq c|x|_2 \leq c'|x|_1$ for all $x \in V$; use the fact that the absolute value on K is non-trivial. It is clear that any two maximum norms, attached to certain K-bases on

S. Bosch, *Lectures on Formal and Rigid Geometry*, Lecture Notes in Mathematics 2105, DOI 10.1007/978-3-319-04417-0,
© Springer International Publishing Switzerland 2014

a finite dimensional K-vector space V, are equivalent. If K is complete, a stronger assertion is possible.

Theorem 1. *Let V be a finite dimensional K-vector space and assume that K is complete. Then all K-vector space norms on V are equivalent. In particular, V is complete under such a norm.*

Proof. Choose a K-basis $v_1, \ldots, v_d$ of V and consider the attached maximum norm $|\cdot|_{\max}$ on V. Let $|\cdot|$ be a second K-vector space norm on V. Then there is a constant $c > 0$ such that $|x| \leq c|x|_{\max}$ for all $x \in V$. In fact, if $x = \sum_{i=1}^{d} \alpha_i v_i$, we have

$$|x| \leq \max_{i=1\ldots d} |\alpha_i||v_i| \leq \max_{i=1\ldots d} |\alpha_i| \max_{i=1\ldots d} |v_i| = c|x|_{\max}$$

for $c = \max_{i=1\ldots d} |v_i|$. Thus, it remains to show that there is a constant $c' > 0$ satisfying $|x|_{\max} \leq c'|x|$ for all $x \in V$.

We want to do this by induction on the dimension d of V. For $d = 0$ the assertion is trivial. Thus, let $d > 0$ and assume that a constant c' as desired does not exist. Then we can construct a sequence $x_n \in V$ such that

$$|x_n|_{\max} = 1 \text{ for all } n \qquad \text{and} \qquad \lim_{n \to \infty} |x_n| = 0.$$

Write $x_n = \sum_{i=1}^{d} \alpha_{ni} v_i$ with coefficients $\alpha_{ni} \in K$ and consider the elements α_{nd} for $i = d$ fixed as a sequence in K. If it is a zero sequence, look at the sequence $x_n' = x_n - \alpha_{nd} v_d$ in $V' = \sum_{i=1}^{d-1} K v_i$. Then, due to the non-Archimedean triangle inequality, $|x_n'|_{\max} = 1$ for almost all indices n and $\lim_{n \to \infty} |x_n'| = 0$. However, this is impossible by the induction hypothesis, since $|\cdot|_{\max}$ and $|\cdot|$ must be equivalent on the subspace $V' \subset V$, which is of dimension $d - 1$. Therefore α_{nd} cannot be a zero sequence. Replacing the x_n by a suitable subsequence, we may assume that there is some $\varepsilon > 0$ satisfying $|\alpha_{nd}| \geq \varepsilon$ for all n. Then

$$y_n = \alpha_{nd}^{-1} x_n = v_d + \sum_{i=1}^{d-1} \alpha_{nd}^{-1} \alpha_{ni} v_i$$

is still a zero sequence in V. Hence, we see that

$$v_d = -\lim_{n \to \infty} \sum_{i=1}^{d-1} \alpha_{nd}^{-1} \alpha_{ni} v_i.$$

In other words, v_d belongs to the closure of V' in V. However, by induction hypothesis, V' is complete and, hence, closed in V. As $v_d \notin V'$, we get a contradiction. $\square$

Corollary 2. *Let* $|\cdot|_1$ *and* $|\cdot|_2$ *be two absolute values on an algebraic field extension* L/K *restricting to the given absolute value* $|\cdot|$ *on* K. *Assume that* K *is complete with respect to* $|\cdot|$. *Then* $|\cdot|_1$ *and* $|\cdot|_2$ *coincide on* L.

Proof. Since L is a union of finite subextensions of L/K, we may assume that the extension L/K is finite. Then, viewing L as a normed K-vector space under $|\cdot|_1$ and $|\cdot|_2$, these norms are equivalent by Theorem 1. Thus, there are constants $c, c' > 0$ such that $|\alpha|_1 \leq c|\alpha|_2 \leq c'|\alpha|_1$ for all $\alpha \in L$. Replacing α by α^n for any integer $n > 0$ and using the multiplicativity of $|\cdot|_1$ and $|\cdot|_2$, we get

$$|\alpha|_1 \leq c^{\frac{1}{n}}|\alpha|_2 \leq c'^{\frac{1}{n}}|\alpha|_1$$

and therefore, by taking limits, $|\alpha|_1 = |\alpha|_2$ for all $\alpha \in L$. $\qquad\qquad\square$

We have just seen that for any algebraic field extension L/K, there is at most one way to extend the given absolute value $|\cdot|$ from K to L, provided K is complete with respect to $|\cdot|$. We want to show now that such an extension will always exist.

Theorem 3. *Let* L/K *be an algebraic extension of fields where* K *is complete with respect to a given absolute value* $|\cdot|$. *Then there is a unique way to extend* $|\cdot|$ *to an absolute value* $|\cdot|'$ *of* L. *In fact,*

$$|\alpha|' = |N_{K(\alpha)/K}(\alpha)|^{\frac{1}{d}}$$

for elements $\alpha \in L$ *where* $N_{K(\alpha)/K}$ *denotes the norm of* $K(\alpha)$ *over* K *and where* d *is the degree of* α *over* K.

If L *is finite over* K, *we see from Theorem 1 that* L *is complete with respect to the absolute value* $|\cdot|'$.

Proof. As $N_{K(\alpha)/K}(\alpha) = \alpha$ for elements $\alpha \in K$, it is clear that $|\cdot|'$ extends $|\cdot|$. To show that $|\cdot|'$ defines a non-Archimedean absolute value on L, let us verify the conditions of 2.1/1. Clearly, $N_{K(\alpha)/K}(\alpha) = 0$ if and only if $\alpha = 0$ and therefore $|\alpha|' = 0$ if and only if $\alpha = 0$. Furthermore, if $\alpha \in L$ is contained in a finite subextension L' of L/K, say of degree n, then we conclude from the definition of norms that $|\alpha|' = |N_{L'/K}(\alpha)|^{\frac{1}{n}}$. Since the norm $N_{L'/K}$ is multiplicative, we see that $|\cdot|'$ is multiplicative as well.

Thus, it remains to show $|\alpha + \beta|' \leq \max\{|\alpha|', |\beta|'\}$ for $\alpha, \beta \in L$. This estimate does not follow right away from properties of the norm, some more work is necessary. First note that for $|\alpha|' \leq |\beta|'$ and $\beta \neq 0$, we can divide by β and thereby are reduced to showing $|1 + \alpha|' \leq 1$ for $\alpha \in L$ satisfying $|\alpha|' \leq 1$. Let $R = \{\alpha \in K \,;\, |\alpha| \leq 1\}$ be the valuation ring of K. With the aid of Hensel's Lemma, see Lemma 4 below, we will show in Lemma 5 that an element $\alpha \in L$ is integral over R if and only if $N_{K(\alpha)/K}(\alpha) \in R$, i.e. if and only if $|\alpha|' \leq 1$. But then the non-Archimedean triangle inequality is easily derived. If $|\alpha|' \leq 1$ for some $\alpha \in L$, then α is integral over R. Hence, the same is true for $1 + \alpha$ and we get $|1 + \alpha|' \leq 1$. $\qquad\qquad\square$

In order to state Hensel's Lemma, let $R = \{\alpha \in K \,;\, |\alpha| \le 1\}$ be the valuation ring of K and let $k = R/\{\alpha \in R \,;\, |\alpha| < 1\}$ be the attached residue field. The canonical projection $R \longrightarrow k$, which will be denoted by $\alpha \longmapsto \tilde{\alpha}$, induces for a variable (or a system of variables) X a projection

$$R[X] \longrightarrow k[X], \qquad f = \sum c_i X^i \longmapsto \tilde{f} = \sum \tilde{c}_i X^i,$$

on the level of polynomial rings.

Hensel's Lemma 4. *Let $f \in R[X]$ be a polynomial in one variable X such that there exists a factorization $\tilde{f} = \tilde{p} \cdot \tilde{q}$ with coprime factors $\tilde{p}, \tilde{q} \in k[X]$, i.e. where $\tilde{p}$ and $\tilde{q}$ are non-zero and their greatest common divisor in $k[X]$ is 1. Then $\tilde{p}, \tilde{q}$ can be lifted to polynomials $p, q \in R[X]$ satisfying*

$$f = p \cdot q, \qquad \deg q = \deg \tilde{q}.$$

Before giving the proof, let us derive the statement on integral dependence that was used in the proof of Theorem 3.

Lemma 5. *As in Theorem 3, let L/K be an algebraic extension of fields and let R be the valuation ring of K. Then, for elements $\alpha \in L$, the following are equivalent:*

(i) α *is integral over R.*
(ii) $N_{K(\alpha)/K}(\alpha) \in R$.

Proof. To begin with, assume condition (i), namely that α is integral over R. Then there is a monic polynomial $h \in R[X]$ satisfying $h(\alpha) = 0$. Let $f \in K[X]$ be the minimal polynomial of α over K. As f must divide h in $K[X]$, there is a decomposition of type $h = fg$ in $K[X]$. We claim that both, f and g belong to $R[X]$. To justify this, consider the Gauß norm on $K[X]$, which is given by

$$\left\| \sum_{i=0}^{n} a_i X^i \right\| = \max_{i=0\dots n} |a_i|.$$

As in Sect. 2.2, one shows that the Gauß norm is multiplicative and this implies $1 = \|h\| = \|f\| \cdot \|g\|$. Since f is a monic polynomial, we have $\|f\| \ge 1$ and there is a constant $c \in K$ such that $|c| = \|f\|^{-1}$. Setting $f' = cf$ and $g' = c^{-1}g$, we get $h = f'g'$ with $\|f'\| = \|g'\| = 1$. In particular, $h = f'g'$ is a decomposition in $R[X]$, which can be transported into $k[X]$, thus implying the decomposition $\tilde{h} = \tilde{f}'\tilde{g}'$. As

$$\deg \tilde{f}' + \deg \tilde{g}' = \deg \tilde{h} = \deg h = \deg f + \deg g,$$

$$\deg \tilde{f}' \le \deg f, \qquad \deg \tilde{g}' \le \deg g,$$

we have necessarily $\deg \tilde{f}' = \deg f$ and $\deg \tilde{g}' = \deg g$. However, since f is monic, $\|f\| > 1$ would imply $\deg \tilde{f}' < \deg f$. Therefore we must have $|c| = 1$ and, hence, $f, g \in R[X]$. Thus, if $f = \sum_{i=0}^{n} c_i X^i \in R[X]$ is the minimal polynomial of α over K, we get $N_{K(\alpha)/K}(\alpha) = (-1)^n c_0 \in R$, which implies condition (ii).

Conversely, assume $N_{K(\alpha)/K}(\alpha) \in R$ as in condition (ii). As before, consider the minimal polynomial $f = \sum_{i=0}^{n} c_i X^i \in K[X]$ of α over K. We want to show that $f(\alpha) = 0$ is, in fact, an integral equation of α over R. Proceeding indirectly, assume that $f \notin R[X]$. Then we have $\|f\| > 1$ and we can choose a constant $c \in K$ such that $|c| = \|f\|^{-1} < 1$. Writing $f' = cf$, we get $\|f'\| = 1$. Since $c_0 = (-1)^n N_{K(\alpha)/K}(\alpha) \in R$ and $c_n = 1$, it follows $0 < \deg \tilde{f}' < \deg f$. Now look at the decomposition $\tilde{f}' = \tilde{p}\tilde{q}$ with $\tilde{p} = 1$ and $\tilde{q} = \tilde{f}'$. Due to Hensel's Lemma 4, we can lift $\tilde{p}$ and $\tilde{q}$ to polynomials $p, q \in R[X]$ such that $cf = f' = pq$ and $\deg q = \deg \tilde{q}$. Since $\deg \tilde{q}$ is strictly between 0 and $\deg f$, we see that $cf = pq$ is a non-trivial decomposition which, however, contradicts the fact that f is irreducible. Therefore we must have $f \in R[X]$, thus, implying condition (i). $\qquad\square$

It remains to do the *proof of Hensel's Lemma*. Starting out from the decomposition $\tilde{f} = \tilde{p}\tilde{q}$, we choose a lifting $q_0 \in R[X]$ of $\tilde{q}$ satisfying $\deg q_0 = \deg \tilde{q}$. Then the highest coefficient of q_0 is a unit in R and, by Euclid's division, there is an equation $f = p_0 q_0 + r_1$ with suitable polynomials $p_0, r_1 \in R[X]$ where $\deg r_1 < \deg q_0$. From this we get $\tilde{f} = \tilde{p}_0 \tilde{q} + \tilde{r}_1$. Since we have

$$\deg \tilde{r}_1 \leq \deg r_1 < \deg q_0 = \deg \tilde{q}$$

and $\tilde{q}$ divides $\tilde{f}$, Euclid's division in $k[X]$ implies $\tilde{r}_1 = 0$. In particular, $\|r_1\| < 1$ and p_0 is a lifting of $\tilde{p}$. Let $m = \deg p_0$ and $n = \deg q_0$. It is now our strategy, to construct polynomials $a, b \in R[X]$ with

$$\|a\|, \|b\| \leq \|r_1\|, \qquad \deg a < m, \qquad \deg b < n,$$

such that

$$f = p_0 q_0 + r_1 = (p_0 + a)(q_0 + b),$$

or, equivalently,

$$bp_0 + aq_0 + ab = r_1. \tag{*}$$

Then the decomposition $f = (p_0 + a)(q_0 + b)$ will be a lifting of $\tilde{f} = \tilde{p}\tilde{q}$, as required.

To do this, we neglect the quadratic term ab in the Eq. $(*)$ for a moment. Let $K[X]_i$ for $i \in \mathbb{N}$ be the R-submodule of $K[X]$ consisting of all polynomials in $K[X]$ of degree $\leq i$. For the valuation ring R and its residue field k the notations $R[X]_i$ and $k[X]_i$ are used in a similar way. Then consider the R-linear map

$$\varphi: R[X]_{m-1} \oplus R[X]_{n-1} \longrightarrow R[X]_{m+n-1}, \qquad (a,b) \longmapsto bp_0 + aq_0,$$

as well as its versions $\varphi \otimes K$ over K and $\varphi \otimes k$ over k. We claim that all of these are isomorphisms. In fact, start with $\varphi \otimes k$. This map is injective, since $\tilde{b}\tilde{p} + \tilde{a}\tilde{q} = 0$ implies that $\tilde{q}$ divides $\tilde{b}$, due to the fact that $\tilde{p}$ and $\tilde{q}$ are coprime. However, since $\deg \tilde{b} < m = \deg \tilde{q}$, we get $\tilde{a} = \tilde{b} = 0$. But then, by reasons of dimensions, $\varphi \otimes k$ is surjective and, hence, bijective. From this we can conclude that φ and $\varphi \otimes K$ are isometric in the sense that

$$\|(\varphi \otimes K)(b,a)\| = \max\{\|a\|, \|b\|\}, \qquad a \in K[X]_{m-1}, \qquad b \in K[X]_{n-1}.$$

In particular, $\varphi \otimes K$ is injective, and the same dimension argument, as used before, shows that $\varphi \otimes K$ is bijective. Furthermore, relying on the fact that $\varphi \otimes K$ is isometric, we finally see that φ is bijective. Now, to lift the decomposition $\tilde{f} = \tilde{p}\tilde{q}$ as stated, let $\varepsilon = \|r\|$. We claim:

There are sequences $p_i \in R[X]_{m-1}, q_i \in R[X]_{n-1}$, and $r_{i+1} \in R[X]_{m+n-1}$, starting with the initial elements p_0, q_0, r_1 constructed above, such that

$$f = \left(\sum_{i=1}^{j} p_i\right)\left(\sum_{i=1}^{j} q_i\right) + r_{j+1}, \qquad j = 0, 1, \dots,$$

where

$$\|p_j\|, \|q_j\| \le \varepsilon^j, \qquad \|r_{j+1}\| \le \varepsilon^{j+1}.$$

Then, as the field K is complete, $p = \sum_{i=1}^{\infty} p_i$ and $q = \sum_{i=1}^{\infty} q_i$ make sense as polynomials in $R[X]$ of degree m, respectively n, and by a limit argument, we get the desired decomposition $f = pq$.

To justify the claim, we proceed by induction on j. So assume that the polynomials p_i, q_i and r_{i+1} have already been constructed, up to some index $j \ge 0$. Then, writing $p' = \sum_{i=1}^{j} p_i$ and $q' = \sum_{i=1}^{j} q_i$ and applying the above properties of the R-linear map φ, now with p', q' in place of p_0, q_0, we can solve the equation

$$r_{j+1} = q_{j+1} p' + p_{j+1} q'$$

for some elements $p_{j+1} \in R[X]_{m-1}$ and $q_{j+1} \in R[X]_{n-1}$ satisfying

$$\|p_{j+1}\|, \|q_{j+1}\| \le \varepsilon^{j+1}.$$

But then we have

$$f = (p' + p_{j+1})(q' + q_{j+1}) + r_{j+2}$$

with $r_{j+2} = -p_{j+1}q_{j+1} \in R[X]_{m+n-1}$ where $\|r_{j+2}\| \le \varepsilon^{2(j+1)} \le \varepsilon^{j+2}$. Thus, our claim is justified, and Hensel's Lemma is proved. $\square$

The problem of extending a non-Archimedean absolute value $|\cdot|$ from a field K to an algebraic extension L has been settled in Theorem 3 for the case where K is complete. If K is not complete with respect to $|\cdot|$, we can pass to its completion $\hat{K}$, which can be constructed as follows. Consider the ring $K^{\mathbb{N}}$ of all infinite sequences in K, addition and multiplication being defined componentwise. The Cauchy sequences define a subring $C(K)$ of $K^{\mathbb{N}}$ and the zero sequences an ideal $Z(K) \subset C(K)$. It is easy to see that the quotient $\hat{K} = C(K)/Z(K)$ is a field and that the canonical map $K \longrightarrow \hat{K}$ sending an element α to the residue class of the constant sequence $\alpha, \alpha, \ldots$ is a homomorphism of fields. In particular, we can view K as a subfield of $\hat{K}$. We can even define an absolute value $|\cdot|'$ on $\hat{K}$ extending the one given on K. Indeed, given any $\alpha \in \hat{K}$, choose a representing Cauchy sequence (α_i) in K. Then the sequence $(|\alpha_i|)$ is a zero sequence or, due to the non-Archimedean triangle inequality, it becomes constant at a certain index i_0. Therefore the limit $c = \lim_{i \to \infty} |\alpha_i|$ exists and is well-defined, and we can set $|\alpha|' = c$. One can show that $\hat{K}$ is complete with respect to $|\cdot|'$ and that it contains K as a dense subfield.

Now if L/K is an algebraic field extension, we can consider the completion $\hat{K}$ of K and its algebraic closure $\hat{K}^{\mathrm{alg}}$. Extending the absolute value of K to $\hat{K}$, as just described, and prolonging it to $\hat{K}^{\mathrm{alg}}$ with the help of Theorem 3, we get a canonical non-Archimedean absolute value on $\hat{K}^{\mathrm{alg}}$, which may be denoted by $|\cdot|$ again. Then we can choose a K-morphism $\tau \colon L \longrightarrow \hat{K}^{\mathrm{alg}}$ and pull back the absolute value from $\hat{K}^{\mathrm{alg}}$ to L via τ. Thereby we obtain an absolute value on L extending the one given on K. However, the latter will not be unique in general, which corresponds to the fact that the K-morphism $\tau \colon L \longrightarrow \hat{K}^{\mathrm{alg}}$ may not be unique.

Taking the algebraic closure of a complete field, we may loose completeness. In particular, the field $\hat{K}^{\mathrm{alg}}$ may not be complete again. However, if we start with an algebraically closed field, its completion will remain algebraically closed. This way it is possible to construct extension fields that are algebraically closed and complete at the same time.

Krasner's Lemma 6. *Let K an algebraically closed field with a non-Archimedean absolute value $|\cdot|$. Then its completion $\hat{K}$ is algebraically closed.*

Proof. Consider an algebraic closure L of $\hat{K}$ and extend the absolute value of $\hat{K}$ to L, using the assertion of Theorem 3. Let $f = \sum_{i=0}^{n} c_i X^i$ be a monic polynomial of degree > 0 in $\hat{K}[X]$. Then f admits a zero $\alpha \in L$, and it is enough to show that α can be approximated by elements in K. To verify this, choose $\varepsilon > 0$ and approximate the coefficients c_i by elements $d_i \in K$ in such a way that the polynomial $g = \sum_{i=0}^{n} d_i X^i \in K[X]$ satisfies $|g(\alpha)| \leq \varepsilon^n$. Assuming $d_n = 1$, write $g = \prod_{i=1}^{n}(X - \beta_i)$ with zeros $\beta_i \in K$. Then $|g(\alpha)| = \prod_{i=1}^{n}|\alpha - \beta_i| \leq \varepsilon^n$ implies that there is an index i such that $|\alpha - \beta_i| \leq \varepsilon$. Consequently, α can be approximated by elements in K. $\square$

The argument used in the proof is referred to as the principle of *continuity of roots*.

Appendix B

Completed Tensor Products

In the following we want to show that the category of affinoid K-algebras admits amalgamated sums where, as usual, K is a field endowed with a non-trivial complete non-Archimedean absolute value. Such amalgamated sums are constructed as completions of ordinary tensor products.

To handle completed tensor products, we need a slightly more general setting. Let R be a ring with a ring norm $|\cdot|$ on it, see 2.3/1, and M a *normed R-module*. Thereby we mean an R-module M together with a map $M \longrightarrow \mathbb{R}_{\geq 0}$, denoted by $|\cdot|$ again, such that for all $x, y \in M$ and $a \in R$ we have

(i) $|x| = 0 \iff x = 0$,
(ii) $|x + y| \leq \max\{|x|, |y|\}$,
(iii) $|ax| \leq |a| \cdot |x|$.

The map $|\cdot|: M \longrightarrow \mathbb{R}_{\geq 0}$ is called a *semi-norm* on M if only conditions (ii) and (iii) are satisfied and (i) possibly not. Furthermore, an R-linear map $\varphi: M \longrightarrow N$ between normed R-modules is called *bounded* if there exists a real constant $\gamma > 0$ such that $|\varphi(x)| \leq \gamma |x|$ for all $x \in M$. In this case γ is referred to as a *bound* for φ.

Looking at topologies that are generated by module norms, we see immediately that bounded morphisms of normed R-modules are continuous. The converse is not always true. However, if there exists a subfield $K \subset R$ such that the norm on R restricts to a non-trivial absolute value on K, then every continuous morphism of normed R-modules is bounded. To justify this, assume that R contains a field K with the stated properties. Then, by restriction of scalars, any R-module M can be viewed as a K-vector space and, in fact, as a normed K-vector space in the sense of 2.3/4. Clearly we have $|ax| \leq |a| \cdot |x|$ for $a \in K$ and $x \in M$, but also

$$|a| \cdot |x| \leq |a| \cdot |a^{-1}ax| \leq |a| \cdot |a^{-1}| \cdot |ax| = |a| \cdot |a|^{-1} \cdot |ax| = |ax|$$

S. Bosch, *Lectures on Formal and Rigid Geometry*, Lecture Notes
in Mathematics 2105, DOI 10.1007/978-3-319-04417-0,
© Springer International Publishing Switzerland 2014

for $a \neq 0$, which shows $|ax| = |a| \cdot |x|$ for all $a \in K$ and $x \in M$. Then, if $\varphi: M \longrightarrow N$ is a continuous morphism of normed R-modules, there exists a constant $\delta > 0$ such that $|\varphi(x)| \leq 1$ for all $x \in M$ satisfying $|x| \leq \delta$. Fixing an element $t \in K$ such that $0 < |t| < 1$, we choose an integer $n \in \mathbb{Z}$ such that $|t|^{n-1} \leq \delta$. Now, considering an arbitrary element $x \in M$, there exists an integer $r \in \mathbb{Z}$ satisfying $|t|^n \leq |t|^r |x| \leq |t|^{n-1}$. Then $|t^r x| \leq \delta$ and, hence, $|\varphi(t^r x)| \leq 1$, as well as $1 \leq |t|^{r-n} |x|$, and we get

$$ |\varphi(x)| = |t|^{-r} \cdot |\varphi(t^r x)| \leq |t|^{-r} \leq |t|^{-r} \cdot |t|^{r-n} \cdot |x| = |t|^{-n} \cdot |x|, $$

which shows that $|t|^{-n}$ is a bound for φ. Thus, we have shown:

Lemma 1. (i) *Any bounded morphism of normed R-modules is continuous.*

(ii) *Conversely, assume that R contains a field K such that the norm on R restricts to a non-trivial absolute value on K. Then every continuous morphism of R-modules is bounded.*

Note that the assumption in (ii) is satisfied if R is a non-zero affinoid K-algebra, for K a field with a non-trivial complete non-Archimedean absolute value. Thus, in this case a morphism of normed R-modules is continuous if and only if it is bounded.

Now let us turn to tensor products and their related bilinear maps. Let M, N, E be normed modules over a normed ring R. An R-bilinear map $\Phi: M \times N \longrightarrow E$ is called *bounded* if there exists a real constant $\gamma > 0$ such that $|\Phi(x, y)| \leq \gamma \cdot |x| \cdot |y|$ for all $x \in M$ and $y \in N$. Again, γ is called a *bound* for Φ. An R-linear or R-bilinear map that is bounded by 1 is called *contractive*.

Proposition 2. *Let M, N be normed modules over a normed ring R. There exists a contractive R-bilinear map $\tau: M \times N \longrightarrow T$ into a complete normed R-module T such that the following universal property holds:*

Given any R-bilinear map $\Phi: M \times N \longrightarrow E$, bounded by some $\gamma > 0$, into a complete normed R-module E, there exists a unique R-linear map $\varphi: T \longrightarrow E$, bounded by γ as well, such that the diagram

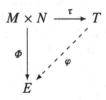

is commutative.

Proof. To construct the map τ, we view the ordinary tensor product $M \otimes_R N$ as a semi-normed R-module using the semi-norm $|\cdot|: M \otimes_R N \longrightarrow \mathbb{R}_{\geq 0}$ given by

$$|z| = \inf\Big(\max_{i=1,\dots,r} |x_i| \cdot |y_i| \Big), \qquad z \in M \otimes_R N,$$

where the infimum runs over all possible representations

$$z = \sum_{i=1}^{r} x_i \otimes y_i, \qquad x_i \in M, \qquad y_i \in N.$$

That we really get a semi-norm on $M \otimes_R N$ is easily verified. Thus, we can define $T = M \,\widehat{\otimes}_R\, N$ as the separated completion of $M \otimes_R N$. It is an R-module again and, in fact, a complete normed R-module, since the semi-norm on $M \otimes_R N$ gives rise to an R-module norm on $M \,\widehat{\otimes}_R\, N$. For elements $x \in M$ and $y \in N$, we write $x \,\widehat{\otimes}\, y$ for the element in $M \,\widehat{\otimes}_R\, N$ that is induced by the tensor $x \otimes y \in M \otimes_R N$. Then it is clear that the map

$$\tau : M \times N \longrightarrow M \,\widehat{\otimes}_R\, N, \qquad (x, y) \longmapsto x \,\widehat{\otimes}\, y,$$

is R-bilinear and contractive. The R-module $M \,\widehat{\otimes}_R\, N$, together with its R-module norm, is called the *completed tensor product* of M and N over R.

Now let us show that the R-bilinear map τ satisfies the universal property of the assertion. So let $\Phi : M \times N \longrightarrow E$ be a bounded R-bilinear map into a complete normed R-module E and let $\gamma > 0$ be a bound for Φ. Using the universal property of ordinary tensor products in terms of the canonical R-bilinear map $\tau' : M \times N \longrightarrow M \otimes_R N$ sending a pair (x, y) to the tensor $x \otimes y$, there is a unique R-linear map $\varphi' : M \otimes_R N \longrightarrow E$ making the following diagram commutative:

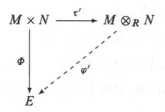

Then consider some element $z = \sum_{i=1}^{r} x_i \otimes y_i \in M \otimes_R N$ where $x_i \in M$ and $y_i \in N$. Since $\varphi'(z) = \sum_{i=1}^{r} \Phi(x_i, y_i)$, we get

$$|\varphi'(z)| \leq \max_{i=1,\dots,r} |\Phi(x_i, y_i)| \leq \gamma \max_{i=1,\dots,r} |x_i| \cdot |y_i|.$$

Taking the infimum over all representations of z as a sum of tensors $\sum_{i=1}^{r} x_i \otimes y_i$ yields $|\varphi'(z)| \leq \gamma |z|$, and we see that φ' is bounded by γ.

Since E is complete, φ' gives rise to an R-linear map $\varphi : M \,\widehat{\otimes}_R\, N \longrightarrow E$ that is bounded by γ as well. Furthermore, we can enlarge the above diagram to obtain the following commutative diagram:

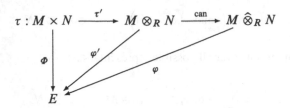

It remains to show that φ is uniquely determined by the relation $\Phi = \varphi \circ \tau$. However, this is clear since φ is unique on the image $\tau(M \times N)$, which generates a dense R-submodule in $M \,\hat{\otimes}_R\, N$.   □

In the situation of Proposition 2, the normed R-module T together with the contractive R-bilinear map $\tau \colon M \times N \longrightarrow T$ is uniquely determined up to isometric isomorphism and will be denoted by $M \,\hat{\otimes}_R\, N$. It is called the *completed tensor product* of M and N over R. For the attached contractive R-bilinear map $\tau \colon M \times N \longrightarrow M \,\hat{\otimes}_R\, N$ we will use the notation $(x, y) \longmapsto x \,\hat{\otimes}\, y$. In other words, we set $x \,\hat{\otimes}\, y = \tau(x, y)$ for $(x, y) \in M \times N$. Note that, independent of the construction in the proof of Proposition 2, there is a canonical R-linear map $M \otimes_R N \longrightarrow M \,\hat{\otimes}_R\, N$, namely the one given by $x \otimes y \longmapsto x \,\hat{\otimes}\, y$. It has a dense image in $M \,\hat{\otimes}_R\, N$, since the closure of this image, just as $M \,\hat{\otimes}_R\, N$, satisfies the universal property of completed tensor products.

As in the case of ordinary tensor products, the universal property defining completed tensor products can be used to derive various standard facts. To list some of them, look at normed R-modules M, N, P. Then there are canonical isometric isomorphisms

$$R \,\hat{\otimes}_R\, M \simeq M,$$

$$M \,\hat{\otimes}_R\, N \simeq N \,\hat{\otimes}_R\, M,$$

$$\left(M \,\hat{\otimes}_R\, N\right) \hat{\otimes}_R\, P \simeq M \,\hat{\otimes}_R\, \left(N \,\hat{\otimes}_R\, P\right),$$

$$\left(M \oplus N\right) \hat{\otimes}_R\, P \simeq \left(M \,\hat{\otimes}_R\, P\right) \oplus \left(N \,\hat{\otimes}_R\, P\right),$$

where the norm on a direct sum like $M \oplus N$ is given by $|x \oplus y| = \max(|x|, |y|)$.

Furthermore, the completed tensor product of two bounded morphisms of normed R-modules can be constructed. Indeed, let $\varphi_i \colon M_i \longrightarrow N_i$ for $i = 1, 2$ be morphisms of normed R-modules that are bounded by constants $\gamma_i > 0$. Then the R-bilinear map

$$M_1 \times M_2 \longrightarrow N_1 \,\hat{\otimes}_R\, N_2, \qquad (x_1, x_2) \longmapsto \varphi_1(x_1) \,\hat{\otimes}\, \varphi_2(x_2),$$

is bounded by $\gamma_1 \gamma_2$ and, thus, gives rise to an R-linear map

$$\varphi_1 \,\hat{\otimes}\, \varphi_2 \colon M_1 \,\hat{\otimes}_R\, M_2 \longrightarrow N_1 \,\hat{\otimes}_R\, N_2, \qquad x_1 \,\hat{\otimes}\, x_2 \longmapsto \varphi_1(x_1) \,\hat{\otimes}\, \varphi_2(x_2),$$

that is bounded by $\gamma_1\gamma_2$ as well. The map $\varphi_1 \hat{\otimes} \varphi_2$ is referred to as the *completed tensor product* of φ_1 and φ_2.

Also note that the associativity isomorphism above admits the following generalization:

Proposition 3. *Let $S \longrightarrow R$ be a contractive homomorphism between normed rings and let M be a normed S-module, as well as N and P normed R-modules. Then there is a canonical isometric isomorphism of normed S-modules*

$$(M \hat{\otimes}_S N) \hat{\otimes}_R P \simeq M \hat{\otimes}_S (N \hat{\otimes}_R P)$$

where $M \hat{\otimes}_S N$ is a normed R-module via the R-module structure of N.

The *proof* is straightforward, see [BGR], 2.1.7/7.

Next let us discuss completed tensor products on the level of *normed algebras*. To do this, fix a normed ring R and consider two normed R-algebras A_1, A_2; by the latter we mean normed rings A_i that are equipped with a contractive ring homomorphism $R \longrightarrow A_i$. In particular, we may view the A_i as normed R-modules, which implies that the completed tensor product $A_1 \hat{\otimes}_R A_2$ exists as a complete normed R-module. We want to show that $A_1 \hat{\otimes}_R A_2$ is, in fact, a normed R-algebra, based on the R-algebra structure of the ordinary tensor product $A_1 \otimes_R A_2$. Using the semi-norm on $A_1 \otimes_R A_2$ as defined in the proof of Proposition 2, we see that the canonical ring homomorphism $R \longrightarrow A_1 \hat{\otimes}_R A_2$ is contractive. Furthermore, for two elements

$$z = \sum_{i=1}^{m} x_i \otimes y_i, \qquad z' = \sum_{j=1}^{n} x'_j \otimes y'_j \qquad \in A_1 \otimes_R A_2,$$

we get

$$|z \cdot z'| = \left| \sum_{i=1}^{m} \sum_{j=1}^{n} x_i x'_j \otimes y_i y'_j \right| \leq \max_{i,j} |x_i x'_j| \cdot |y_i y'_j|$$

$$\leq \max_{i=1,\dots,m} |x_i| \cdot |y_i| \cdot \max_{j=1,\dots,n} |x'_j| \cdot |y'_j|,$$

which yields

$$|zz'| \leq |z| \cdot |z'|.$$

when taking the infimum over all representations of z and z' as sums of tensors. Thus, passing from $A_1 \otimes_R A_2$ to its completion, it follows that, indeed, the completed tensor product $A_1 \hat{\otimes}_R A_2$ is a normed R-algebra where the multiplication is characterized by

$$(x \mathbin{\hat{\otimes}} y) \cdot (x' \mathbin{\hat{\otimes}} y') = xx' \mathbin{\hat{\otimes}} yy'$$

and the structural morphism $R \longrightarrow A_1 \mathbin{\hat{\otimes}}_R A_2$ by $a \longmapsto a \mathbin{\hat{\otimes}} 1 = 1 \mathbin{\hat{\otimes}} a$.

We want to characterize $A_1 \mathbin{\hat{\otimes}}_R A_2$ in terms of a universal property for normed R-algebras.

Proposition 4. *Let R be a normed ring and A_1, A_2 normed R-algebras. Then the contractive R-algebra homomorphisms*

$$\sigma_1 \colon A_1 \longrightarrow A_1 \mathbin{\hat{\otimes}}_R A_2, \qquad a_1 \longmapsto a_1 \mathbin{\hat{\otimes}} 1,$$

$$\sigma_2 \colon A_2 \longrightarrow A_1 \mathbin{\hat{\otimes}}_R A_2, \qquad a_2 \longmapsto 1 \mathbin{\hat{\otimes}} a_2,$$

admit the following universal property of amalgamated sums:

Let $\varphi_1 \colon A_1 \longrightarrow D$ and $\varphi_2 \colon A_2 \longrightarrow D$ be two homomorphisms of normed R-algebras that are bounded by constants $\gamma_1 > 0$ and $\gamma_2 > 0$ and assume that D is complete. Then there is a unique R-algebra homomorphism $\varphi \colon A_1 \mathbin{\hat{\otimes}}_R A_2 \longrightarrow D$, bounded by $\gamma_1 \gamma_2$, such that the diagram

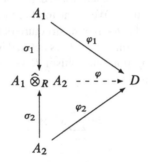

is commutative.

Proof. Consider homomorphisms of normed R-algebras $\varphi_1 \colon A_1 \longrightarrow D$ as well as $\varphi_2 \colon A_2 \longrightarrow D$ where D is complete and assume that φ_1 and φ_2 are bounded by constants $\gamma_1 > 0$ and $\gamma_2 > 0$. Then

$$A_1 \times A_2 \longrightarrow D, \qquad (a_1, a_2) \longmapsto \varphi_1(a_1) \cdot \varphi_2(a_2),$$

is an R-bilinear map that is bounded by $\gamma_1 \gamma_2$. Thus, by the universal property of completed tensor products in Proposition 2, it gives rise to an R-linear map

$$\varphi \colon A_1 \mathbin{\hat{\otimes}}_R A_2 \longrightarrow D, \qquad a_1 \mathbin{\hat{\otimes}} a_2 \longmapsto \varphi_1(a_1) \cdot \varphi_2(a_2),$$

that is bounded by $\gamma_1 \gamma_2$. Furthermore, φ satisfies

$$\varphi\big((a_1 \,\hat{\otimes}\, a_2) \cdot (a_1' \,\hat{\otimes}\, a_2')\big) = \varphi(a_1 a_1' \,\hat{\otimes}\, a_2 a_2') = \varphi_1(a_1 a_1') \cdot \varphi_2(a_2 a_2')$$
$$= \varphi_1(a_1) \cdot \varphi_2(a_2) \cdot \varphi_1(a_1') \cdot \varphi_2(a_2')$$
$$= \varphi(a_1 \,\hat{\otimes}\, a_2) \cdot \varphi(a_1' \,\hat{\otimes}\, a_2')$$

for $a_1, a_1' \in A_1$ and $a_2, a_2' \in A_2$. This shows that φ is multiplicative on the image of $A_1 \otimes_R A_2$ in $A_1 \,\hat{\otimes}_R A_2$ and, hence, by continuity, on $A_1 \,\hat{\otimes}_R A_2$ itself. Since

$$\varphi(a_1 \,\hat{\otimes}\, a_2) = \varphi\big((a_1 \,\hat{\otimes}\, 1) \cdot (1 \,\hat{\otimes}\, a_2)\big) = \varphi_1(a_1) \cdot \varphi_2(a_2)$$

for $a_1 \in A_1$ and $a_2 \in A_2$, it is clear by a continuity argument as before that φ is unique on $A_1 \,\hat{\otimes}_R A_2$.   □

If $\psi_i : A_i \longrightarrow B_i$, $i = 1, 2$, are bounded morphisms of normed R-algebras, their completed tensor product

$$\psi_1 \,\hat{\otimes}\, \psi_2 : A_1 \,\hat{\otimes}_R A_2 \longrightarrow B_1 \,\hat{\otimes}_R B_2, \qquad a_1 \,\hat{\otimes}\, a_2 \longmapsto \psi_1(a_1) \,\hat{\otimes}\, \psi_2(a_2),$$

is defined as a bounded R-linear map, but can also be obtained within the context of normed R-algebras using the universal property of Proposition 4; both versions coincide.

Next we want to study the behavior of restricted power series under completed tensor products. To do this, let A be a complete normed ring and $\zeta = (\zeta_1, \ldots, \zeta_n)$ a set of variables. Then, as usual, the A-algebra of *restricted power series* in ζ with coefficients in A is given by

$$A\langle\zeta\rangle = \Big\{ \sum_{\nu \in \mathbb{N}^n} a_\nu \zeta^n \in A[\![\zeta]\!] \; ; \; a_\nu \in A, \lim_{\nu \in \mathbb{N}^n} a_\nu = 0 \Big\}.$$

It is a complete normed A-algebra under the *Gauß norm*

$$\Big| \sum_{\nu \in \mathbb{N}^n} a_\nu \zeta^n \Big| = \max_{\nu \in \mathbb{N}^n} |a_\nu|.$$

Proposition 5. *Let R be a complete normed ring, A a complete normed R-algebra, and $\zeta = (\zeta_1, \ldots, \zeta_n)$ a set of variables. Then, using the Gauß norm on $R\langle\zeta\rangle$ and $A\langle\zeta\rangle$, there is a canonical isometric isomorphism of normed R-algebras*

$$A \,\hat{\otimes}_R R\langle\zeta_1, \ldots, \zeta_n\rangle \xrightarrow{\;\sim\;} A\langle\zeta_1, \ldots, \zeta_n\rangle.$$

Proof. We want to show that the canonical maps

$$\sigma_1 : A \longrightarrow A\langle\zeta\rangle, \qquad \sigma_2 : R\langle\zeta\rangle \longrightarrow A\langle\zeta\rangle,$$

which are contractive, satisfy the universal property mentioned in Proposition 4. To do this, consider two morphisms of R-algebras

$$\varphi_1 : A \longrightarrow D, \qquad \varphi_2 : R\langle \zeta \rangle \longrightarrow D$$

into a complete normed R-algebra D such that φ_1 and φ_2 are bounded by constants $\gamma_1, \gamma_2 > 0$. Then there is a well-defined R-algebra homomorphism

$$\varphi : A\langle \zeta \rangle \longrightarrow D, \qquad \sum_{\nu \in \mathbb{N}^n} a_\nu \zeta^\nu \longmapsto \sum_{\nu \in \mathbb{N}^n} \varphi_1(a_\nu) \cdot \varphi_2(\zeta^\nu).$$

Indeed, if the a_ν form a zero sequence in A, their images form a zero sequence in D since $|\varphi_1(a_\nu)| \le \gamma_1 |a_\nu|$. Furthermore, we have $|\varphi_2(\zeta^\nu)| \le \gamma_2$ for all ν so that the infinite sums of type $\sum_\nu \varphi_1(a_\nu) \cdot \varphi_2(\zeta^\nu)$ are converging. Hence, φ is well-defined, and it is bounded by $\gamma_1 \gamma_2$, as shown by the estimate

$$\left| \sum_{\nu \in \mathbb{N}^n} \varphi_1(a_\nu) \cdot \varphi_2(\zeta^\nu) \right| \le \gamma_1 \gamma_2 \cdot \max_\nu |a_\nu| = \gamma_1 \gamma_2 \cdot \left| \sum_{\nu \in \mathbb{N}^n} a_\nu \zeta^\nu \right|.$$

By continuity, φ is even a homomorphism of R-algebras and, in fact, the unique bounded homomorphism making the diagram

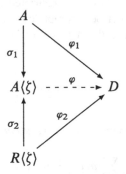

commutative. Thus, we are done.   □

For the remainder of this section, we want to look at affinoid K-algebras where, as usual, K is a field with a complete non-Archimedean absolute value that is non-trivial. Any such algebra A may be viewed as a complete normed K-algebra by choosing a residue norm on it. Furthermore, we know from 3.1/20 that any two residue norms $|\cdot|$ and $|\cdot|'$ on A are equivalent in the sense that they induce the same topology on A. In particular, the identity map $(A, |\cdot|) \longrightarrow (A, |\cdot|')$ and its inverse are bounded due to Lemma 1.

Now let $\tau_1 : R \longrightarrow A_1$ and $\tau_2 : R \longrightarrow A_2$ be two homomorphisms of affinoid K-algebras. In order to construct the completed tensor product $A_1 \hat{\otimes}_R A_2$, we need to specify appropriate norms on R, A_1, and A_2 in such a way that τ_1 and τ_2 are contractive. We do this in terms of residue norms. In fact, choosing epimorphisms $\alpha: T_m \longrightarrow R$ and $\alpha_i: T_{n_i} \longrightarrow A_i$, $i = 1, 2$, we can use 3.1/19 in conjunction with 3.1/7 and 3.1/9 to construct commutative diagrams

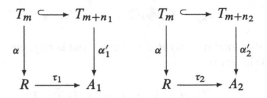

where α_1' and α_2' are extensions of α_1 and α_2 and, hence, are surjective. Considering the residue norms associated to α, α_1', and α_2' on R and the A_i, it is clear that the maps τ_1 and τ_2 are contractive and, hence, that the completed tensor product $A_1 \,\widehat{\otimes}_R\, A_2$ can be constructed. If we consider a second set of residue norms on R, A_1, and A_2 such that τ_1 and τ_2 are contractive, then the resulting semi-norms on $A_1 \otimes_R A_2$ that are used to construct the completed tensor product, are seen to be equivalent. As a result, the attached completions can canonically be identified and it follows that, indeed, the completed tensor product $A_1 \,\widehat{\otimes}_R\, A_2$ is well-defined, up to a set of equivalent ring norms on it, just as is the case for affinoid K-algebras and their possible residue norms on them. We will keep this in mind and talk about "the" completed tensor product of A_1 and A_2 over R. However, when it comes to particular norms on $A_1 \,\widehat{\otimes}_R\, A_2$, we have to be more specific.

Our main objective for the remainder of this section is to show:

Theorem 6. *Let $\tau_1: R \longrightarrow A_1$ and $\tau_2: R \longrightarrow A_2$ be homomorphisms of affinoid K-algebras. Then the completed tensor product $A_1 \,\widehat{\otimes}_R\, A_2$ is an affinoid K-algebra as well. In other words, the category of affinoid K-algebras admits amalgamated sums.*

To prepare the proof of the theorem, we start with some consequences of Proposition 5.

Proposition 7. *Let $\xi_1, \ldots, \xi_m$ and $\zeta_1, \ldots, \zeta_n$ be sets of variables, and K' an extension field of K with a complete absolute value extending the one given on K. Then there are canonical isometric isomorphisms*

$$K\langle \xi_1, \ldots, \xi_m \rangle \,\widehat{\otimes}_K\, K\langle \zeta_1, \ldots, \zeta_n \rangle \overset{\sim}{\longrightarrow} K\langle \xi_1, \ldots, \xi_m, \zeta_1, \ldots, \zeta_n \rangle,$$

$$K' \,\widehat{\otimes}_K\, K\langle \zeta_1, \ldots, \zeta_n \rangle \overset{\sim}{\longrightarrow} K'\langle \zeta_1, \ldots, \zeta_n \rangle,$$

with respect to the Gauß norm on the occurring Tate algebras.

Proposition 8. *Let A_1 and A_2 be affinoid K-algebras. Then $A_1 \,\widehat{\otimes}_K\, A_2$ is an affinoid K-algebra as well. Similarly, if K' is an extension field of K with a complete absolute value extending the one given on K, then $K' \,\widehat{\otimes}_K\, A_i$ is an affinoid K'-algebra.*

More specifically, choose epimorphisms of K-algebras $\alpha_i \colon T_{n_i} \longrightarrow A_i$ *for* $i = 1, 2$, *and consider the attached residue norms on* A_1 *and* A_2. *Then the canonical morphism of K-algebras*

$$\alpha \colon T_{n_1+n_2} = T_{n_1} \,\widehat{\otimes}_K\, T_{n_2} \longrightarrow A_1 \,\widehat{\otimes}_K\, A_2$$

is surjective and its kernel is generated by $\ker \alpha_1$ *and* $\ker \alpha_2$, *thus giving rise to an isomorphism of K-algebras*

$$T_{n_1+n_2}/(\ker \alpha_1, \ker \alpha_2) \xrightarrow{\sim} A_1 \,\widehat{\otimes}_K\, A_2.$$

The latter is an isometric isomorphism if we consider on $T_{n_1+n_2}/(\ker \alpha_1, \ker \alpha_2)$ *its canonical residue norm. Likewise, the homomorphisms of K-algebras*

$$\alpha_i' \colon K'\langle \zeta_1, \dots, \zeta_{n_i} \rangle = K' \,\widehat{\otimes}_K\, T_{n_i} \longrightarrow K' \,\widehat{\otimes}_K\, A_i, \qquad i = 1, 2,$$

are surjective, and their kernels are generated by $\ker \alpha_i$, *thus giving rise to isometric isomorphisms*

$$\left(K' \,\widehat{\otimes}_K\, T_{n_i}\right)/(\ker \alpha_i) \xrightarrow{\sim} K' \,\widehat{\otimes}_K\, A_i, \qquad i = 1, 2.$$

Proof. We show that $T_{n_1+n_2}/(\ker \alpha_1, \ker \alpha_2)$ and, likewise, $K'\langle \zeta_1, \dots, \zeta_{n_i} \rangle/(\ker \alpha_i)$ satisfy the universal property of completed tensor products. To do this, consider a commutative diagram of type

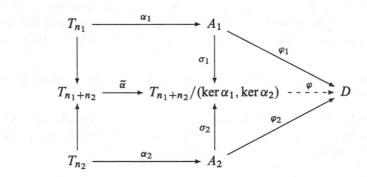

where σ_i is induced by the inclusion $T_{n_i} \hookrightarrow T_{n_1+n_2}$, $i = 1, 2$, and where $\tilde{\alpha}$ is the canonical projection. Concerning the right part of the diagram, D is a complete normed K-algebra and the $\varphi_i \colon A_i \longrightarrow D$, $i = 1, 2$, are homomorphisms that are bounded by constants $\gamma_1, \gamma_2 > 0$. Using Proposition 7 and interpreting $T_{n_1+n_2}$ as the completed tensor product $T_{n_1} \,\widehat{\otimes}_K\, T_{n_2}$, there exists a canonical homomorphism of K-algebras $T_{n_1+n_2} \longrightarrow D$ that is bounded by $\gamma_1 \gamma_2$ and that, apparently, will factor through the quotient $T_{n_1+n_2}/(\ker \alpha_1, \ker \alpha_2)$ via a unique homomorphism of K-algebras

$$\varphi: T_{n_1+n_2}/(\ker\alpha_1, \ker\alpha_2) \longrightarrow D$$

making the above diagram commutative. Let us equip now the affinoid K-algebra $T_{n_1+n_2}/(\ker\alpha_1, \ker\alpha_2)$ with its residue norm via $\tilde{\alpha}$. Then, by the definition of residue norms, we see that the maps σ_1 and σ_2 are contractive since the canonical inclusions of T_{n_i} into $T_{n_1+n_2}$ preserve Gauß norms. Furthermore, by the definition of residue norms again, φ is bounded by $\gamma_1\gamma_2$ since the same is true for the composition $\varphi\circ\tilde{\alpha}$; one may also use the fact that for every $\overline{f} \in T_{n_1+n_2}/(\ker\alpha_1, \ker\alpha_2)$ there is an inverse image $f \in T_{n_1+n_2}$ satisfying $|f| = |\overline{f}|$, cf. 3.1/5. Altogether we conclude that $T_{n_1+n_2}/(\ker\alpha_1, \ker\alpha_2)$ along with the contractions σ_1, σ_2 satisfy the universal property of a completed tensor product $A_1 \widehat{\otimes}_K A_2$. Thus, we are done with the first part of the assertion. The completed tensor products of type $K' \widehat{\otimes}_K A_i$ are dealt with similarly. □

Proposition 9. *Let* $\sigma: S \longrightarrow R$ *as well as* $\tau_1: R \longrightarrow A_1$ *and* $\tau_2: R \longrightarrow A_2$ *be homomorphisms of affinoid K-algebras. Then there is a canonical homomorphism of normed K-algebras* $A_1 \widehat{\otimes}_S A_2 \longrightarrow A_1 \widehat{\otimes}_R A_2$, *and the latter is an epimorphism.*

More specifically, consider residue norms on R, S, A_1, and A_2, and assume that σ and the τ_i are contractive. Then the norm on $A_1 \widehat{\otimes}_R A_2$ coincides with the residue norm derived from the norm on $A_1 \widehat{\otimes}_S A_2$.

Proof. We proceed similarly as in the proof of Proposition 8 and consider a commutative diagram of type

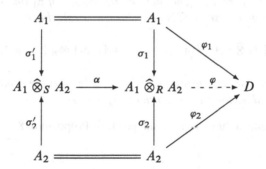

where D is a complete normed R-algebra and the $\varphi_i: A_i \longrightarrow D$, $i = 1, 2$, are homomorphisms of R-algebras that are bounded by constants $\gamma_1, \gamma_2 > 0$. Furthermore, φ is the unique homomorphism of R-algebras, bounded by $\gamma_1\gamma_2$, that is derived from the universal property of $A_1 \widehat{\otimes}_R A_2$. It follows that $\varphi\circ\alpha$ is the unique homomorphism of S-algebras derived from the universal property of $A_1 \widehat{\otimes}_S A_2$; it is bounded by $\gamma_1\gamma_2$ as well. Now consider the factorization

$$\alpha: A_1 \widehat{\otimes}_S A_2 \longrightarrow (A_1 \widehat{\otimes}_S A_2)/\ker\alpha \hookrightarrow A_1 \widehat{\otimes}_R A_2$$

where $\ker\alpha$ is a closed ideal in $A_1 \widehat{\otimes}_S A_2$ since α is contractive and, hence, continuous. Thus, proceeding in the manner of 3.1/5 (i) and (ii), we can equip the quotient $(A_1 \mathbin{\widehat{\otimes}}_S A_2)/\ker \alpha$ with the residue norm derived from the norm on $A_1 \mathbin{\widehat{\otimes}}_S A_2$. Clearly, the homomorphisms σ_1 and σ_2 factor through contractive homomorphisms of R-algebras $\tilde{\sigma}_i \colon A_i \longrightarrow (A_1 \mathbin{\widehat{\otimes}}_S A_2)/\ker \alpha$, $i = 1, 2$, and it is easily seen that $(A_1 \mathbin{\widehat{\otimes}}_S A_2)/\ker \alpha$ along with $\tilde{\sigma}_1$ and $\tilde{\sigma}_2$ satisfy the universal property of the completed tensor product $A_1 \mathbin{\widehat{\otimes}}_R A_2$. Thus, we are done. $\square$

Now the *Proof of Theorem* 6 can be carried out without problems. We assume that $\tau_1 \colon R \longrightarrow A_1$ and $\tau_2 \colon R \longrightarrow A_2$ are contractive homomorphisms of affinoid K-algebras, the latter being equipped with suitable residue norms. Then the completed tensor product $A_1 \mathbin{\widehat{\otimes}}_K A_2$ is an affinoid K-algebra by Proposition 8 and so is the completed tensor product $A_1 \mathbin{\widehat{\otimes}}_R A_2$, since it is a quotient of $A_1 \mathbin{\widehat{\otimes}}_K A_2$ by Proposition 9.

Finally, we want to mention the following generalization of the first part of Proposition 8:

Proposition 10. *Let* $\tau_1 \colon R \longrightarrow A_1$ *and* $\tau_2 \colon R \longrightarrow A_2$ *be homomorphisms of affinoid K-algebras, and consider ideals* $\mathfrak{a}_1 \subset A_1$ *as well as* $\mathfrak{a}_2 \subset A_2$. *Furthermore, fix residue norms on* R, A_1, *and* A_2 *such that* τ_1 *and* τ_2 *are contractive, and provide the quotients* $A_1/\mathfrak{a}_1$ *and* $A_2/\mathfrak{a}_2$ *with the residue norms derived from the given residue norms on* A_1 *and* A_2 *via the canonical projections* $\alpha_i \colon A_i \longrightarrow A_i/\mathfrak{a}_i$. *Then*

$$\alpha_1 \mathbin{\widehat{\otimes}} \alpha_2 \colon A_1 \mathbin{\widehat{\otimes}}_R A_2 \longrightarrow (A_1/\mathfrak{a}_1) \mathbin{\widehat{\otimes}}_R (A_2/\mathfrak{a}_2)$$

is surjective and its kernel is generated by the images of $\mathfrak{a}_1$ *and* $\mathfrak{a}_2$ *in* $A_1 \mathbin{\widehat{\otimes}}_R A_2$. *This way* $\alpha_1 \mathbin{\widehat{\otimes}} \alpha_2$ *gives rise to an isomorphism of R-algebras*

$$\left(A_1 \mathbin{\widehat{\otimes}}_R A_2\right)/(\mathfrak{a}_1, \mathfrak{a}_2) \overset{\sim}{\longrightarrow} (A_1/\mathfrak{a}_1) \mathbin{\widehat{\otimes}}_R (A_2/\mathfrak{a}_2),$$

which is isometric if we consider on $(A_1 \mathbin{\widehat{\otimes}}_R A_2)/(\mathfrak{a}_1, \mathfrak{a}_2)$ *the residue norm derived from the completed tensor product norm on* $A_1 \mathbin{\widehat{\otimes}}_R A_2$.

Proof. Use the same arguments as in the proof of Proposition 8. $\square$

References

[EGR]   A. Abbes, *Éléments de géométrie rigide*, vol. I. Construction et étude géométrique des espaces rigides (Birkhäuser, Basel, 2010)

[A]   M. Artin, *Grothendieck Topologies*. Notes on a Seminar by M. Artin, Harvard University, Cambridge, 1962

[Be1]   V. Berkovich, *Spectral Theory and Analytic Geometry over Non-Archimedean Fields*. Mathematical Surveys and Monographs, vol. 33 (American Mathematical Society, Providence, 1990)

[Be2]   V. Berkovich, *Étale cohomology for non-Archimedean analytic spaces*. Publ. Math. IHES **78** (1993)

[Bo]   S. Bosch, *Algebraic Geometry and Commutative Algebra*. Universitext (Springer, London, 2013)

[BGR]   S. Bosch, U. Güntzer, R. Remmert, *Non-Archimedean Analysis*. Grundlehren, Bd. 261 (Springer, Heidelberg, 1984)

[F I]   S. Bosch, W. Lütkebohmert, Formal and rigid geometry, I. Rigid spaces. Math. Ann. **295**, 291–317 (1993)

[F II]   S. Bosch, W. Lütkebohmert, Formal and rigid geometry, II. Flattening techniques. Math. Ann. **296**, 403–429 (1993)

[F III]   S. Bosch, W. Lütkebohmert, M. Raynaud, Formal and rigid geometry, III. The relative maximum principle. Math. Ann. **302**, 1–29 (1995)

[F IV]   S. Bosch, W. Lütkebohmert, M. Raynaud, Formal and rigid geometry, IV. The reduced fibre theorem. Invent. Math. **119**, 361–398 (1995)

[AC]   N. Bourbaki, *Algèbre Commutative*, Chap. I–IV (Masson, Paris, 1985)

[EVT]   N. Bourbaki, *Espaces Vectoriels Topologiques*, Chap. I (Hermann, Paris, 1953)

[C]   B. Conrad, Deligne's notes on Nagata compactifications. J. Ramanujan Math. Soc. **22**, 205–257 (2007); Erratum. J. Ramanujan Math. Soc. **24**, 427–428 (2009)

[FC]   G. Faltings, C.-L. Chai, *Degeneration of Abelian Varieties*. Ergebnisse der Mathematik, 3. Folge, Bd. 22 (Springer, Heidelberg, 1990)

[F]   K. Fujiwara, Theory of tubular neighborhood in étale topology. Duke Math. J. **80**, 15–57 (1995)

[FK]   K. Fujiwara, F. Kato, *Foundations of rigid geometry*, I (2013) [arxiv:1308.4734]

[Go]   R. Godement, *Théorie des Faisceaux* (Herrmann, Paris, 1964)

[GR]   H. Grauert, R. Remmert, Über die Methode der diskret bewerteten Ringe in der nicht-Archimedischen Analysis. Invent. Math. **2**, 87–133 (1966)

[Gr]   A. Grothendieck, Sur quelques points d'algèbre homologique. Tôhoku Math. J. **9**, 119–221 (1957)

S. Bosch, *Lectures on Formal and Rigid Geometry*, Lecture Notes   249
in Mathematics 2105, DOI 10.1007/978-3-319-04417-0,
© Springer International Publishing Switzerland 2014

[EGA I]   A. Grothendieck, J.A. Dieudonné, *Éléments de Géométrie Algébrique I*. Grundlehren, Bd. 166 (Springer, Heidelberg, 1971)

[EGA II]   A. Grothendieck, J.A. Dieudonné, Éléments de Géométrie Algébrique II. Publ. Math. **8** (1961)

[EGA III]   A. Grothendieck, J.A. Dieudonné, Éléments de Géométrie Algébrique III. Publ. Math. **11, 17** (1961/1963)

[EGA IV]   A. Grothendieck, J.A. Dieudonné, Éléments de Géométrie Algébrique IV. Publ. Math. **20, 24, 28, 32** (1964/1965/1966/1967)

[H]   R. Huber, *Étale Cohomology of Rigid Analytic Varieties and Adic Spaces*. Aspects of Mathematics, vol. E 30 (Vieweg, Braunschweig, 1996)

[K1]   R. Kiehl, Theorem A und Theorem B in der nichtarchimedischen Funktionentheorie. Invent. Math. **2**, 256–273 (1967)

[K2]   R. Kiehl, Der Endlichkeitssatz für eigentliche Abbildungen in der nichtarchimedischen Funktionentheorie. Invent. Math. **2**, 191–214 (1967)

[Kö]   U. Köpf, Über eigentliche Familien algebraischer Varietäten über affinoiden Räumen. Schriftenr. Math. Inst. Univ. Münster, 2. Serie, Heft 7 (1974)

[L]   W. Lütkebohmert, Formal-algebraic and rigid-analytic geometry. Math. Ann. **286**, 341–371 (1990)

[M1]   D. Mumford, An analytic construction of curves with degenerate reduction over complete local rings. Compos. Math. **24**, 129–174 (1972)

[M2]   D. Mumford, An analytic construction of degenerating abelian varieties over complete local rings. Compos. Math. **24**, 239–272 (1972)

[R1]   M. Raynaud, Géométrie analytique rigide d'apres Tate, Kiehl,.... Table ronde d'analyse non archimedienne. Bull. Soc. Math. Fr. Mém. **39/40**, 319–327 (1974)

[R2]   M. Raynaud, Variétés abéliennes et géométrie rigide. Actes du congrès international des Mathématiciens (Nice 1970), tome 1 (1971), pp. 473–477

[RG]   M. Raynaud, L. Gruson, Critères de platitude et de projectivité. Invent. Math. **13**, 1–89 (1971)

[S]   J.-P. Serre, Géométrie algébrique et géométrie analytique. Ann. Fourier **6**, 1–42 (1956)

[T]   J. Tate, *Rigid Analytic Spaces*. Private Notes (1962) [Reprinted in Invent. Math. **12**, 257–289 (1971)]

[Te]   M. Temkin, On local properties of non-Archimedean analytic spaces. Math. Ann. **318**, 585–607 (2000)

[W]   T. Wedhorn, *Adic Spaces*. Lecture Script, Institut für Mathematik, Universität Paderborn, 19 June 2012

Index

LECTURE NOTES IN MATHEMATICS   🐎 Springer

Edited by J.-M. Morel, B. Teissier; P.K. Maini

Editorial Policy (for the publication of monographs)

1. Lecture Notes aim to report new developments in all areas of mathematics and their applications - quickly, informally and at a high level. Mathematical texts analysing new developments in modelling and numerical simulation are welcome.

   Monograph manuscripts should be reasonably self-contained and rounded off. Thus they may, and often will, present not only results of the author but also related work by other people. They may be based on specialised lecture courses. Furthermore, the manuscripts should provide sufficient motivation, examples and applications. This clearly distinguishes Lecture Notes from journal articles or technical reports which normally are very concise. Articles intended for a journal but too long to be accepted by most journals, usually do not have this "lecture notes" character. For similar reasons it is unusual for doctoral theses to be accepted for the Lecture Notes series, though habilitation theses may be appropriate.

2. Manuscripts should be submitted either online at www.editorialmanager.com/lnm to Springer's mathematics editorial in Heidelberg, or to one of the series editors. In general, manuscripts will be sent out to 2 external referees for evaluation. If a decision cannot yet be reached on the basis of the first 2 reports, further referees may be contacted: The author will be informed of this. A final decision to publish can be made only on the basis of the complete manuscript, however a refereeing process leading to a preliminary decision can be based on a pre-final or incomplete manuscript. The strict minimum amount of material that will be considered should include a detailed outline describing the planned contents of each chapter, a bibliography and several sample chapters.

   Authors should be aware that incomplete or insufficiently close to final manuscripts almost always result in longer refereeing times and nevertheless unclear referees' recommendations, making further refereeing of a final draft necessary.

   Authors should also be aware that parallel submission of their manuscript to another publisher while under consideration for LNM will in general lead to immediate rejection.

3. Manuscripts should in general be submitted in English. Final manuscripts should contain at least 100 pages of mathematical text and should always include

   - a table of contents;
   - an informative introduction, with adequate motivation and perhaps some historical remarks: it should be accessible to a reader not intimately familiar with the topic treated;
   - a subject index: as a rule this is genuinely helpful for the reader.

   For evaluation purposes, manuscripts may be submitted in print or electronic form (print form is still preferred by most referees), in the latter case preferably as pdf- or zipped ps-files. Lecture Notes volumes are, as a rule, printed digitally from the authors' files. To ensure best results, authors are asked to use the LaTeX2e style files available from Springer's web-server at:

   ftp://ftp.springer.de/pub/tex/latex/svmonot1/ (for monographs) and
   ftp://ftp.springer.de/pub/tex/latex/svmultt1/ (for summer schools/tutorials).

Additional technical instructions, if necessary, are available on request from lnm@springer.com.

4. Careful preparation of the manuscripts will help keep production time short besides ensuring satisfactory appearance of the finished book in print and online. After acceptance of the manuscript authors will be asked to prepare the final LaTeX source files and also the corresponding dvi-, pdf- or zipped ps-file. The LaTeX source files are essential for producing the full-text online version of the book (see http://www.springerlink.com/openurl.asp?genre=journal&issn=0075-8434 for the existing online volumes of LNM). The actual production of a Lecture Notes volume takes approximately 12 weeks.

5. Authors receive a total of 50 free copies of their volume, but no royalties. They are entitled to a discount of 33.3 % on the price of Springer books purchased for their personal use, if ordering directly from Springer.

6. Commitment to publish is made by letter of intent rather than by signing a formal contract. Springer-Verlag secures the copyright for each volume. Authors are free to reuse material contained in their LNM volumes in later publications: a brief written (or e-mail) request for formal permission is sufficient.

Addresses:
Professor J.-M. Morel, CMLA,
École Normale Supérieure de Cachan,
61 Avenue du Président Wilson, 94235 Cachan Cedex, France
E-mail: morel@cmla.ens-cachan.fr

Professor B. Teissier, Institut Mathématique de Jussieu,
UMR 7586 du CNRS, Équipe "Géométrie et Dynamique",
175 rue du Chevaleret
75013 Paris, France
E-mail: teissier@math.jussieu.fr

For the "Mathematical Biosciences Subseries" of LNM:

Professor P. K. Maini, Center for Mathematical Biology,
Mathematical Institute, 24-29 St Giles,
Oxford OX1 3LP, UK
E-mail: maini@maths.ox.ac.uk

Springer, Mathematics Editorial, Tiergartenstr. 17,
69121 Heidelberg, Germany,
Tel.: +49 (6221) 4876-8259

Fax: +49 (6221) 4876-8259
E-mail: lnm@springer.com